PHYLLOTAXIS

Phyllotaxis, the study of the patterns exhibited by leaves and other organs of plants, raises some of the deepest questions of plant morphogenesis. What principles of biological organization produce these dynamical geometric systems? The constant occurrence of the Fibonacci sequence in such systems is a phenomenon that has fascinated botanists and mathematicians for centuries.

In this book, the many facets of phyllotaxis are dealt with in an integrated manner for the first time. The author describes a unified concept of phyllotaxis based on experimental, anatomical, cellular, physiological, and paleontological observations. The book provides a framework for formal analyses of botanical data, and emphasizes the relevance of the phyllotactic paradigm in the study of other structures, such as crystals and proteins. It is of interest to crystallographers and physicists as well as to botanists and mathematicians.

PHYLLOTAXIS

A systemic study in plant morphogenesis

ROGER V. JEAN
University of Québec at Rimouski

CAMBRIDGE UNIVERSITY PRESS
Cambridge, New York, Melbourne, Madrid, Cape Town, Singapore, São Paulo, Delhi

Cambridge University Press
The Edinburgh Building, Cambridge CB2 8RU, UK

Published in the United States of America by Cambridge University Press, New York

www.cambridge.org
Information on this title: www.cambridge.org/9780521104692

First published 1994
Reprinted 1995
This digitally printed version 2009

A catalogue record for this publication is available from the British Library

ISBN 978-0-521-40482-2 hardback
ISBN 978-0-521-10469-2 paperback

To all of those who contributed to the understanding of the phenomenon of phyllotaxis, and in particular to Bravais and Bravais who laid down the foundations of its mathematical treatment, and to Church who envisioned the importance of the systemic approach to it.

And first and foremost to P., K.Y., and I.-M.

Here we are in the kingdom of numbers, and each number is a flower. It joins another one at the right place to bloom into a bouquet. The numbers and the forms authorizing the slow elaboration and the skillful development of flowers and crystals always generate the same designs. These designs represent another face of light which is total architecture; they are the consequence of a subtle, prodigious planetary geometry wherein randomness means aberration.

Contents

Acknowledgments

I express my gratitude to John Palmer (developmental plant physiologist, Department of Botany, University of New South Wales, Australia), to Rolf Sattler (plant morphologist, Department of Biology, McGill University, Montreal), and to Joachim Vieth (plant morphologist, Institut de Recherche en Biologie Végétale, University of Montreal). I hope I have adequately incorporated their stimulating suggestions on an early form of the present monograph. Many thanks are due to the team at the computer graphics service of the University of Québec at Rimouski for their help in redrawing some seven figures; to my student Martine Houde for providing good reprints of many of the figures; to Francine Bélanger, at the library of the same university; and most of all to Marcelle Martin, also at the library, for her competent assistance over the years. My gratitude also goes to the people at Cambridge University Press, in particular to Lauren Cowles, mathematics and computer science editor, to Janet Polata, editorial assistant, and to Sophia Prybylski, production editor. The author's research is funded by Grant A6240 from the National Sciences and Engineering Research Council of Canada.

Prologue

1. Subject and aims of the book

This book deals with a fascinating subject in plant morphogenesis called **phyllotaxis**. In the various areas of botany, phyllotaxis is often considered to be the most striking phenomenon and the toughest subject, raising the most difficult questions. Phyllotaxis studies the symmetrical (asymmetrical) constructions determined by organs and parts of organs of plants, their origins, and their functions in the environment. These constructions are the **phyllotactic patterns**, and their building blocks, in their young stage, are called the **primordia**. The primordia differ in number, size, position, rate of formation, and shape, thus giving considerable diversity to phyllotactic patterns. Yet the phenomenon of phyllotaxis is simple, insofar as all the phyllotactic systems showing spirality belong to Fibonacci-type sequences of integers, characterized by the rule that every term in it is the sum of the preceding two terms, as in the Fibonacci sequence ⟨1, 1, 2, 3, 5, 8, 13, ...⟩.

In a well-known chapter on phyllotaxis, Coxeter (1969) calls the appearance of the Fibonacci sequence in botany a "fascinatingly prevalent tendency," called by Cook (1914) "the Law of Wiesner." The prevalence of the Fibonacci sequence in phyllotactic patterns is often referred to as "the mystery of phyllotaxis," and "the bugbear of botanists." These Fibonacci patterns are one of the most puzzling products of the activity of apical meristems, buds, and shoots. Their study has brought about new ideas and considerable progress in our knowledge of the organization of vegetative shoots and reproductive structures in plants. A large part of this knowledge is based on induction and comes from intuition. Therefore hypothetico-deductive constructs, theoretical models and methods are needed that can serve as guides for botanists and experimentalists.

Wardlaw (1965a) defined the **plant apex** as "*a dynamical geometrical system possessing a biological organization.*" It thus is of interest to the mathematician as well as to the biologist. One of my objectives in this book is to show that phyllotaxis has become a multidisciplinary subject of research. The history of **observational phyllotaxis** goes back to Bonnet (1754) and even before to Kepler (1571–1639) who observed the frequent occurrence of the number 5 in plants, and to Leonardo da Vinci (1452–1519) who anticipated Bonnet's observations on the spiral arrangement of leaves. The history of **mathematical phyllotaxis** starts in the first half of the nineteenth century with pioneers such as Braun (1831), Schimper (1836), Bravais and Bravais (1837) and Lestiboudois (1848). History shows that the many mathematical and botanical facets of phyllotaxis have hardly ever been dealt with in an integrated manner. There is a need to fill the unacceptable gap generated by scattered knowledge on phyllotaxis, and to explain with mathematical modeling, based on experimental, anatomical, cellular, physiological, and paleontological observations, what determines the observed regularity. In this book I propose frameworks for formal analyses of botanical data, and clearly show that mathematical modeling is concerned not only with a posteriori descriptions but also with the causes inducing the phenomenon. Phyllotaxis cannot be considered anymore as a scientific curiosity upon which to improvise our competency. This monograph shows that phyllotaxis has the status of a discipline with a strategic position in the main trends of mathematical biology and plant morphogenesis.

This book stresses the importance of making a systemic approach to phyllotaxis, so that earlier concepts can harmoniously take their place. My intention is to provide a **unified concept of phyllotaxis**, incorporating its phyletic, ontogenetic, and functional aspects. The phyletic aspect concerns the problem of the origins of phyllotactic patterns and the question of morphogenetical parallelism. The functional aspect includes teleonomic and interpretative questions [e.g., packing efficiency, self-similarity (Section 9.1), compositional hierarchy (Section 11.4), and entropy model (Chapter 6)], and deals with plant patterns in their environment. The ontogenetic aspect is the one most commonly dealt with, by some kind of mechanistic hypothesis at the level of the plant apex.

Part I shows how to describe and characterize phyllotactic patterns, in order to be able to recognize patterns in a practical fashion. It deals with **pattern recognition**, and proposes a model by which we can easily and adequately do so. Chapter 1 is an introductory account of what is commonly known as the phenomenon of phyllotaxis. Two important tools for pattern recognition are the Fundamental Theorem of Phyllotaxis (Chapter 2) and the Pattern Determination Table (Chapter 5). Part II deals with the problem of **pattern generation** in phyllotaxis through mathematical modeling, and I introduce a model and a theory of phyllotaxis. This is a systemic and evolutionary theory, in which **comparative morphology** and morphogenetical parallelism, dealt with in Part III, plays a role. In Part III the fundamental question of the functions and **origins of phyllotactic patterns** is developed. Part IV of the book contains selected topics that do not readily find places in the main text.

2. The problem of origins of phyllotactic patterns

The question of the origins of phyllotaxis is given a major place in this book which delves into the foundations of phyllotaxis, rooted in the physicochemical basis of the universe. In their studies of natural phenomena in the nineteenth and twentieth centuries, scientists adopted the philosophy of reductionism, dividing these phenomena into separate parts. Consequently in phyllotaxis the studies are mainly limited to ontogenetic and mechanistic approaches. The word phyllotaxis itself seems very restrictive (from the Greek *phullon,* leaf, and *taxis,* arrangement), meaning the study of arrangements of leaves. It shows neither the extensiveness of the domain it covers, nor the need to reach for general principles able to deal with the problems in an appropriate universal framework. A phyletic and synthetic viewpoint on the phenomenon of phyllotaxis considered as part of the environment can give the most important keys for its understanding, as the book will show.

The science of phyllotaxis was born as a branch of botany when naturalists tried to work out how to describe the newly observed regularity in the arrangements of leaves on stems. Since then, the description of phyllotaxis has attained a high degree of elegance, represented, for example, by Prusinkiewicz and Lindenmayer's (1990) reproduction of colorful and very realistic sunflowers, using L-systems and computer graphics. But imitating nature is not necessarily understanding nature. Hence this book stresses the necessity for making systemic models that take into consideration all that has been learned about the phenomenon over the past 170 years in order to be able to establish causality and origins, to make predictions, and to develop a potential for practical applications.

Phyllotaxis has been intensively studied during the last twenty years. In recent years crystallographers have entered the field, and botanists have begun to use crystallographic terms to describe the subject. The discovery of quasicrystals in 1984 has shown that fivefold symmetry, characterizing phyllotactic patterns in particular, is not absent from inorganic material as we used to think. The transition zones observed in the flowers of the common daisy, where phyllotaxis is seen to rise from one pair of Fibonacci numbers to the next, can be said to have a quasicrystalline nature. The patterns displayed by buds and organs of plants are found to a certain extent in other biological and physical phenomena. Part III shows that phyllotaxislike patterns occur in other areas of research, in particular in molecular biology. These patterns are found in the structures of polymers, viruses, jellyfishes, proteins, resonating metal plates, and even hurricane cloud bands. They point to the necessity of redefining the botanical concept of phyllotaxis, with consequences for the future mathematical modeling of the phenomenon.

The word *systemic* in the title of this book is meant to reflect the fact that general comparative morphology, which encompasses biological comparative morphology, can throw light on the understanding of the fundamental mechanisms and organization of phyllotaxis. Despite the reluctance to compare structures with different substrata, we will be driven to the conclusion that objects sharing a common structure reflect the operation of a common principle. Trying to explain the same structure by referring to one context only, botanical or not, seems illogical. The process of explaining the structure – phyllotactic patterns in particular – becomes a systemic quest.

My theory of phyllotaxis brings together aspects of five theories that have been proposed by morphologists, botanists, paleontologists, and geneticists. These theories are: Church's evolutionary theory; the Lestiboudois–Bolle law of duplication and induction lines; Zimmermann's telome theory on the origins of land plants; Meyen's morphogenetical parallelism or nomothetics, according to which intersectorial comparisons will bring a revolution into biology; and Lima-de-Faria's still polemical but fascinating autoevolutionism. They provide five levels of investigation and understanding that can help to unlock the mysteries of the origins of phyllotaxis, dealt with here in an unprecedented manner.

3. The level of presentation

I agree with biologists who express the view that mathematical books in biology are often divorced from the underlying biological processes, so that they are of relatively low interest to biologists. It is understandable and desirable that biology suggests mathematical theories that can be dealt with for themselves. A very good example is the theory of L-systems invented by a botanist. But the viewpoint adopted here is that the mathematical developments are tools able to give a better insight into the biological phenomena. Since this monograph is biomathematical, mathematicians and botanists should be able to follow the concepts.

Given moreover that phyllotaxis has now evolved into a multidisciplinary field of research, the style of presentation takes into consideration the variety of potential readers. Consequently the book has no special prerequisites, except a general interest in mathematics and botany. The reader not acquainted with the field is given the necessary explanations. This is especially true for Chapters 1 and 2, where the presentation is progressive, and where simple problems solved in Appendix 2 are proposed. Chapter 2 contains an intuitive presentation of the most important theorem of phyllotaxis, which I called **the fundamental theorem** of phyllotaxis. Given that some of the difficulties of the subject come from the lack of understanding of the special words used to describe it, particular care has been taken in their definitions, and the book contains an extensive glossary (Appendix 1).

Although this book can thus be used as an introduction to phyllotaxis, at the same time it is an advanced textbook addressed to all those who contributed to the field. It presents the status of the field, and the general perspective in which it should be studied. The reference list is a comprehensive guide to authors and their works, which goes back to the 1830s. The list should be particularly useful for the freshman who wishes to study the field. This monograph can be used as a personal teaching tool for an amazing subject, and it will also be useful for advisory workers and graduate students in the plant sciences. The first two chapters in particular can be used to update the generally very inadequate teaching of phyllotaxis in undergraduate courses of botany, and the usually rudimentary presentation of the subject in advanced textbooks of botany. This monograph is also intended for teaching phyllotaxis to mathematics students.

4. Related works in the field

As the references show, many articles have been written on phyllotaxis, but there are surprisingly very few books, and no really alternative texts. There are also a few books containing a single chapter on phyllotaxis. In chronological order, the book by Schwendener (1878) is obsolete and of historical interest only. The one by van Iterson (1907) is also mainly of historical interest, with the exception that his studies on packing of spheres on cylinders are still in use today. D. W. Thompson's classical and important pioneering book on general biomathematics, published in 1917 and reprinted twice, has a chapter on phyllotaxis which is obsolete and even wrong in parts.

Church (1904b, 1920b) and Richards (1948, 1951), are very important pioneering authors whose endeavors greatly influenced the development of phyllotaxis. Church's contributions have been ignored for decades. I take pride in resuscitating them. Church's works are recalled mainly in Chapters 3 and 8 of the present monograph. His books are important sources of botanical information, gold mines of illustrations of phyllotactic patterns, and major contributions to addressing the fundamental issues. The monographs by Richards are also still very important. Richards is responsible for the introduction of a mathematical treatment of descriptive phyllotaxis in the centric representation. Some of his contributions are described in Chapter 4 and deduced from Jean's more general approach. Chapter 5 compares Jean's Pattern Determination Table to Richards's famous phyllotaxis index.

Dealing with the experimental, anatomical, and physiological aspects of phyllotaxis, there is a compilation by Cutter (1965), some general monographs by Wardlaw (e.g., 1965a,b, where phyllotaxis is devoted a few chapters), and a book by Loiseau (1969). The latter is mainly centered on the polemical (outside France) work of Plantefol in which mathematics are eliminated from the discussion. The book by Williams (1975) contains excellent drawings. It is limited almost entirely to observational data with little discussion of theory. *Phytomathématique* by Jean (1978b) is an intuitive initiation to phyllotaxis, presented as an object of fascination and as an introduction to the literature on the subject. The more recent literature offers presentations of selected topics on the subject, in articles or collective books, such as those by Erickson (1983), Schwabe (1984), and Thornley and Johnson (1990); they have the deficiencies of highly selective bibliographies.

The set of contributions to the subject published before 1983 are reported for the first time altogether in a book by Jean (1983c). The adapted English version (Jean, 1984a) was translated into Chinese in 1990. Adler's contribution in particular is given therein an important place. His model makes an extensive use of continued fractions, a tool almost unmentioned in the present monograph. One of the reasons is that for the purpose of dealing with the cylindrical lattice at the descriptive level there are more direct and efficient algorithms (presented in Chapter 2). The 1983 book can be described as a museum of ideas from everybody.

The author's models and his systemic perspective and theory of phyllotaxis constitute the main thread of the present monograph. Around them are harmonized the various contributions to phyllotaxis. Thus the book integrates the works published before 1983, especially those works that can be considered modern landmarks in the history of mathematical phyllotaxis: the contributions of Coxeter (1972), Adler (1974, 1977a), Thornley (1975a), Ridley (1982a), and Marzec and Kappraff (1983). The book also deals with all developments that occurred after 1983, including the work of crystallographers and botanists using crystallographic methods (Chapters 9 and 10). It gives a description (in Chapter 5) of the method used by Erickson (1983) and co-workers for evaluating phyllotactic patterns, and shows the accuracy, reliability, and ease of a method (Jean, 1987a) based on the allometry-type model (Jean, 1983a) presented in Chapter 4. Chapter 9 shows convergences of Adler's (1977a) and Marzec's (1987) models, with a recent model (Jean, 1990) called the τ-model. Chapter 12 introduces systemic-like approaches such as Niklas's (1988), whose computer simulation takes into account the various parameters that could influence light interception by leaves; Green's (1992), looking for the immediate cause of phyllotactic patterns with minimal surface energy concepts on the tunica compared to a resonating metal plate; and Douady and Couder's (1992), who compared the primordia of plants to droplets of ferrofluid in a magnetic field. Appendix 5 is devoted to Williams and Brittain's model (1984), which is correlated with the author's Pattern Determination Table of Chapter 5. Appendix 8 is devoted to Meinhardt's (1982, 1984) treatment of diffusion–reaction equations that are still of special interest for modelers in phyllotaxis. Recent refined attempts with diffusion equations in phyllotaxis include Schwabe and Clewer's (1984) and Chapman and Perry's (1987), discussed in Chapter 12.

Part I
Pattern recognition

Introduction

Behind the untold diversity of plant architecture we can find mathematical constants – many great minds such as Goethe and Leonardo da Vinci were aware of that. It was however only in the first half of the nineteenth century that naturalists developed coherent accounts of phyllotaxis. In the 1830s the brothers L. and A. Bravais presented the first mathematical treatment of the phenomenon, which is known today as the cylindrical representation of phyllotaxis (introduced in Chapter 2). The centric representation of phyllotaxis used later by Church and Richards, also known as the spiral lattice, is dealt with in Chapters 1 and 2.

Chapter 1 gives an elementary **description of phyllotaxis**. The reader will learn to recognize phyllotactic patterns; he will be introduced to the terminology, parameters, and concepts used for their description. Among the concepts we find the divergence angle d, the parastichy pair (m, n), and the plastochrone ratio R. This chapter gives a geometrical description of the spiral pattern of florets or seeds in the sunflower head and of the cross section of terminal buds under the microscope. The theorem of Chapter 2, which is called here the **fundamental theorem of phyllotaxis**, relating d and (m, n), will be given particular formulations and many applications. Among the applications we find the interpretation of a type of puzzling pattern known as spiromonostichy observed in *Costus* (Chapter 2) and the study of proteins (Chapter 10). This theorem gives an interesting insight into the historical development of phyllotaxis (Chapter 2).

Chapter 3 deals with the **hierarchical representation of phyllotaxis**, introduced by Jean (1976a), on which the interpretative model of Part II is based. Phyllotactic hierarchies will be seen to be generated by L-systems and growth matrices. Zimmermann's theory and Bolle's theory mentioned in the Prologue and the translocation of substances in plants provide some of the biological bases of the hierarchical representation, which has also formal bases in Hermant's fractals for example. Gnomonic growth and branching will be shown to be very important processes in phyllotaxis.

In Chapter 4 the **allometry-type model** is derived in the cylindrical representation and used to give new mathematical relations involving Church's bulk ratio B, Richards's area ratio A, and the plastochrone P. Richards's phyllotaxis index (the P.I. used by botanists) and a generalized Coxeter formula will be deduced from the model. The main utility of the model is the **Pattern Determination Table** which is abundantly applied in Chapter 5 where we learn to practically characterize spiral patterns. This chapter also discusses the difficulties involved in practical pattern recognition.

The first three appendices of Part IV are meant to promote an understanding of the concepts developed in Part 1. Appendix 1 is by far the most complete **glossary** on phyllotaxis ever; Appendix 2 gives the **answers to the problems** at the ends of Chapters 1 to 4; and Appendix 3 presents numerous review **questions**. Appendix 4 extends the mathematical developments around the fundamental theorem of phyllotaxis (Chapter 2), and contains algorithms to obtain d from (m, n) and vice versa.

1
The centric representation

1.1. Parastichy pairs (m, n)

1.1.1. Patterns in plants

Pattern formation in organisms is one of the commonest phenomena observed in nature. Virtually all animals and plants possess symmetries that result in pattern formation. Most animals seem to be bilaterally symmetrical; others show radial symmetry. For a whole plant, radial symmetry is more common, while such plant organs as leaves usually exhibit bilateral symmetry. Such symmetrical morphologies are perceived as patterns and are usually regarded as being attractive or beautiful. These patterns appear in many ways in plants: in the arrangements of leaves or branches on the stem, in the venation of leaves, and particularly in flowers with their patterns of shapes and colors. Of course humans are not the only ones to appreciate such patterns; pollinating insects are also attracted to flowers by this means.

In spite of the overwhelming diversity of plant architecture, there are common patterns that link a wide range of species. These patterns can be seen macroscopically, as when viewing shoot tips or adult plants from above, and microscopically (using scanning electron micrographs), at an early stage of development, in sectioned or in dissected shoot apices. An outstanding example of such patterns is the spiral arrangement of florets in the capituli of sunflowers and daisies.

This book is not concerned with all types of patterns. It is restricted to phyllotactic patterns and to the study of **phyllotaxis**, a central subject in plant morphogenesis dealing with the arrangements of plant organs such as leaves, bracts, branches, petals, florets, and scales, called in their young stages **primordia**. Phyllotactic patterns can be described with just a few mathematical constants and principles as we will see.

1.1.2. Whorled and spiral patterns

What are the common patterns of leaf arrangements seen in nature? Although they are related, two large categories of patterns can be recognized, the whorled and the spiral patterns. In a number of common species the leaves are arranged in **whorls** at a level of the stem. The number n of leaves in a whorl varies from species to species, in the same species, and can even vary in the same specimen. The horsetails (*Equisetum*) are a good example; there n may vary from 6 to 20. It is characteristic of whorled patterns that the leaves at any one node are generally inserted above the gaps of the preceding one, thus the term **alternating whorls**. When the leaves in each whorl are just above the ones in other whorls, we have **superposed whorls**, as in *Ruta* and *Primula*. Whorled patterns are also known as "verticillate patterns."

In transverse sections of shoots the consecutive whorls form concentric rings. Figure 1.1 shows an alternating whorl where five primordia of the same size (the whorl) begin to grow in the five gaps between the five slightly larger, older primordia themselves inserted in the five previous gaps. The case $n=2$, called **decussate**, presents the same idea but with two primordia. Successive pairs of opposite leaves (opposite with respect to the center of the apical bud) are arranged at right angles to each other.

In another pattern a single leaf is inserted at each node, and in this case the leaf at the next node is positioned at 180° to the previous one, thus determining two vertical rows of leaves along the stem. This pattern, which is very common especially in the grasses, is termed **distichous**. The magnificent fan palm shows a distichous arrangement of its leaves, all in the same vertical plane, consecutively born leaves alternating on each side of the central axis.

The most common pattern, the **spiral pattern**, also involves the insertion of a single primordium; but in this case it is possible to trace two sets or families of spirals round the stem, which run in opposite directions and which appear to cross one another. In the botanical literature these spirals are called **parastichies**. Two families of parastichies constitute a **parastichy pair**. They are formed, for example, by leaves on the stems of many plants with a cylindrical surface. The spirals made by the scales on the pineapple fruit are another example. They are perhaps even more obvious when seen from above, especially in species that show closely

Figure 1.1. An alternating whorled system observed in a transverse section of a shoot of *Lycopodium selago* (fir club moss) showing five leaf primordia in a whorl. The dots in the middle of five primordia serve to identify the second whorl. The primordia arise at the center of the shoot, and they move outward by growth, while increasing in size. The system is denoted by (5, 5). There are *Lycopodium selago* with (5, 6), (4, 5), (4, 4) and (3, 3) parastichy pairs. (From Church, 1904b.)

packed leaves or florets in the inflorescence or capitulum of composites, such as the common daisy. When we look at transverse sections of apical buds of shoots under the microscope, we obtain a surface view of the primary patterns, which are later extended or stretched. Surface views of spiral apices are clearly seen in Figure 1.2 at the macroscopic level and in Figure 1.3 at the microscopic level.

1.1.3. Contact parastichies

How can one characterize these spiral patterns? We can start characterizing the patterns by counting the numbers of spirals visible to the naked eye. For example, by taking a look at the surface of the common pineapple fruit we very easily notice that there are generally three families of spirals, due to the generally hexagonal form of the scales in contact side by side. The hexagons have three pairs of opposite sides that determine three directions in which the scales are aligned. One can easily count the numbers x, y, and z of spirals in each direction. In 1933 a member of the

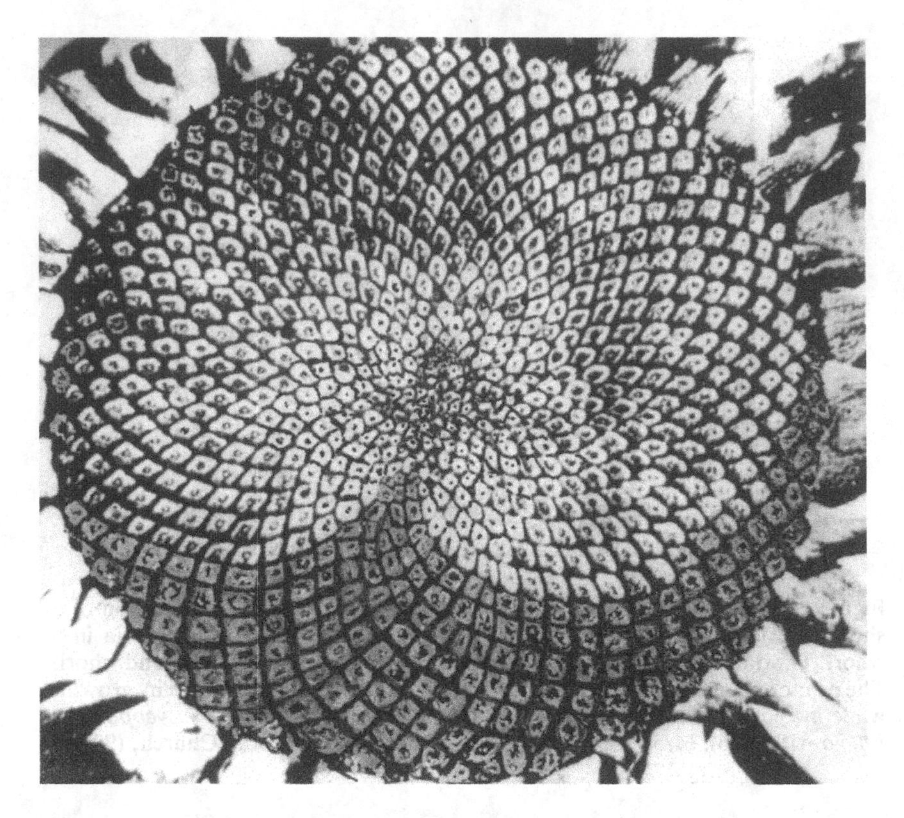

Figure 1.2. Capitulum of a sunflower (*Helianthus annuus*) from which the flowers have been removed to reveal quadrilateral seeds in contact side by side. These seeds are arranged in a pattern of two sets of parastichies running in opposite directions (called a contact parastichy pair), which number 34 and 55 respectively in the outer parts. Seeds have been removed to show the direction of the curves. The numbers of parastichies decline towards the center, with pairs coalescing into single spirals. (From Church, 1904b.)

Hawaiian society that published the *Pineapple Quarterly* did just that. He noticed that x, y, and z take the respective values 5, 8, and 13, and more often the values 8, 13, and 21 for larger fruits. No exceptions are reported. When the scales of the pineapple fruit are quadrilateral we have

Caption for Figure 1.3
(1) Transverse section of an apical bud of *Araucaria excelsa* (Norfolk Island pine) showing quadrilateral leaf primordia in contact side by side. The black dots on three primordia can be used to number them according to the Bravais–Bravais theorem (see the text). (2) Cross section of an apical bud of *Pinus pinea* showing the contact parastichy pair (5, 8) in the Fibonacci sequence. (Both figures from Church, 1904b.)

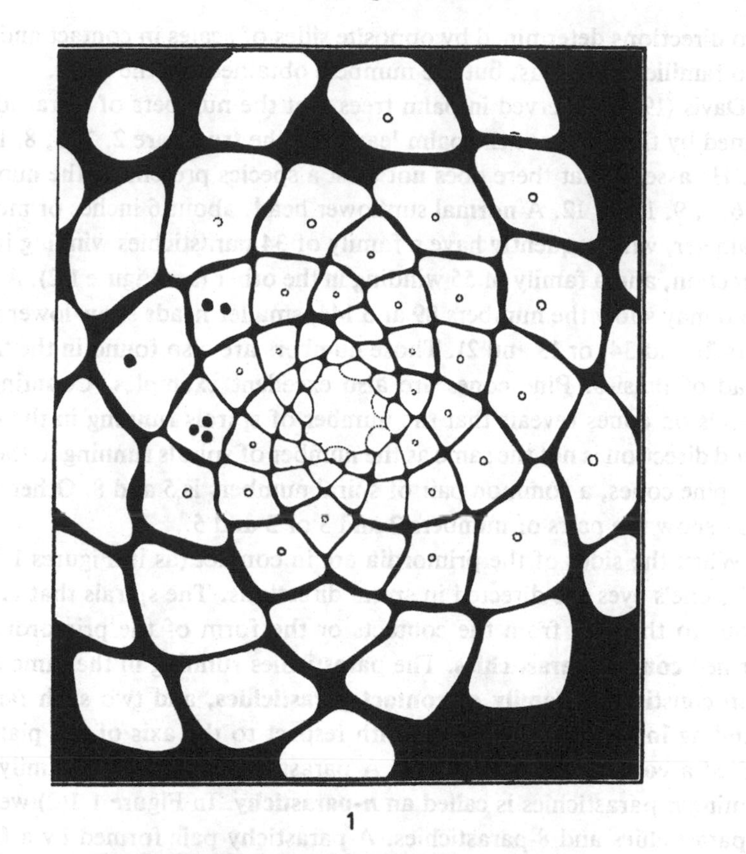

1

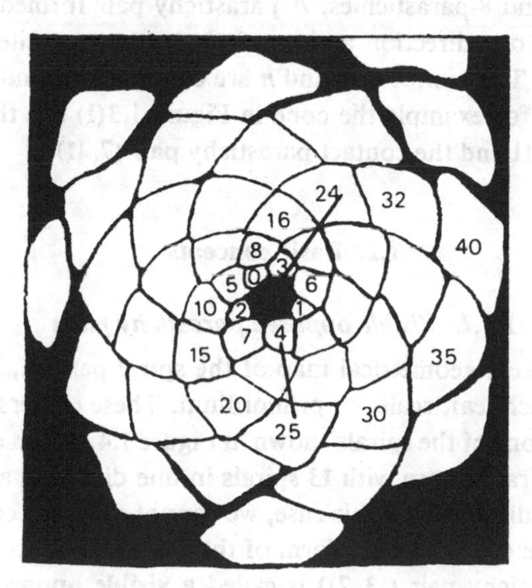

2

two directions determined by opposite sides of scales in contact and thus two families of spirals, but the numbers obtained are the same.

Davis (1971) observed in palm trees that the numbers of spirals determined by the traces of the palm leaves on the trunk are 2, 3, 5, 8, 13, or 21. He asserts that there does not exist a species presenting the numbers 4, 6, 7, 9, 11, or 12. A normal sunflower head, about 6 inches or more in diameter, will frequently have a family of 34 parastichies winding in one direction, and a family of 55 winding in the other (see Figure 1.2). A large head may show the numbers 89 and 144; smaller heads show lower numbers 21 and 34, or 13 and 21. Those numbers are also found in the flower head of daisies. Pine cones are also excellent examples. Counting the spirals on cones reveals that the number of spirals running in the right-hand direction is not the same as the number of spirals running to the left. On pine cones, a common pair of spiral numbers is 5 and 8. Other cones may show the pairs of numbers 2 and 3 or 3 and 5.

When the sides of the primordia are in contact (as in Figures 1.2 and 1.3), one's eyes are directed in spiral directions. The spirals that are obvious to the eye, from the contacts or the form of the primordia, are termed **contact parastichies**. The parastichies running in the same direction constitute a **family of contact parastichies**, and two such families winding in opposite directions with respect to the axis of the plant are called a **contact parastichy pair**. A parastichy (spiral) in a family containing n parastichies is called an **n-parastichy**. In Figure 1.3(2) we have 5-parastichies and 8-parastichies. A parastichy pair formed by a family of m spirals in one direction and n spirals in the opposite direction is denoted (m, n). The numbers m and n are called the secondary numbers of the pair. So for example the cone in Figure 1.3(1) has the secondary numbers 7 and 11 and the contact parastichy pair (7, 11).

1.2. Basic concepts

1.2.1. Visible opposed parastichy pairs

In order to make a geometrical map of the spiral pattern, we mark out the center of each leaf, scale, or primordium. These centers are the dots at the intersections of the spirals shown in Figure 1.4, which depicts in the outer part a spiral pattern with 13 spirals in one direction and 21 spirals in the opposite direction. In this case, we cannot speak of contact parastichies, since we do not see the form of the primordia in contact side by side. The parastichy pair (13, 21) is called a **visible opposed parastichy**

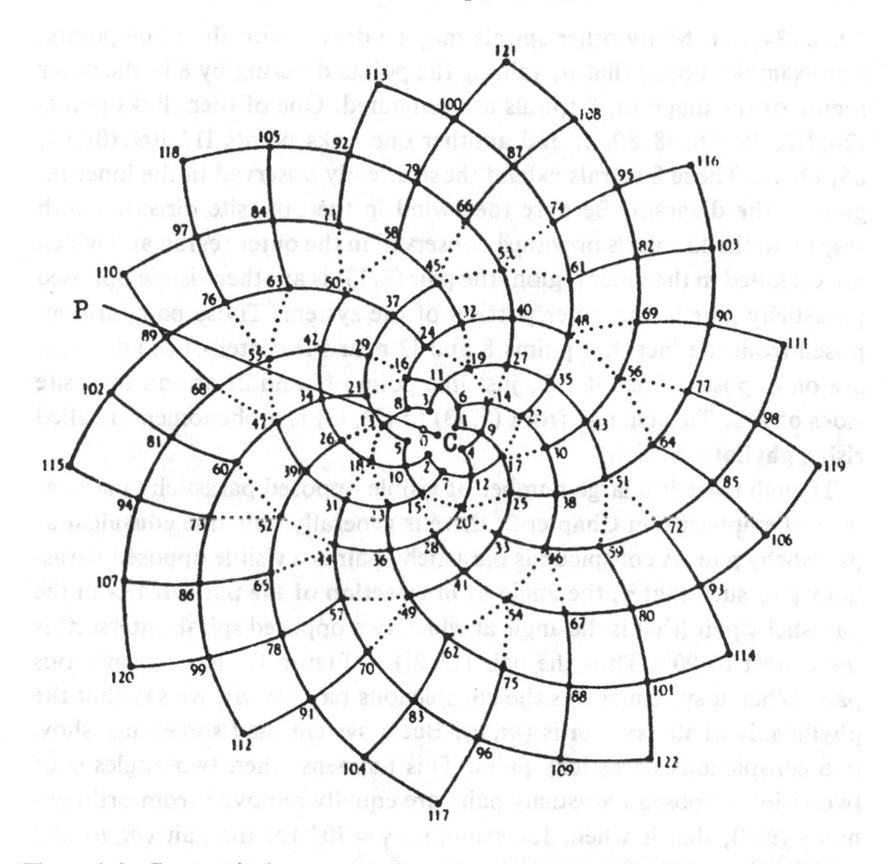

Figure 1.4. Geometrical representation of a plant apex. The diagram illustrates how the phenomenon of rising phyllotaxis (also observed in Figure 1.2) occurs. Here it is clear that the phyllotaxis rises along the Fibonacci sequence; in the inner part we have a (8, 13) visible opposed parastichy pair, and in the outer part a (21, 13) visible opposed pair. This is due to the fact that in such a system the Fibonacci numbers are clustered around, and alternate on each side of, the segment *PC*, where *C* represents the center of the apex. This center is the common pole of all the spirals; it is also where all the primordia originate. The divergence angle of the system is the Fibonacci angle. (From Richards, 1948.)

pair. This term does not mean "visibly" opposed parastichy pair. It is not the opposition of the spiral pair that is said to be visible but the pair, from the simple fact that at every intersection of two spirals there is a primordium, as one can see in the figure.

In the inner part of the diagram of Figure 1.4 we can observe the visible opposed parastichy pair (8, 13) (follow the integers 45, 37, 29, 21, 13, and 5 to find one of the 8 spirals; another one goes through the points

50, 42, 34, ...). Many other spirals may be drawn with the same points. For example, notice that by joining the points differing by 8 in the outer region of the diagram, 8 spirals are generated. One of them links points 120, 112, 104, 96, 88, 80, ... and another one links points 117, 109, 101, 93, 85, 77, ... Those 8 spirals extend those already observed in the inner region of the diagram. Because they wind in the opposite direction with respect to the 13 spirals previously observed in the outer region and which are extended in the inner region, the pair (8, 13) is another visible opposed parastichy pair in the outer portion of the system. Those pairs are opposed from the fact that points 8 and 13 near the center of the diagram are on opposite sides of *PC*, just like points 13 and 21 are on opposite sides of *PC*. This change from (8, 13) to (21, 13) is a phenomenon called **rising phyllotaxis**.

Though there is a large number of visible opposed parastichy pairs, as we will emphasize in Chapter 2, there is generally only one **conspicuous parastichy pair**. A conspicuous parastichy pair is a visible opposed parastichy pair such that γ, the **angle of intersection of the parastichies** in the parastichy pair (that is the angle at which two opposed spirals intersect) is the closest to 90°. Thus the pair (13, 21) in Figure 1.4 is a conspicuous pair. When a system shows the conspicuous pair (m, n), we say that the **phyllotaxis of the system** is (m, n). But a system may sometimes show two conspicuous parastichy pairs. This happens when two angles γ of two visible opposed parastichy pairs are equally removed from orthogonality (90°), that is when, for example, $\gamma = 70°$ for the pair (m, n) and 110° for the pair $(m, m+n)$.

Thus we know of two particular types of visible opposed parastichy pairs: the conspicuous parastichy pair and the contact parastichy pair. Practically speaking, when observing a plant specimen, they will generally be the same pair. Notice that in Figures 1.2 and 1.3 the shape of the primordia is quadrilateral. Thus we have in each case two families of contact parastichies and, consequently, one contact parastichy pair. But, there may be (for example, in pineapples with hexagonal scales) three families of contact parastichies and thus two contact parastichy pairs. Considering that there is generally one conspicuous parastichy pair and often two contact parastichy pairs, it may thus happen that a contact parastichy pair is not a conspicuous pair, though the contact pair is most visible (or conspicuous, in the common sense of the word) to the eyes attracted in directions determined by the shape of the primordia. Again, a contact parastichy pair is defined by the shape of the primordia in contact, while a conspicuous parastichy pair is a visible opposed parastichy pair where the opposed spirals show an angle of intersection γ closest to 90°.

Consult the Glossary (Appendix 1) for definitions of the various parastichy pairs, starting with the terms "parastichy," "family of parastichies," and "parastichy pair."

1.2.2. Genetic spiral and the Bravais–Bravais theorem

The numbering of the primordia in Figures 1.3(2) and 1.4 is in agreement with the Bravais–Bravais theorem, named by the author after the botanist and the mathematician–crystallographer who discovered it. In Figure 1.4 the primordia are numbered according to their age, that is according to the order in which they arise on the plant apex, with 0 being the youngest. Notice that on each of the 13 spirals, a 13-parastichy, the points are successively 13 primordia apart in terms of age (for example, the spiral linking the primordia 13, 26, 39, 52, 65, 78, 91, 104, 117), and on any of the 21 spirals (a 21-parastichy) they are 21 primordia apart.

Referring to Figure 1.3(2), once we have counted 5 spirals in one direction, 8 in the other, and if a primordium is given a number such as 0, then the primordia adjacent to 0 are numbered 5 and 8 and all other primordia can be numbered. The **Bravais–Bravais theorem** (1837) states that on an n-parastichy of a phyllotactic spiral pattern, the numbers on two adjacent primordia differ by n. Thus on any one of the 5-parastichies the numbers differ by 5.

The **genetic spiral** (Schimper, 1836, Braun, 1835), also called the generative, fundamental, or ontogenetic spiral, is a continuous line going through the consecutively borne primordia from the oldest (near the rim) to the youngest (near the center of the apex) by the shortest path around this center. In Figure 1.4 the genetic spiral winds counterclockwise. To visualize a portion of it just follow the consecutive numbers 111, 110, 109, 108, ... Of course you can follow them clockwise or counterclockwise (chirality). But point 111 for example is closer to 110 by going counterclockwise from 111 (older than 110) to 110. You can check that the genetic spiral is counterclockwise in Figures 1.5 and 2.1 (be careful with the reverse numbering of the primordia in the latter case), and clockwise in Figures 1.6 and 4.2. Some authors prefer the terms right- and left-handed spirals. Bravais and Bravais showed that whenever the secondary numbers in the contact parastichy pair (m, n) are relatively prime, there is only one genetic spiral, and it can be drawn by using the theorem above.

Davis (1963), and Davis and Davis (1987) studied the direction of the genetic spiral on 71,640 palm-trees in 42 regions of the globe. The analysis of the results (by the chi-square test and by the test of Wilcoxon) led them to conclude that the proportion of left-handed spirals significantly decreases from north to south, with left-handed spirals outnumbering right-handed spirals in the Northern Hemisphere, and right-handed outnumbering left-handed in the Southern Hemisphere. Must we conclude that the direction of the genetic spiral depends on the geomagnetic latitude? Does it depend on the species? Are there as many left-handed spirals as there are right-handed spirals? The subject is open for investigation. The numerous references on it include Imai (1927), Allard (1951), Wardlaw (1965b), Loiseau (1969), Gomez-Campo (1970), Gregory and Romberger (1972), Davis and Mark (1981), Tennakone, Dayatilaka and Ariyaratne (1982), and XuDong, JuLin and Bursill (1988) (the latter two references propose a convention for chirality).

1.2.3. Divergence angle d and plastochrone ratio R

No matter which visible opposed parastichy pairs we see (or how many), if we draw lines from any two successively numbered primordia (n and $n+1$) to the center of the apex, we define an angle called the **divergence angle** or simply the divergence. The divergence is denoted by d and is always smaller than 180°. It is very often represented by a number smaller than 1/2 (180/360), such as $d=2/5$ or 144° (360° × 2/5) which is common. In nature the divergence angle tends to be relatively constant for spiral arrangements, and the value generally measured is approximately $137\frac{1}{2}°$. For example, this particular divergence can be measured in Figure 1.3(2) between points 24 and 25 or 2 and 3.

When we want to describe a pattern by taking measurements on primordia in a plant bud, the divergence angle is not the only parameter required. We also need to know the distances of the centers of the primordia to the center of the apex. If we take the ratio of the distances for two successively numbered primordia, we obtain what is called the **plastochrone ratio**, which is denoted by R and is a number larger than 1. Of course, due to irregularities in patterns, in order to estimate this ratio, we have to measure a number of such paired distances and calculate an average. With this new parameter, we have all the information needed to construct a diagram reflecting the actual pattern at the apex.

1.3. Mathematical constants

1.3.1. Fibonacci and Lucas sequences

The secondary numbers in the parastichy pairs reported above on pineapples, cones, and sunflowers are consecutive **Fibonacci numbers**. The first few Fibonacci numbers are 1, 1, 2, 3, 5, 8, 13, 21, 34, where the sum of two consecutive numbers is the next number. A recursive application of this rule gives the very simple but important mathematical sequence called the **Fibonacci sequence**, also known as the **main sequence**, which is the ordered set of all the Fibonacci numbers:

$$\langle 1, 1, 2, 3, 5, 8, 13, 21, 34, \ldots \rangle, \tag{1.1}$$

The symbols ⟨ and ⟩ mean that the numbers between them constitute an infinite sequence. The sequence was named in the nineteenth century by the French mathematician Edouard Lucas after Leonardo Fibonacci, the famous mathematician from Pisa who described the sequence in the twelfth century.

Fibonacci numbers occur with great regularity (see Section 1.1.3), as is apparent from the results of an investigation (Brousseau, 1968) using 4,290 cones from ten species of pine found in California, from which it was established that only 74 cones (1.7%) deviated from the Fibonacci pattern. Another investigation (Jean, 1992a) based on 12,750 observations on 650 species found in the literature over the last 150 years shows that the Fibonacci sequence arises in more than 92% of all the possible spiral cases.

When visible opposed parastichy pairs do not contain numbers that are consecutive values in the Fibonacci sequence, they often contain those from another famous sequence, the Lucas sequence, derived in the same manner as the Fibonacci sequence, but with initial terms 1 and 3. Among the 12,750 observations mentioned above, the Lucas sequence was found in about 2%. Figure 1.3(1) shows precisely one of those relatively rare specimens in which the visible opposed pair (7, 11) can be seen. Figure 1.5 is a diagrammatic representation of a spiral pattern having the visible opposed pair (11, 18). These three numbers are Lucas numbers and belong to the Lucas sequence. The next chapter will help us to understand why the divergence angle corresponding to Lucas sequence is approximately equal to 99.5°. Sometimes, but only very rarely, consecutive terms of yet another Fibonacci-type sequence can be observed, the sequence ⟨1, 4, 5, 9, 14, 23, ...⟩, for which the divergence angle is approximately equal to 78°.

Let us move on, now, to ways in which the various concepts that have been introduced tie in with the Fibonacci sequence. Denote by F_k the kth Fibonacci number. Thus $F_5 = 5$ and $F_9 = 34$. We have

$$F_{k+1} = F_k + F_{k-1}, \quad \text{for } k = 2, 3, 4, \ldots, \tag{1.2}$$

with $F_1 = F_2 = 1$. The kth term of the Lucas sequence is denoted by L_k, $L_{k+1} = L_k + L_{k-1}$ for $k = 2, 3, 4, \ldots$, and $L_1 = 1$, $L_2 = 3$. This sequence

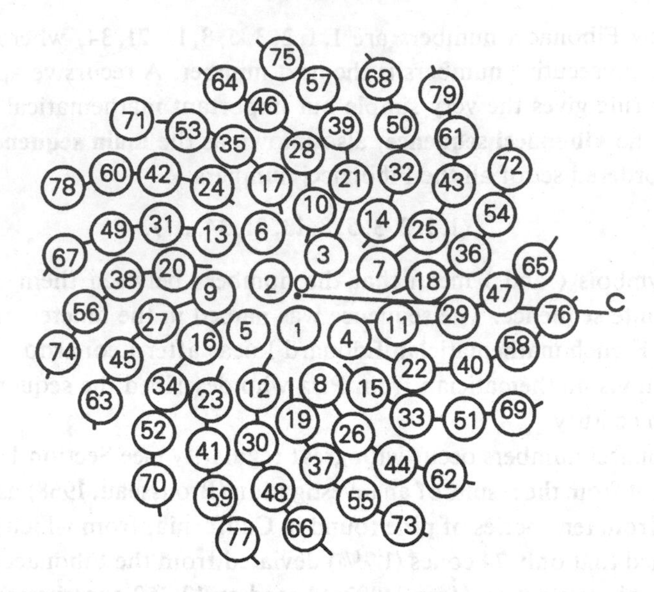

Figure 1.5. Diagrammatic representation of a spiral pattern, showing a family of 18 spirals. Along each spiral the numbers on the primordia differ by 18. A family of 11 spirals linking the primordia by 11 can easily be drawn (one of these spirals links primordia 77, 66, 55, 44, 33, 22, and 11). The divergence can be measured and is seen to be approximately equal to 99.5°. Lucas numbers are seen to alternate on each side of *OC* while getting closer to *OC*.

is also known as Schoute's accessory sequence and the first accessory sequence.

1.3.2. The golden ratio τ

Closely related to the Fibonacci sequence is the golden ratio, denoted here by τ (the Greek letter tau), whose value is

$$\tau = (\sqrt{5}+1)/2. \tag{1.3}$$

This is an irrational number approximately equal to 1.618. It has also been denoted by the Greek letter ϕ (phi) after Phidias, the greatest sculptor of antiquity. Denoted by ϕ, the golden ratio is a symbol of harmony. The Greeks were fascinated by the golden ratio because of the aesthetic proportions it generates in geometry, art, and architecture. For example, tests have shown that among the many possible rectangles, the one for which the ratio of the base to the height is the golden ratio is the most pleasant to the eye. This number arises in many ways in the regular pentagon (five equal sides), and fivefold symmetry is one of the main characteristics of the phenomenon of phyllotaxis. The absence of fivefold symmetry from inorganic matter is known as the crystallographic restriction.

The symbol τ adopted throughout the book has a deeper meaning. Not only does it symbolize ratio and proportion, but it also represents one of the oldest symbols of life well adapted to the form, the symbol known to ancient Egyptians (the ankh) and ancient Hindus. This book is about one of the most fascinating expressions of life, harmony, and fivefold symmetry. Tau is also the symbol generally used in purely mathematical literature.

The ratio τ may be defined as follows: If a straight line segment (or the circumference of a circle) of length 1, is divided in two parts, so that the ratio of the larger portion (x) to the smaller ($1-x$), equals the ratio of the whole (1) to the larger portion (x), then the segment (or circumference) has been divided in the golden ratio: $x/(1-x)=1/x$, $x=(\surd 5-1)/2=1/\tau$, $1/x=\tau$ (see Problem 1.4.3).

1.3.3. Relationships between the constants

The golden ratio is a number whose decimal expansion does not end, and does not show periodicity. How then is such a number related to the simple integers in the Fibonacci sequence? To see it, just take the ratios of consecutive Fibonacci numbers: $1/1, 2/1, 3/2, 5/3, 8/5, 13/8, 21/13, \ldots$. By using a calculator we notice that these ratios get closer to the golden ratio as the consecutive Fibonacci numbers in the ratios become larger. In other words we have

$$\tau = \text{limit}\, F_{k+1}/F_k, \tag{1.4}$$

meaning that τ is the value that the ratio F_{k+1}/F_k of two consecutive Fibonacci numbers approaches as k becomes very large.

The importance of the golden ratio appears when we consider the divergence angle d of a plant specimen showing a contact parastichy pair (F_k, F_{k+1}). For large values of k the corresponding value of d will be around $137\frac{1}{2}°$. It can be verified that the following relations hold between this angle, the golden ratio, and the Fibonacci pair:

$$137\tfrac{1}{2}° \approx (360/\tau^2)° \approx 360[1/(F_{k+1}/F_k+1)]° \approx 360[1-1/(F_{k+1}/F_k)]°.$$

The value $(360/\tau^2)°$ and, by extension τ^{-2}, and their approximations $137\frac{1}{2}°$ and 0.381, are called the **Fibonacci angle**.

Coming back to the definition of the golden ratio (end of Section 1.3.2), we have that the smaller part of the circumference is subtended by an angle at the center, which is equal to the Fibonacci angle. The reflex angle has the value $360°/\tau$ since $\tau+1=\tau^2$ and $(\tau-1)/\tau=1/\tau^2$ so that $360/\tau+360/\tau^2=360$. The angle $360(3+1/\tau)^{-1}° \approx 99.5°$, or $(3+1/\tau)^{-1} \approx 0.276$, corresponding to the Lucas sequence, is called the Lucas angle. The study of the correspondences between divergence angles and Fibonacci-type sequences is developed in Chapter 2.

The primordial fact about phyllotaxis is that in the early development of a phyllotactic pattern, the plastochrone ratio R decreases, the divergence angle d generally converges rapidly towards the Fibonacci angle, and the secondary numbers are consecutive terms of the Fibonacci sequence.

1.4. A model for pattern analysis

1.4.1. Geometry of the spiral lattice

In order to make a mathematical study of plant patterns such as those observed previously so as to obtain meaningful relations among the various parameters in the patterns, we must make realistic but mathematically interesting assumptions. We will assume that the spirals in each family are evenly spaced, identical, logarithmic spirals going through the centers of the primordia (represented as points) and that the divergence and plastochrone ratio are constants. This is called the **centric representation** of the naturally occurring spiral patterns, a spiral lattice of points. The points are distributed along a genetic spiral. The flat flowering head of the sunflower represents the equivalent of the centric projection.

The constant angle of a logarithmic spiral is the angle made at any point of the curve by the tangent to the curve at that point and the radius linking the pole (point C in Figure 1.6) to the point. The constant angles of the logarithmic spirals in the families of m and n spirals are denoted by ϕ_m and ϕ_n (assumed to be of opposite signs: one positive, one negative), respectively. It can be proved that if (m,n) is a visible opposed parastichy pair in the centric representation of a plant pattern, with divergence d, and plastochrone ratio R, and if u and v are respectively the integers nearest to md and nd, then the following relations hold (see Problem 1.7):

$$\Delta_m = 2\pi(u-md), \qquad \Delta_n = 2\pi(v-nd), \tag{1.5}$$

$$mv-nu = \pm 1, 0 < \gamma = \phi_n - \phi_m \quad (\text{or } \phi_m - \phi_n), \tag{1.6}$$

$$m \ln R = \Delta_m \cot \phi_m, \tag{1.7}$$

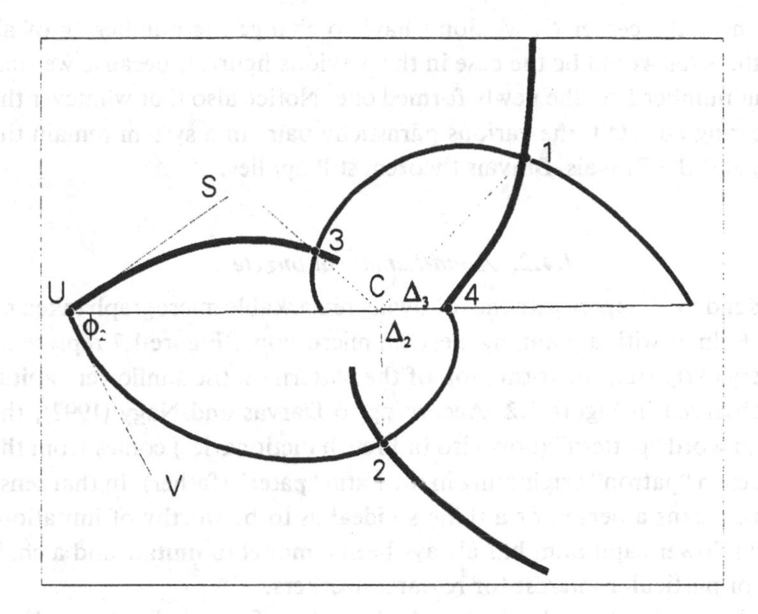

Figure 1.6. Centric representation of a plant pattern showing the spiral pair $(m, n) = (3, 2)$. We have, for example, $\Delta_3 =$ (angle at the center between consecutive primordia on any spiral in the family of 3 spirals) = angle $1C4$ = angle $UC3$, $\phi_2 =$ angle CUV, and $\Delta_2 =$ angle $2C4$ = angle $3C1$.

$$n \ln R = \Delta_n \cot \phi_n, \tag{1.8}$$

$$\cot \phi_m / \cot \phi_n = m\Delta_n / n\Delta_m. \tag{1.9}$$

To better understand the meaning of the symbols used, refer to Figure 1.6 where they are illustrated and use the following definitions. The Δ_m and Δ_n are the angles (in radians, of opposite signs between $-\pi$ and $+\pi$) made by the vectors issued from the pole of the spirals to the centers of two consecutive primordia on any spiral in the family of m and n spirals, respectively. Also γ (equal to angle VUS in Figure 1.6) is the angle of intersection between any two opposed spirals. We will come back to these relations from the point of view of the cylindrical representation of the spiral patterns in Chapters 2 and 4.

Notice that the numbering of the primordia in Figure 1.6 is different from the numbering in previous figures. In Figure 1.6, the oldest primordium (the farthest away from the center C) is numbered 1, and the youngest is numbered 4. This notation is more convenient developmentally and is now becoming generally adopted. Indeed, when a new primordium arises near the center C, we don't have to change the numbering of all the others (as would be the case in the previous figures), because we simply put number 5 on the newly formed one. Notice also that whatever the numbering adopted, the various parastichy pairs in a system remain the same, and the Bravais–Bravais theorem still applies.

1.4.2. A mathematical puzzle

Let us end this chapter with the following remarkable micrograph taken by J. H. Palmer with a scanning electron microscope. Figure 1.7 represents the very early stage of formation of the pattern on the sunflower capitulum observed in Figure 1.2. According to Darvas and Nagy (1992), the English word "pattern" (now also in French dictionaries) comes from the old French "patron" originating in the Latin "pater" (father). In that sense pattern means a person or a thing so ideal as to be worthy of imitation. The sunflower capitulum has always been a **model to imitate** and a challenge of particular interest for research workers.

Modern technology shows that the intricacy of the challenge has been underestimated. In particular, the primordial florets are initiated in rapid succession from the periphery to the center of the apex, and not the other way around as modelers often think. Even in this early stage, two families of contact parastichies can already be clearly observed. The primordia being quadrilateral, the parastichies made by the florets can be easily counted to reveal a (55, 89) contact parastichy pair.

Palmer's work on the sunflower shows that during the period of active floret production and the generation of the floret pattern, there is no central area or position with organizing ability, equivalent to the apex in a vegetative shoot. The center of the receptacle appears to have no controlling role since it can be isolated or cut out without changing the floret pattern or rate of floret production, whereas destruction of the apex in the vegetative shoot invariably arrests pattern production and formation of leaf primordia. A special problem in the phyllotaxis of the sunflower capitulum is to explain how the pattern-forming influences can interact together with sufficient rapidity to accommodate the high rate of floret production (2 or 3 per hour at peak stages), when the distances in the capitulum over which they must travel, diffuse, or operate are vast in developmental terms (2 mm or more).

Palmer and Steer (1985) have emphasized the dynamics of the floret-generation process by showing how the site of floret production, called

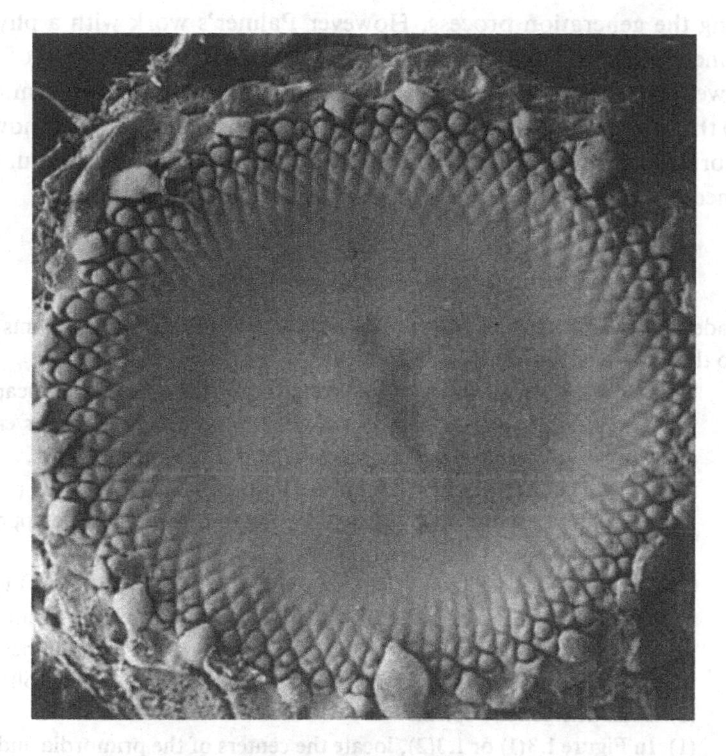

Figure 1.7. Scanning electron micrograph of a very young capitulum of *Helianthus annuus,* showing the process of floret initiation proceeding toward the center on the generative front with a remarkable degree of symmetry. The front surrounds the generative area which occupies the central region of the receptacle. The receptacle is about 2.5 mm in diameter and is covered by floret primordia for about 50% of its area. The 89 short contact parastichies contain four primordia and the 55 long parastichies, six or seven. (From J. H. Palmer, personal communication.)

the generative front, remains roughly constant in circumference and diameter for much of the floret generation period. Palmer regards the generative front as being equivalent to a standing wave, where the florets and associated tissue are formed as tissue moves through the front and is carried away outward by the production of new tissue on the inside of the front in the generative area.

An inherent assumption in approaches to phyllotaxis is that the generation of the lateral leaf or flower is a consequence of interacting influences at a position. This has lead mathematicians and botanists investigating phyllotaxis to study pattern production as though this was equivalent to studying the generation process. However Palmer's work with a phytohormone called benzyladenine (BA) clearly establishes the ability of the sunflower to generate florets at random positions in the capitulum. To add to the mystery, the surgical experiments of Hernandez (1988) showed that normal florets with irregular spacing can be produced without the influence of positional information from older parts of the shoot.

1.5. Problems

The reader who wants to check his or her ability with the following problems can refer to the solutions given in Appendix 2.

1.1 Count the spirals on cones of different conifers or on several pineapple fruits. Establish the numbers in the contact parastichy pair for each, and draw a frequency distribution diagram for the various pairs.

1.2 (1) In Figure 1.3(1) count the numbers in the families of contact parastichies and use the appropriate notation to record a visible opposed spiral pair.

(2) Use the Bravais–Bravais theorem to number the primordia in Figure 1.3(1). Hint: Put 1 on the primordium with a dot on it. Count the number of contact spirals in the two directions and put numbers on the two contacts of 1 marked with dots. Then the numbering follows.

(3) Draw the genetic spiral.

1.3 (1) In Figure 1.3(1) or 1.3(2), locate the centers of the primordia and the center of the shoot apex, and measure, using a millimeter ruler, the plastochrone ratio for successive pairs of as many consecutive primordia as possible (using Problem 1.2). Then calculate the mean value for this ratio. You may find it useful to work on enlarged figures made on an enlarging photocopier.

(2) Using a protractor, measure the divergence angle for as many primordia as you can, and calculate the mean value.

(3) As all plant material is variable, it is essential to know the degree of variability around a mean; this is achieved by calculating the standard deviation of the mean value. Use the individual determinations to calculate the estimated standard deviation

$$S=\sqrt{[\Sigma(x_i-X)^2/(N-1)]},$$

for each of the two mean values above, where x_i is the individual value of any determination, $i=1$ to N, X the mean value and N the total number of determinations that have been made.

1.4 (1) Determine F_{16} and F_{21}.

(2) Write out the first few terms of the Lucas sequence.

(3) A definition of the golden ratio uses the circumference of a circle. Suppose that the length of the smaller portion of the circumference

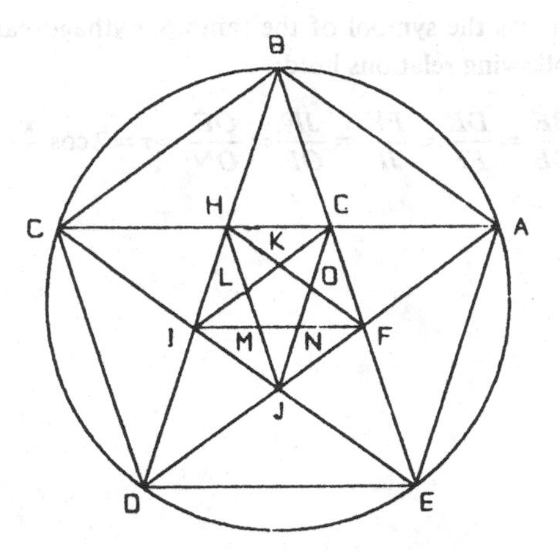

Figure 1.8. Representations of the regular pentagon and the regular pentagram.

is equal to 1 and that the larger portion has the value x. Perform the operations given in the definition to obtain that $x=\tau$.

1.5 Use the calculator to obtain results that verify relation (1.4) between the golden ratio τ and the Fibonacci sequence; conclude that $(F_k/F_{k+1})^2 \approx 1/\tau^2$, for large k. Notice that as k gets very large the approximation gets better.

1.6 Assuming that a plant shows the parastichy pair (F_k, F_{k+1}) and a divergence angle of $360(1/\tau^2)°$, prove the following relation between these observations: $1/\tau^2 = \lim[1-1/(F_{k+1}/F_k)] = \lim F_{k-1}/F_{k+1}$.

1.7 In Figure 1.6 measure the divergence angle d, the plastochrone ratio R, the various angles, and verify that (1.5) through (1.9) hold.

1.8 With various Fibonacci pairs $(m,n)=(F_k, F_{k+1})$, verify that the relation $mv-nu=\pm 1$ holds in the centric representation when $d=\tau^{-2}$. Do the same with consecutive Lucas pairs and the angle $(3+1/\tau)^{-1}$.

1.9 (1) Prove that $m\Delta_n - n\Delta_m = \pm 2\pi$.

(2) Use part 1 to prove that $\Delta_n\Delta_m(\cot\phi_m - \cot\phi_n) = \pm 2\pi \ln R$.

(3) Use part 2 to prove the following classical relation:

$$mn\ln^2 R - 2\pi\ln R\cot\gamma + \Delta_n\Delta_m = 0. \tag{1.10}$$

(This relation will be obtained in Chapter 4 from another point of view.)

(4) Put $\gamma = 90°$ in (1.10) to obtain what I called **Richards's formula**:

$$\ln R = (-\Delta_n\Delta_m/mn)^{1/2}. \tag{1.11}$$

1.10 Figure 1.8 is extremely rich in golden ratios. It represents the regular pentagram (starred pentagon) inscribed in the regular pentagon. The pentagram was the symbol of the famous Pythagorean Society. Show that the following relations hold:

$$\frac{BE}{DE}=\frac{DE}{FE}=\frac{FE}{JF}=\frac{JF}{OF}=\frac{OF}{ON}=\tau=2\cos\frac{\pi}{5}$$

2
The fundamental theorem and its applications

2.1. A cornerstone in phyllotaxis – insight into history

This chapter develops, intuitively and formally, a theorem of fundamental importance in phyllotaxis, and shows applications of this theorem. Other applications will be given in subsequent chapters. The theorem establishes the complete geometrical relation between two concepts presented in Chapter 1: the visible opposed parastichy pairs (m, n) and the divergence angle d of a system. From it I devised four algorithms to obtain easily one parameter from the other. The theorem expresses essential properties of the lattices used in most of the explanatory models and attempts to understand phyllotaxis. Chapter 3 presents an illustration of the theorem, which throws light on the Lestiboudois–Bolle theory of vascular phyllotaxis. This illustration opens the way to a third representation of phyllotaxis, the hierarchical representation that plays an important role in the interpretative model of phyllotaxis, the subject of Chapter 6. We will see in Chapters 5 and 7 that the theorem is involved in practical pattern assessment.

The lack of such a theorem has been a major reason for errors in mathematical and biological phyllotaxis, because of the confusion generated by the large number of families of spirals observed on a single plant. A family of parastichies can be generated simply by linking two points in the lattice representing the pattern on a plant, and the family partitions all the points of the lattice. From the very beginning of the investigations in phyllotaxis (Bonnet, 1754; Goethe, 1790), the multiplicity of spirals puzzled its observers. The fundamental theorem and its particular forms clarify the meaning of these families of parastichies, settling a two-century-old debate.

Braun (1831) used to say that only the genetic spiral exists, all the others being illusory. He stuck to that belief and built around it the genetic spiral theory, which has been abandoned today. Church (1920) chose among the many pairs of families the orthogonal family (angle of intersection γ of parastichies $\approx 90°$). He stuck to his choice too, and built around it his equipotential theory which fell into oblivion. (Some aspects of this theory are still very relevant as we will see in Chapter 12.) D'Arcy Thompson (1917) proclaimed that there was no reason to prefer one parastichy or family of parastichies, and concluded, under the influence of the botanist Sachs (1882), that there was an irreducible subjectivity transforming the whole subject into mystical idealism and fantastic speculations.

As an answer to Thompson, Plantefol (1948) looked for a set of parastichies having a biological reality. How to select it among all the possibilities is, according to LeGuyader (1985), the question that preoccupied Plantefol. According to LeGuyader, the main achievement of Plantefol was to propose a family of parastichies, which he called foliar helices, having an ontological reality that the other families do not have, and to rectify the error made by Schimper and Braun who confused the ideas of chronological and ontological spirals. Plantefol's hypothesis of foliar helices, of their contiguity, and of their generative centers is still very popular in France. The notion of the generative center is, whatever its merits and biological support, a hypothesis. Many say that Plantefol's views considerably retarded the progress of morphology in France. It is certain that his hypothesis distracted attention from the biomathematical advances made by the French pioneers Bravais and Bravais (1837) and Lestiboudois (1948). Bravais and Bravais introduced the cylindrical representation of phyllotaxis, and recognized the relevance of continued fractions in the area. Following their lead, many workers in the field brought biomathematical insights into the problems of phyllotaxis. (Coxeter 1972; Adler, 1974, 1977a; Ridley, 1982a; Marzec & Kappraff, 1983; Jean, 1983a, 1987a, 1988a, see Chapters 4 and 5, Appendix 4; Rothen & Koch, 1989a.)

We must recognize that the genetic spiral has a reality (to use Plantefol's word), that the orthostichies (particular sets of parastichies parallel to the axis of a plant from which phyllotactic fractions are obtained) have a biological foundation, and that the measurements from which the Fibonacci angle is obtained have a strong reality. For Larson (1983) the genetic spiral and the orthostichies have real structural meanings in vascular phyllotaxis (see his statements in Section 3.2.3). We must accept that all these families of parastichies, which have puzzled so many investigators, have, de facto, a reality from which we cannot escape. At least from a geometrical point of view, there is no reason to stick to one spiral or another. Thompson was correct when he chose all of them, though this choice drove him out of the discussion. The idea is to have each and every family of parastichies contribute whatever information they can to the phyllotactic system as a whole. The fundamental theorem of phyllotaxis is a major step in that direction. It says for example that a contact parastichy pair gives an indication of the value of the divergence angle of the system, and that some visible opposed parastichy pairs give better indications of this value.

Although the theorem handles the many families of parastichies, some confusion remained even after the publication by Adler (1974) of an important particular case of it. Moreover, most of the subsequent reviews concerned with the state of development of the subject, mainly on biological phyllotaxis (such as Rutishauser, 1981; Erickson, 1983; Schwabe, 1984; Carr, 1984; Steeves & Sussex, 1989, pp. 109–23), ignore the result. A good reason seems to be that Adler's presentation of the theorem is intertwined with the complexities of continued fraction, and is hidden in an argumentation meant to support his model of contact pressure for pattern generation. The result was rediscovered independently (Jean, 1986a) as a consequence of the proof of what I called the Bravais–Bravais formula for approximating the divergence angle of a system, a formula developed in Section 2.5.2. A new proof of the theorem was then proposed using elementary properties of similar triangles in the cylindrical representation of phyllotaxis, a representation equivalent to the centric representation of Chapter 1 and defined in Section 2.5.1.

In mathematics many proofs are frequently given to theorems which are fundamental in the sense that they ramify in the very tissue of mathematics and its applications. The theorem object of the present chapter is fundamental in the full meaning of the word. It relates mathematical and biological phyllotaxis, and is compulsory to master the field. A general proof of it is sketched in Appendix 4 in the context of Farey sequences. Goodal (1991) uses the result in his eigenshape analysis of the cut-grow algorithm in phyllotaxis (presented in Chapter 12), and refers to it as the Adler–Jean theorem.

2.2. Introduction to the theorem

2.2.1. Visible opposed parastichy pairs for the Fibonacci angle

Chapter 1 suggests that a geometrical relation exists between the visible opposed parastichy pairs (F_k, F_{k+1}) and the Fibonacci angle $1/\tau^2$ ($\approx 137.5°$) representing the divergence angle between successive primordia. To put

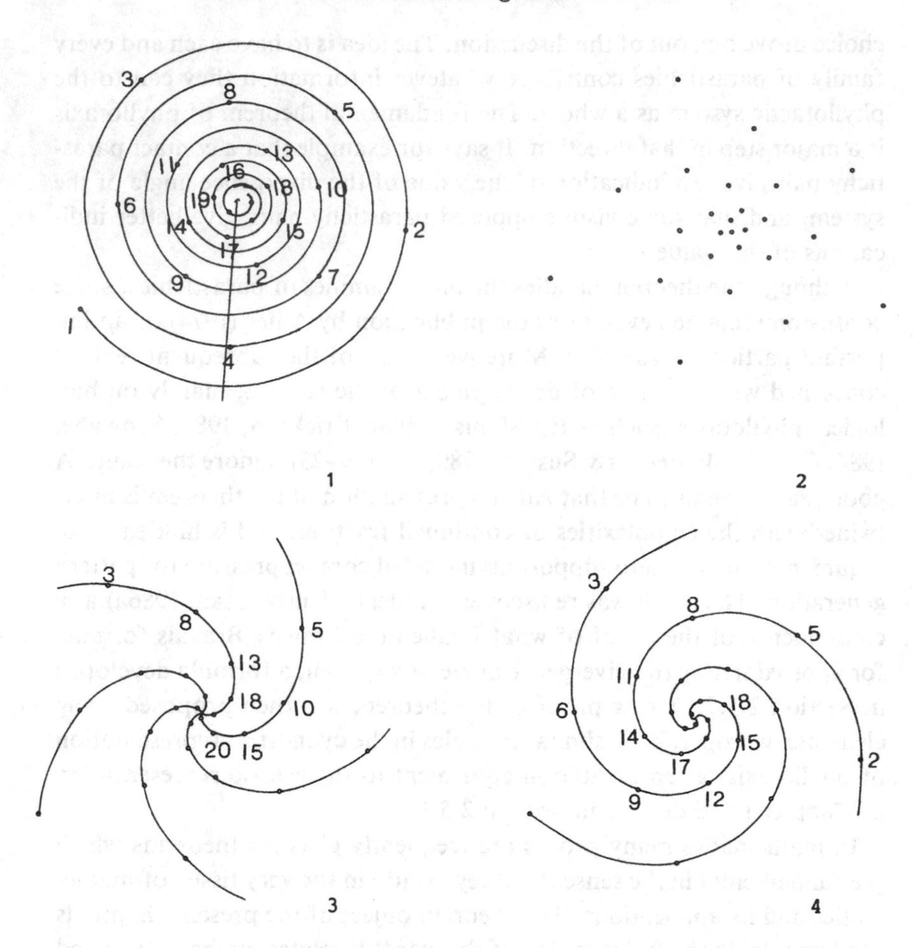

Figure 2.1. (1) A spiral on which points are distributed at $137\frac{1}{2}°$ with respect to the pole. (2) Cluster of points obtained by erasing the spiral. (3) and (4) Opposed families of parastichies containing respectively 5 and 3 spirals, drawn with this set of points. In the first case, the points are linked from 5 to 5; in the second from 3 to 3, in agreement with the Bravais–Bravais theorem.

this relation into better light, we draw a spiral having a small pitch, like the one in Figure 2.1(1). Then we put points around it – let us say 30 – at $137\frac{1}{2}°$ from one another, and number the points consecutively. These points could represent the loci of insertion of the leaves on a cylindrical leafy stem, transformed into a cone and viewed from above, that is projected in the plane. Or alternatively they could represent the florets in a spiral lattice representing a daisy capitulum. Now we erase the spiral and make 12 or 15 copies of the set of points on transparencies. On each copy we draw a family of x spirals with the same pole, going through the points and making a partition of the set of points. This can be done easily by linking the points so that the numbers on them differ by 2, 3, ..., Figures 2.1(3) and 2.1(4) give two examples with $x=3$ and $x=5$, where the points on any spiral are seen to differ by 3 and 5 respectively, giving 3-parastichies and 5-parastichies.

It can be seen in this way that x can only have the following values: 1 [the genetic spiral: Fig. 2.1(1)], 2, 3, 5, 7, 8, 11, 12, 13, 14, 17, ... A number N is excluded from this ordered list of numbers, when it is a multiple of a number x in the list and when it determines a set of parastichies winding in the same direction as the set of x parastichies. Now we superimpose two copies of the set of points, in which the families of spirals wind in opposite directions, and we look for the parastichy pairs such that at every intersection of any two spirals there is a point of the set. This gives us the visible opposed parastichy pairs. It is seen that the only possible visible opposed pairs are (2, 1), (2, 3), (5, 3), (5, 8), (13, 8), ..., all made with consecutive Fibonacci numbers.

2.2.2. Phyllotactic fractions associated with the Fibonacci angle

Now let us consider Figure 2.1(1) from another point of view. We take a point, say 4, and another near the line joining 4 to the pole, say 7. Following the genetic spiral, we go from 4 to 7. We have to make about 1 turn around the pole and we meet 3 points, excluding point 4. We say that 1/3 is a phyllotactic fraction of the system. Take again point 4 and another point near the line, say 9. To go from 4 to 9 along the genetic spiral we must make about 2 turns and we meet 5 points. The value 2/5 is another phyllotactic fraction. The same process made with point 4 and point 12, and with point 4 and point 17, which lie more and more closely to the straight line connecting point 4 and the pole, gives the phyllotactic fractions 3/8 and 5/13, respectively. Notice that using point 2, instead of 4, and the consecutive points 7, 10, 15, 23 that get closer and closer to the segment linking 2 to the pole give the same fractions, plus 8/21. On a long leafy stem of willow, for example, with a lot of buds on it, we can even obtain phyllotactic fractions such as 13/34.

Phyllotactic fractions are approximations of the divergence angle of a system. The simple process described above is the way to derive them. We just have to look for two leaves on the cylindrical stem or for two primordia that appear to be approximately one above the other with respect to the center of the apex, and we go from one to the other by proceeding along the genetic spiral and by registering the number of turns and the number of primordia met, excluding the first one. In trees the commonly occurring phyllotactic fractions belong to the sequence ⟨1/2, 1/3, 2/5, 3/8, 5/13, ...⟩. Of course the growth of the specific branch should not have distorted too much the initial relations in the bud that gave rise to that particular branch. The sequence is generated by adding two consecutive numerators and two consecutive denominators, so that the next fraction in the sequence is 8/21. It is easily proved that this sequence converges toward $1/\tau^2 \approx 0.381$, corresponding to the divergence angle of 137.5° used to build the spiral lattice of Figure 2.1.

The particular fractions in this example are alternately smaller and greater than τ^{-2}. Going around the pole of the spiral lattice in Figure 2.1(1) to obtain the consecutive phyllotactic fractions produces an oscillatory movement of decreasing amplitude around the line joining, for example, point 4 to the pole. This means that each one of the closed intervals [1/3, 2/5], [3/8, 2/5], [3/8, 5/13], [8/21, 5/13], brackets the divergence τ^{-2}. In degrees these intervals are [120, 144], [135, 144], [135, 138.46], [137.14, 138.46], nested around $137\frac{1}{2}°$ and squeezing that angle more and more. If a spiral with a larger number of points is built, more nested intervals are obtained, their lengths tending to become zero and their intersection tending to contain only the value τ^{-2}.

To summarize, the preceding two processes lead to a sequence of nested intervals and to a sequence of visible opposed parastichy pairs. The numbers in the pairs are the denominators of the end points of the intervals. For example, the visible opposed parastichy pair (3, 5) corresponds to the interval [1/3, 2/5]. The theorem in the next section says that there is a one-to-one correspondence between the sequence of intervals and the sequence of visible opposed parastichy pairs.

2.3. The fundamental theorem of phyllotaxis

2.3.1. Various forms

General form (Jean 1988a)

Let (m, n) be a parastichy pair, where m and n are relatively prime, in a system with divergence angle d. The following properties are equivalent:

(1) There exists unique integers $0 \le v < n$, and $0 \le u < m$ such that $|mv - nu| = 1$, and $d \le \frac{1}{2}$ is in the closed interval whose end points are u/m and v/n;

(2) The parastichy pair (m, n) is visible and opposed.

The proof of this theorem is based on an elegant result on visible parastichy pairs. The result (Jean, 1988a) is that the pair (m, n) is visible if and only if $|m(nd)-n(md)|=1$, where (xd) is the integer nearest to xd, for $x=m$ and n. It suggests an alternative more manageable definition, presented in Section 2.5.1, of a visible parastichy pair, which generalizes the notion of visible point in a lattice introduced by Hardy and Wright (1945).

Particular forms

1. For the **sequence of normal phyllotaxis** (Adler, 1974) that is the Fibonacci-type sequence

$$J\langle 1, t, t+1, 2t+1, 3t+2, 5t+3, \ldots\rangle, \tag{2.1}$$

where $t \geq 2$ and $J \geq 1$ are integers, J multiplying all the terms of the sequence, and where $N_{t,k}=F_k t+F_{k-1}$ is the general term:

the parastichy pair $(JN_{t,k}, JN_{t,k+1})$ is visible and opposed,
if and only if
d is in the interval whose endpoints are $F_k/JN_{t,k}$ and $F_{k+1}/JN_{t,k+1}$.

It follows that for any integers $t \geq 2$, $J \geq 1$, we have in particular that

the divergence is in the interval $[1/J(t+1), 1/Jt]$
if and only if
the parastichy pair $J(t, t+1)$ is visible and opposed.

The divergence is in the interval $[1/J(t+1), 2/J(2t+1)]$ if and only if $J(2t+1, t+1)$ is a visible and opposed parastichy pair. Continuing to add numerators and denominators in the last interval, so that the denominators are always consecutive terms of the sequence, we establish that the divergence is in the interval $[3/J(3t+2), 2/J(2t+1)]$ if and only if $J(2t+1, 3t+2)$ is a visible opposed parastichy pair, and so on.

The Fibonacci sequence corresponds to $t=2$ and $J=1$. It gives the following theorem which will be referred to as **Adler's theorem**:

The parastichy pair (F_k, F_{k+1}) is visible and opposed
if and only if
d is equal to F_{k-2}/F_k or to F_{k-1}/F_{k+1} or is between these two values.

We already know that the two ratios in Adler's theorem come very close to $1/\tau^2$ as k increases. The latter number is in the interval defined by the ratios. The theorem says in particular that:

$(8, 5)$ is a visible opposed pair if and only if $d \in [3/8, 2/5]$,
$(13, 8)$ is a visible opposed pair if and only if $d \in [3/8, 5/13]$.

That is why in Section 1.2 we said that a given visible opposed parastichy pair conveys the knowledge that the divergence is to be found in a certain range of values, a result already experimentally known to Richards (1948). As another example, if the visible opposed parastichy pair (21, 34) is observed for the florets or seeds in a sunflower head, then we know that the divergence angle for that sunflower head is very close to $360(1/\tau^2)°$, given that it is in the interval [8/21, 13/34], that is [137.14, 137.65°]. Conversely, if measured values of the divergence angle cluster around $137\frac{1}{2}°$, then the visible opposed pairs of the system will consist of consecutive Fibonacci numbers. That is why we observed only visible opposed parastichy pairs having secondary numbers equal to consecutive Fibonacci numbers when we worked on Figure 2.1.

A result similar to Adler's theorem can be formulated for the Lucas sequence, corresponding to $t=3$ and $J=1$ in the sequence of normal phyllotaxis. Notice that the values $t=1$ and $J=1$ give the angle of $222\frac{1}{2}°$ which is the complement of the Fibonacci angle over 360°. The botanical meaning of $J>1$ is in the existence of multijugate and multimerous patterns introduced in Sections 6.4.2 and 8.1, respectively.

2. For the **sequence of anomalous phyllotaxis**, that is the Fibonacci-type sequence

$$\langle 2, 2t+1, 2t+3, 4t+4, 6t+7, \ldots \rangle, \tag{2.2}$$

where $t \geq 2$ is an integer:

the pair $(2t+1, 2t+3)$ is visible and opposed
if and only if
d is in the interval $[t/(2t+1), (t+1)/(2t+3)]$;

the pair $(4t+4, 2t+3)$ is visible and opposed if and only if d is in the interval $[(2t+1)/(4t+4), (t+1)/(2t+3)]$, etc.

The general form of the end points of the intervals for normal and anomalous phyllotaxis is $F_k/JN_{t,k}$ and $N_{t,k}/(2F_kt+F_{k+1}+F_{k-1})$, respectively. Taking the limit as k tends towards infinity gives the **limit-divergence angles** of normal and anomalous phyllotaxis, that is $360(J(t+\tau^{-1}))^{-1}$ and $360(2+(t+\tau^{-1})^{-1})^{-1}$ (see Table 2.1). In Part II we will learn how sequences (2.1) and (2.2) arise.

2.3.2. Useful algorithms relating d and (m, n)

Four algorithms may be derived from the fundamental theorem that allow us to determine an interval for d from the visible opposed parastichy pair

Table 2.1. *Examples of limit-divergence angles corresponding to patterns of normal and anomalous phyllotaxis*

Sequence	Limit divergence angle
⟨1, 2, 3, 5, 8, 13, ...⟩	137.51° (normal, $J=1$, $t=2$)
⟨1, 3, 4, 7, 11, 18, ...⟩	99.50° (normal, $J=1$, $t=3$)
⟨1, 4, 5, 9, 14, 23, ...⟩	77.96° (normal, $J=1$, $t=4$)
⟨1, 5, 6, 11, 17, 28, ...⟩	64.08° (normal, $J=1$, $t=5$)
⟨1, 6, 7, 13, 20, 33, ...⟩	54.40° (normal, $J=1$, $t=6$)
⟨1, 7, 8, 15, 23, 38, ...⟩	47.25° (normal, $J=1$, $t=7$)
⟨2, 5, 7, 12, 19, 31, ...⟩	151.14° (anomalous, $t=2$)
⟨2, 7, 9, 16, 25, 41, ...⟩	158.15° (anomalous, $t=3$)
⟨2, 9, 11, 20, 31, 51, ...⟩	162.42° (anomalous, $t=4$)
2⟨1, 2, 3, 5, 8, 13, ...⟩	68.75° (normal, $J=2$, $t=2$)
2⟨1, 3, 4, 7, 11, 18, ...⟩	49.75° (normal, $J=2$, $t=3$)
3⟨1, 2, 3, 5, 8, 13, ...⟩	45.84° (normal, $J=3$, $t=2$)

(m,n), and visible opposed parastichy pairs (m,n) from a value of d. This section presents two of them. Appendix 4 presents the other two. The knowledge of the computational algorithm in particular will be of help in following the discussion on practical pattern recognition in Chapters 5 and 7. The algorithms are:

Computational algorithm: given d, all the visible pairs are obtained.
Diophantine algorithm: given any visible opposed pair, an interval for d is obtained (Appendix 4).
Contraction algorithm: given a visible opposed pair made with secondary numbers in the sequences of normal and anomalous phyllotaxis, an interval for d is obtained.
Graphical algorithm: given d the visible pairs are obtained (Appendix 4).

Let us work again with $d=\tau^{-2}$, to show how it is possible to derive elegantly and easily the same visible opposed parastichy pairs as in Section 2.2, this time using a calculator instead of geometry.

1. **Computational algorithm** (for the determination of the visible opposed parastichy pairs from a value of the divergence angle). Let us suppose that the divergence angle of a regular lattice is $d=\tau^{-2}$. For consecutive values of $k=1,2,3,4,\ldots$, determine the sequence A of fractions $(kd)/k$ in their lowest terms, where (kd) is the integer nearest to kd. If $kd=I+0.5$ where I is an integer, then choose both I and $I+1$. The

sequence A up to $k=21$ is the following (if in the process fractions are repeated, we take the first occurrence only):

$$\frac{0}{1}, \frac{1}{2}, \frac{1}{3}, \frac{2}{5}, \frac{3}{7}, \frac{3}{8}, \frac{4}{11}, \frac{5}{12}, \frac{5}{13}, \frac{5}{14}, \frac{6}{17}, \frac{7}{18}, \frac{7}{19}, \frac{8}{21}.$$

Consider now the integer x which is the first one giving $(xd)=1$. Here $x=2$ given that 2 times d is approximately equal to 0.76 and that (0.76) is equal to 1. Now order the fractions above by increasing values, by inserting them consecutively between 0/1 and $(xd)/x=1/2$, taking into consideration that the next fraction to be placed (here 1/3, then 2/5, etc.) is the Farey sum of two fractions already ordered [the Farey sum of the fractions p/q and s/t is the fraction $(p+s)/(q+t)$]. For example we have $1/3=(0+1)/(1+2)$, and $2/5=(1+1)/(3+2)$, etc. It is known that the Farey sum of two fractions is between those fractions. We obtain:

$$\frac{0}{1} \;\Big|\; \frac{1}{2},$$

$$\frac{0}{1}\;\frac{1}{3} \;\Big|\; \frac{1}{2},$$

$$\frac{0}{1}\;\frac{1}{3} \;\Big|\; \frac{2}{5}\;\frac{1}{2},$$

$$\frac{0}{1}\;\frac{1}{3} \;\Big|\; \frac{2}{5}\;\frac{3}{7}\;\frac{1}{2},$$

$$\frac{0}{1}\;\frac{1}{3}\;\frac{3}{8} \;\Big|\; \frac{2}{5}\;\frac{3}{7}\;\frac{1}{2},$$

$$\vdots$$

$$\frac{0}{1}\;\frac{1}{3}\;\frac{6}{17}\;\frac{5}{14}\;\frac{4}{11}\;\frac{7}{19}\;\frac{3}{8}\;\frac{8}{21} \;\Big|\; \frac{5}{13}\;\frac{7}{18}\;\frac{2}{5}\;\frac{5}{12}\;\frac{3}{7}\;\frac{1}{2}.$$

The consecutive denominators m and n of the two fractions of A on both sides of the vertical segment representing the value of d, at any step of the process, give the visible opposed pair (m, n). The visible opposed pairs are thus here (1, 2), (3, 2), (3, 5), (8, 5), (8, 13), and (21, 13). Consecutive denominators at every step give visible pairs [e.g., (3, 8), (5, 12)]. If d is the rational number p/q, there are two sequences of visible opposed pairs, given that point p/q is on the vertical axis. If d is an irrational number, as in the example here, there is only one sequence of visible opposed pairs, each pair in the sequence being made, in the example, with consecutive Fibonacci numbers.

2. **Contraction algorithm** (for the determination of an interval for the divergence from a visible opposed parastichy pair in the cases of normal or anomalous phyllotaxis). Let us call the **contraction of the visible opposed parastichy pair** (m, n), the pair $(m, n-m)$ if $n>m$, and the pair $(m-n, n)$ if $m>n$ (Adler, 1974). It is known that the contraction of a visible opposed parastichy pair is a visible opposed parastichy pair (see Appendix 4).

Suppose that the pair (19, 31) is a visible opposed parastichy pair. We are looking for an interval whose end points have 19 and 31 as denominators. Take the consecutive contractions of that pair up to the point that a pair of the type $J(t, t+1)$ or $(2t+1, 2t+3)$, given in the particular forms of the theorem, is obtained:

$$
\begin{array}{lcr}
(19, 31) & \leftrightarrow & [13/31, 8/19]. \\
\downarrow (19, 12) & \leftrightarrow & [5/12, 8/19] \uparrow \\
\downarrow (7, 12) & \leftrightarrow & [5/12, 3/7] \uparrow \\
\multicolumn{3}{l}{\downarrow (7, 5), \text{ then } t = 2 \text{ and } d \text{ is in } [2/5, 3/7] \uparrow}
\end{array}
$$

We notice that the secondary numbers in the given pair are consecutive terms in the sequence of anomalous phyllotaxis $\langle 2, 5, 7, 12, 19, 31, \ldots \rangle$, and that the pair (7, 5) is of the form $(2t+3, 2t+1)$ for $t = 2$. Then we write the corresponding interval, that is $[t/(2t+1), (t+1)/2t+3)] = [2/5, 3/7]$ given by the theorem, and then take the Farey sums of the end points as above. It follows that the divergence is in [13/31, 8/19], that is between the values 150.97° and 151.58° enclosing the limit divergence of anomalous phyllotaxis with $t = 2$, that is 151.14° (see Table 2.1). If the secondary numbers in the visible opposed pair lead to the contraction $J(t, t+1)$, then we take Farey sums as above, starting with the interval $[1/J(t+1), 1/Jt]$ given by the theorem for normal phyllotaxis.

2.4. Interpretation of spiromonostichy in *Costus* and *Tapeinochilus*

Spiromonostichy, also called heliomonostichy, is a type of phyllotactic pattern observed in the shoots of plants such as *Costus* and *Tapeinochilus*. It is illustrated in Figure 2.2. This type of pattern is a genuine puzzle in that it violates **Hofmeister's axiom** on which most of the models for pattern generation are based. This axiom states that each leaf arises in the largest gap between the existing leaves or primordia, and as far away from them as possible. In spiromonostichy, on the contrary, the primordia are initiated close to one another along a single (mono) spiral (spiro).

Another problem raised by *Costus* was the great variability of the divergence angle in a single specimen. This angle appears to be between 1/8 and 1/3, that is, in the interval [45°, 120°]. But Snow (1952) showed that these fluctuations were based on transverse sections made too far below the apex. He measured the divergence angles in two apices through the level of insertion of leaf #1. The results were as follows (in degrees):

apex 1: 45, 50, 46, 49, 49 – mean value 47.8,
apex 2: 50, 47, 48, 46, 47 – mean value 47.6.

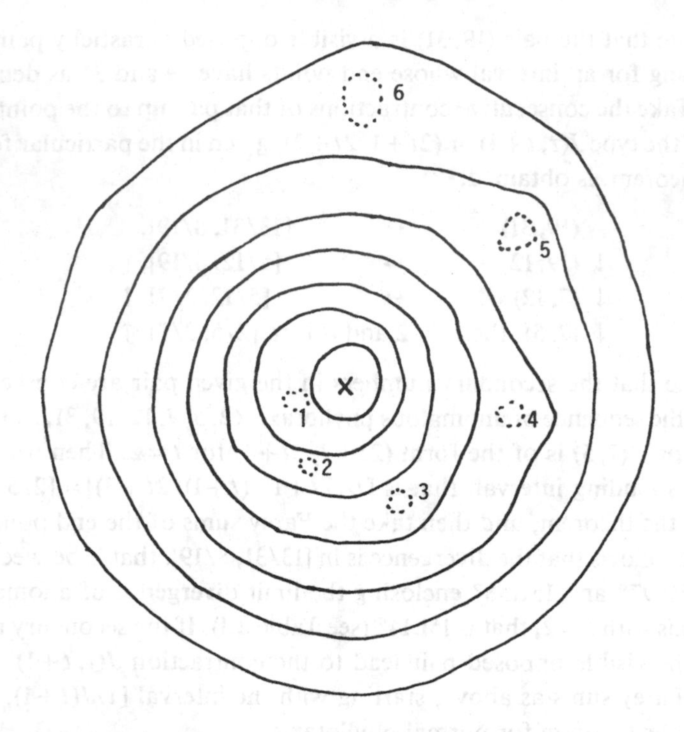

Figure 2.2. View of *Costus* shoot from above showing six leaf primordia.

The mean value of all measurements is $d = 47.7°$. There is a unique integer t such that $360/(t+1) < d < 360/t$ (this is an interval for d corresponding to a case of normal phyllotaxis). We find $t = 7$. Notice that each one of the ten angles are in the interval $[360°/8, 360°/7] = [45°, 51.4°]$. According to the particular case of the fundamental theorem for normal phyllotaxis, the pair $(t+1, t) = (8, 7)$ is a visible opposed parastichy pair. The mean value is also seen to be in the interval $[360/8°, 360(2/15)°] = [45°, 48°]$, so that the pair $(8, 15)$ is also visible and opposed. The angle corresponding to the sequence $\langle 1, 7, 8, 15, 23, 38, \ldots \rangle$ is $47.26°$ (see Table 2.1), a value close to the mean value of all the measurements.

Consequently, I propose that spiromonostichy among the leaves in shoots of *Costus* and *Tapeinochilus* is a type of normal phyllotaxis defined by the sequence $\langle 1, 7, 8, 15, 23, 38, \ldots \rangle$. The reasons why this has not been recognized earlier might be the following: (1) the interval of time between the initiation of two consecutive leaves is so long that there is always only one leaf primordium at the shoot apex, the others, being well differentiated are no longer part of the apical meristem (Smith, 1941).

(2) The genetic spiral is the most visible spiral to the eye, given that the leaves are only 47° apart. (3) As many as 15 primordia are needed for the system (8, 7) to be conspicuous. According to Smith, the effect of the long time interval is to remove the hypothetical competitive factor that may govern the relative positions of the leaf primordia at the apex. The new primordium would just tend to arise in the part of the apical meristem most favorable for its growth, irrespective of the positions of the previous primordia. We will come back to this case in Section 8.6, where recent data will be analyzed.

2.5. Bravais–Bravais approximation formula

2.5.1. The cylindrical lattice

Bravais and Bravais (1837) intuitively found an approximation formula for the divergence angle. The formula, expressing a particular phyllotactic fraction, is:

$$d \approx (tu+sv)/(tm+sn), \tag{2.3}$$

where u, v, m, and n are the integers given in the general form of the theorem, and t and s are particular integers obtainable from the Bravais–Bravais lattice for the study of phyllotaxis. This cylindrical lattice will now be defined.

The spruce cone depicted in Figure 2.3(1) may be thought of as a cylinder. The surface of the cylinder is generated by a line, called the generator, moving parallel to itself in the three-dimensional space along the circumference of a circle which is the base of the cylinder. Cutting a cylindrical cone along a generator, represented by the Y axis and the line $X=1$ in Figure 2.3(2), and unrolling it in the plane, gives the cylindrical representation of the cone. In this representation the scales are represented by their centers as points. The lattice is assumed to be regular, that is the coordinates (d,r) of point 1 are sufficient to build the whole lattice of points. The divergence angle d is constant and is the horizontal distance between any two consecutive points on the genetic spiral. The parameter r is called the **rise** and corresponds to the plastochrone ratio in the centric representation. It is the vertical distance between any two consecutive points n and $n+1$ on the genetic spiral. In the cylindrical lattice the primordia are evenly distributed within a vertical strip of width equal to 1, and the parastichies are represented as parallel, evenly spaced straight lines. The coordinates of point n are $(nd-(nd), nr)$, where (nd) is the

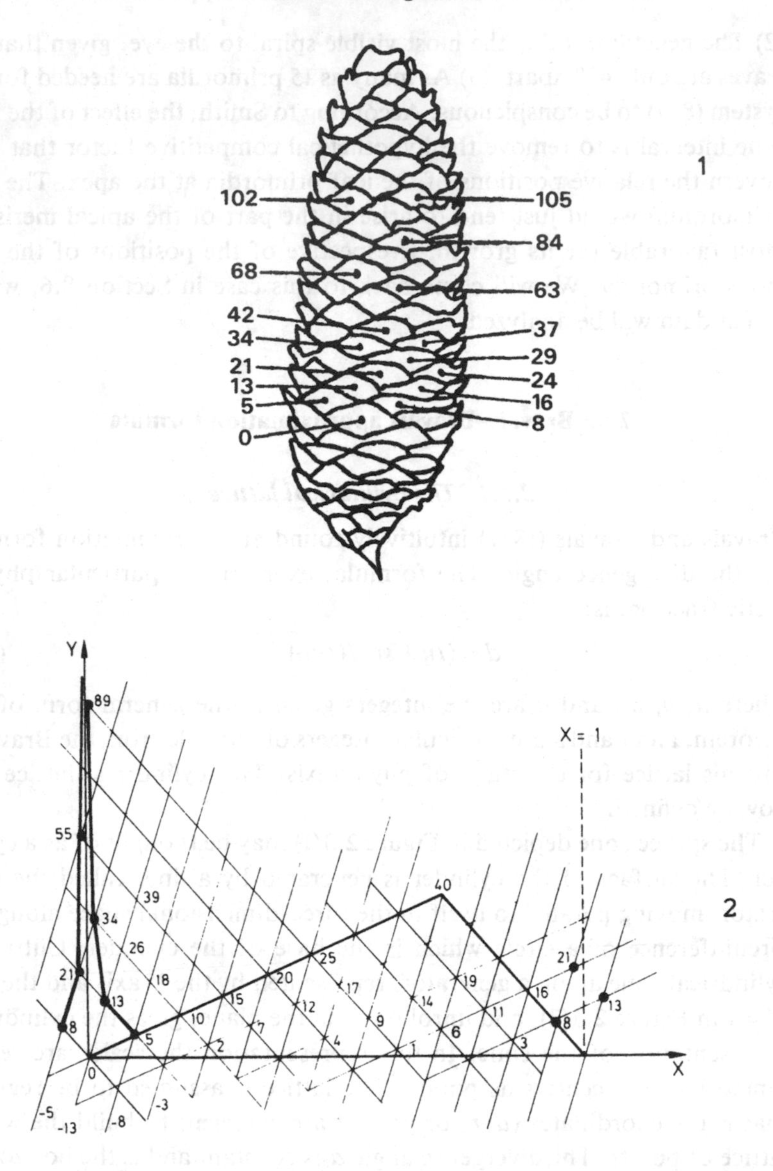

Figure 2.3. (1) Spruce cone showing (8, 5) contact parastichy pair and numbered according to the Bravais–Bravais theorem. (2) Planar representation of the cone, called the cylindrical representation or the Bravais–Bravais lattice. The horizontal unit measure is the distance between two representations of the same point. Because of the value of the divergence angle used to build the lattice (τ^{-2}), the consecutive neighbors of the Y axis alternate on each side of the axis while approaching it.

integer nearest to nd. The abscissa of n is called the **secondary divergence** of point n, and is a value in the closed interval $[-0.5, 0.5]$. (Working with secondary divergences in the range $[0, 1]$, as in past analyses, is a source of potential confusion in the determination of the visible parastichy pairs.)

In this representation a parastichy pair (m, n) is opposed if points m and n are on different sides of the Y axis. Working in the vertical strip of points with abscissas between $-1/2$ and $+1/2$, **a pair** (m, n) **is visible** if the triangle made with points 0, m, and n does not contain another point of the lattice on and inside it (Jean, 1988a). This definition of a visible parastichy pair is equivalent to the one given in Chapter 1 (see also the Glossary definition for this term). The two points m and n that are nearest to the origin of coordinates determine the conspicuous pair (m, n). This pair is visible and opposed, and the angle of intersection γ of the opposed parastichies in the pair is closer to 90° than the angle of intersection in any other visible opposed parastichy pair of the system.

2.5.2. *Derivation of the formula*

On a cone like the one in Figure 2.3(1), we cannot see the internodes between the scales, as we can see the internodes between leaves on a stem. But nevertheless we can determine a phyllotactic fraction in the following way. Choose a scale at the bottom of the cone and put 0 on it. Then number the scales according to the Bravais–Bravais theorem presented in Chapter 1 using the knowledge that the cone has a (8, 5) contact parastichy pair, $m = 8$, $n = 5$. The consecutive scales on an 8-parastichy differ by 8 and on a 5-parastichy they differ by 5. To obtain a phyllotactic fraction we must proceed from scale 0 to another scale which seems approximately above 0 with respect to the axis of the cone, such as scale 21.

To better follow the line of reasoning, let us move on to Figure 2.3(2) showing the parastichy pair (8, 5). In order to reach scale 21 (for example), we have to go through $tm + sn$ $(= 21)$ scales, excluding 0, and we make $((tm + sn)d)$ turns around the cone (cylinder), where the outer parentheses mean the integer nearest to $(tm + sn)d$. In the case of scale 21 we have $t = 2$ and $s = 1$. Indeed, the cylindrical lattice shows that we must make $s = 1$ step in the direction of the 5-parastichies, from 0 to 5, and $t = 2$ steps in the direction of the 8-parastichies, from 5 to 13 and from 13 to 21, to proceed from 0 to 21. More generally it follows that the phyllotactic fraction obtained is $((tm + sn)d)/(tm + sn)$. But according to the fundamental theorem, d is between u/m and v/n, so that $((tm + sn)d) =$

$(tmd+snd) \approx (tm(u/m)+sn(v/n)) = (tu+sv) = tu+sv$. Thus we obtained the approximation formula.

We know how to determine u and v from the diophantine algorithm (Appendix 4) or from the contraction algorithm above, starting from the relation $mv-nu = 8v-5u = \pm 1$ or from the pair (8, 5). But from Figure 2.3(2) we can obtain that $v=2$ and $u=3$. Indeed, considering that $mv-nu=\pm 1$ (see the fundamental theorem), then if 1 in the equation is scale number one, then it is noticed that to go from 0 to 1, $v=2$ steps must be made in the direction of the 8-parastichies, and $u=3$ steps in the direction of the 5-parastichies. There is obviously another way to go from 0 to 1, giving $u=5$ and $v=3$, but then $u/m=5/8$ and $v/n=3/5$ are greater than $\frac{1}{2}$, so that the divergence angle cannot be between these two fractions as required by the theorem. Finally, the approximation given by the formula is $(3t+2s)/(8t+5s)=8/21$, a value in the interval $[3/8, 2/5]$ given by Adler's theorem for the pair (8, 5). If we had worked with point 34 instead of point 21 we would have obtained the approximation 13/34 ($t=3$, $s=2$).

2.6. Problems

2.1 (1) Perform the geometrical construction of spirals on the transparencies mentioned in Section 2.2, and try to draw sets of spirals that would contain values of x other than 1, 2, 3, 5, 7, 8, 11, 12, 13, 14, 17, … . Notice, for example, that 4 is excluded, since the family linking the points by 2 also has to link them by 4, because of the value of the divergence angle (i.e., τ^{-2}). The same can be said for the spirals linking the points by 3 and by 6. But notice also that 8 is not excluded from the list, even though it is a multiple of 2, because the spirals linking the points by 8 wind in the opposite direction to the spirals linking them by 2.

(2) Now superimpose any two transparencies with spirals winding in opposite directions in order to verify that the only visible opposed pairs are made with secondary numbers which are consecutive Fibonacci numbers.

2.2 (1) See that by performing the geometrical operations in Problem 2.1 with the divergence angle of $360(3+1/\tau)^{-1\circ} \approx 99\frac{1}{2}°$, we obtain the visible opposed pairs made of consecutive Lucas numbers. What, then, is the sequence of nested intervals containing the value $(3+1/\tau)^{-1}$? Compare the denominators in the consecutive intervals with the consecutive visible opposed pairs.

(2) State a theorem similar to Adler's theorem, for the Lucas sequence, denoting by L_k the kth Lucas number ($L_1=1$, $L_2=3$, $L_3=4$,

$L_4=7,\ldots$). Conclude in particular that the pair (4, 3) is visible and opposed if and only if the divergence is in [90°, 120°], that the pair (4, 7) is visible and opposed if and only if the divergence angle is in [1/4, 2/7], and that the pair (11, 7) is visible and opposed if and only if the divergence is in [98.18°, 102.85°]. (The larger the consecutive numbers in the Lucas sequence are, the smaller the interval around $99\frac{1}{2}$° is.)

(3) Prove that $(3+1/\tau)^{-1}$ is in all the intervals mentioned in (1). Hint: Use the fact that $L_k=F_{k-1}+F_{k+1}$.

2.3 In a regular lattice where $d=105°$ or 7/24, determine the first six visible opposed parastichy pairs by using the computational algorithm.

2.4 If (47, 29) is a visible opposed parastichy pair, determine an interval of values for the divergence angle by using either the answer to Problem 2.2, the diophantine algorithm (Appendix 4), or the contraction algorithm. Hint: For the diophantine algorithm, Appendix 2 at the answer to Problem 2.4 gives the general method for solving diophantine equations.

2.5 Considering the theorem for the sequence of normal phyllotaxis with $J=1$, show that when the visible opposed parastichy pair is made with very large consecutive numbers in the sequence then the divergence angle is around the value $1/(t+\tau^{-1})$. This is called the limit divergence of normal phyllotaxis for $t\geq 2$. Notice that as t increases, the angle approaches 0. Notice that the sum of the two angles for $t=1$ and $t=2$ is equal to 1 (complementary angles). Show that the divergence of anomalous phyllotaxis is given by $(2+(t+\tau^{-1})^{-1})^{-1}$, and that as t increases, the angle approaches $\frac{1}{2}$, that is 180°.

3

Hierarchical control in phyllotaxis

3.1. Lestiboudois–Bolle theory of duplications

The fundamental theorem (Chapter 2), in particular its special form called the Adler theorem, can be linked to early results in phyllotaxis and illustrated in the following way. Let us consider again Figure 2.3(2). In this representation of a phyllotactic pattern we noticed that the Fibonacci numbers get closer to the Y-axis as we move up in the diagram, because of the value of the divergence angle. (We observed a similar phenomenon in Figure 1.4 around the line PC.) The phyllotactic fractions that can be obtained from this system are for example, 2/5, 3/8, 5/13, 8/21, corresponding to cycles of 5, 8, 13, 21 leaves, respectively.

Now for each of the cycles, let us project the points of the lattice on the X-axis. For a cycle of 8 points for example, the ordered list of points is then, from right to left, 8-3-6-1-4-7-2-5, the same as in Figure 3.1 for that cycle. In both figures, following the points in their natural order along the genetic spiral, we must go almost 3 times to the right, starting at 1, to meet 8 leaves. We can thus learn from Figure 3.1 that 3/8 is a phyllotactic fraction of the system. The numbers m and n in two consecutive cycles constitute the visible opposed parastichy pair (m, n), and the corresponding fractions are the end points of the intervals given by the Adler theorem.

The levels in Figure 3.1 illustrate Bolle's (1939) phyllotactic law in his theory of *Antriebe,* called induction lines by Vieth (1965). These induction lines, as imagined by Bolle, travel through the cycles and induce the patterns of leaves. The law is that "the induction lines from a cycle bifurcate for the lower leaves of the cycle, while for the upper leaves they continue their paths without bifurcating." This is exactly the Lestiboudois law of duplication (1848, p. 84). For example, considering the cycle 3, 1, 4, 2, 5

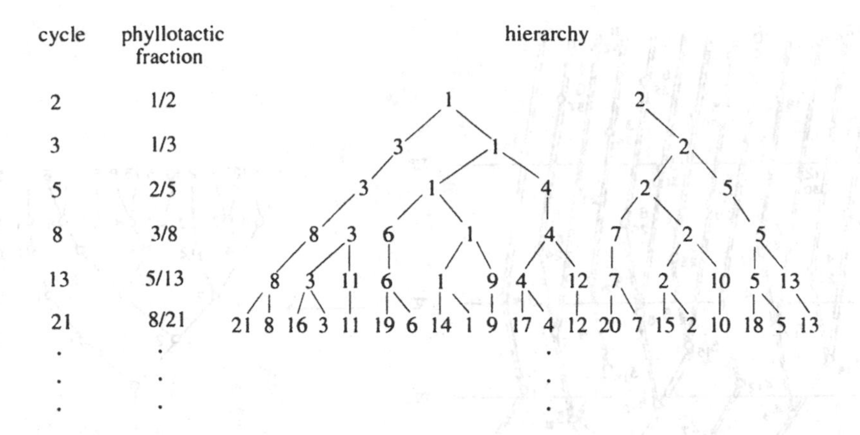

Figure 3.1. A leafy stem or the cone shown in Figure 2.3(2) characterized by the main sequence is a hierarchical system of simple and bifurcated induction lines. This is the hierarchy ⟨2, 3⟩ expressed by the numbers of nodes in levels 0 and 1 of the hierarchy. The order of the points in each row expresses a phyllotactic fraction. Each row is linked to adjacent rows according to the Lestiboudois–Bolle law of induction.

in Figure 3.1, there are bifurcations from leaves 1, 2, and 3 which are double nodes and "the lower leaves" found in the preceding cycle. Moreover, 8 leaves are present in the next cycle given that leaves 4 and 5 "continue their paths without bifurcating." In other words, if there are n leaves in a cycle, the leaves $x = 1, 2, 3, \ldots, p$, $p < n$, of that cycle that were present in the previous cycle, bifurcate to give leaves $x + n$ in the next cycle. We thus have a hierarchy of simple and bifurcated induction lines. The hierarchy illustrates the fundamental theorem of Chapter 2 and represents the system. It is denoted by ⟨2, 3⟩, 2 and 3 being the numbers of nodes in the first two levels, respectively.

Irrespective of the manner in which leaves are displayed, an internal vascular system must nonetheless serve them. Bolle discovered an amazing coincidence between the vascular skeleton linking the foliar traces and the induction lines for the capitulum of *Cephalaria* (Dipsacaceae) shown in Figure 3.2(1). "Bolle thinks that in almost all dicotyledonous plants, one of the two cotyledons (a seed leaf containing food material) is the point of departure of a bifurcated induction line . . . which induces the first pair of leaves . . . . If the two branches of a bifurcated induction line have different lengths and slopes (overtopping), there will be an alternate system of leaves . . . . The two branches continue, the one having a smaller slope staying simple, the other one bifurcating again, until the

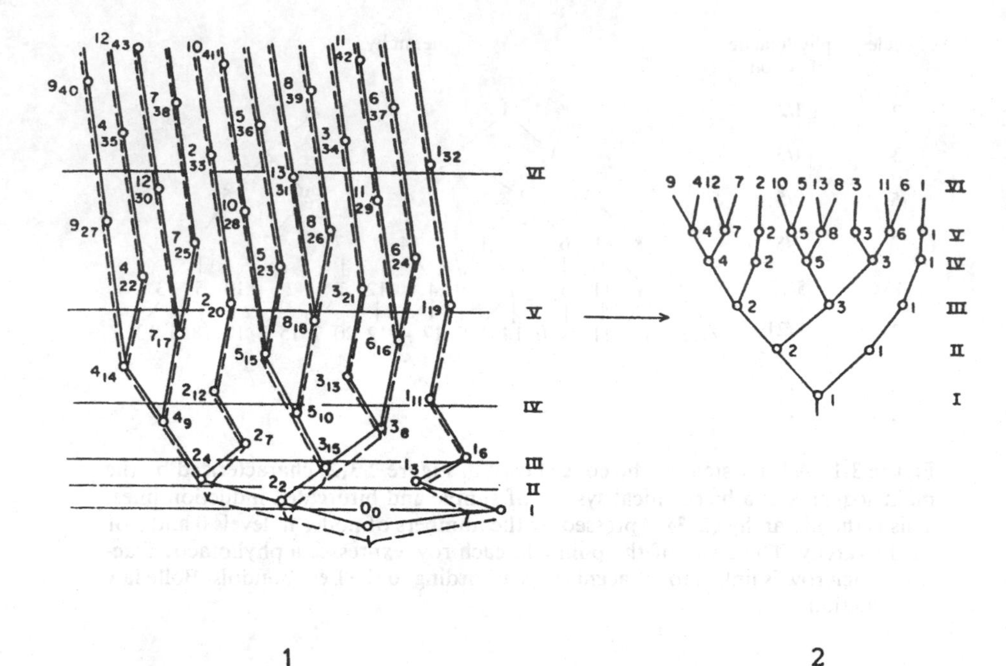

Figure 3.2. (1) Dotted lines drawn by Bolle to link the foliar traces. Heavy lines are induction lines for the system of alternate leaves of a capitulum of *Cephalaria*. The smaller numbers serve to count the points. (After Bolle, 1939, p. 173, redrawn by Vieth, 1965.) (2) Diagram depicting scheme in (1); it is essentially identical to the diagram in Figure 3.1 to a cyclic permutation.

number of leaves by cycle (3, 5, 8, ...), characterizing the plant under construction, is realized" (from Vieth, 1965, p. 77). This reminds Clos (1861) who presented a theory of partition based on a phenomenon of partition which he said occurs generally, and according to which the axis of an inflorescence splits in two branches, one staying simple while the other divides again and again.

Szabo (1930), apparently at the origin of Bolle's theory, reveals morphological and anatomical data on which the theory is based (see Fig. 3.3). Vieth confirmed this data: "in Bolle's theory the arrangement of the induction lines, particularly the distribution of simple and double lines, is of fundamental importance. For Szabo the connections of the vascular bundles constitute a first important criterion." Bolle considers Szabo's work (1930) as a proof of the validity of his deduction. Szabo numbered the bracts and foliar traces of Dipsacaceae capituli, and studied the manner in which the bundles are assembled: some bracts are linked by their

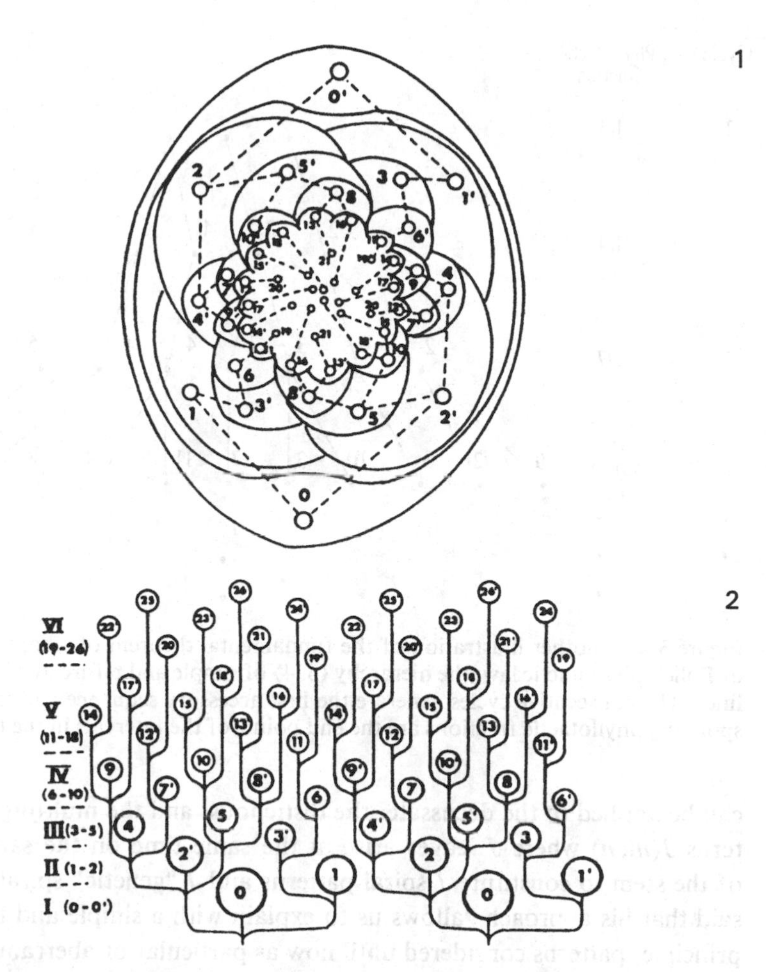

Figure 3.3. (1) Bispiraled $2(2,3) = (4,6)$ capitulum. (From Szabo, 1930, p. 21, redrawn by Vieth, 1965.) (2) Diagram representing the links between the primordia of the capitulum. (From Szabo, 1930, p. 29.)

foliar traces to only one bract, while others are linked to two. The simple and double junctions of the foliar traces are distributed in the way proposed by Bolle for the distribution of his bifurcated and simple induction lines.

The theory of duplications can be applied to any type of spiral phyllotaxis. A hierarchy of simple and double nodes is obtained. The hierarchy $\langle 3,4 \rangle$ generating the sequence $\langle 1,3,4,7,11,18,\ldots \rangle$, called the first accessory or the Lucas sequence, is illustrated in Figure 3.4 (where there are three and four nodes in the first two levels). Bolle explains that the theory

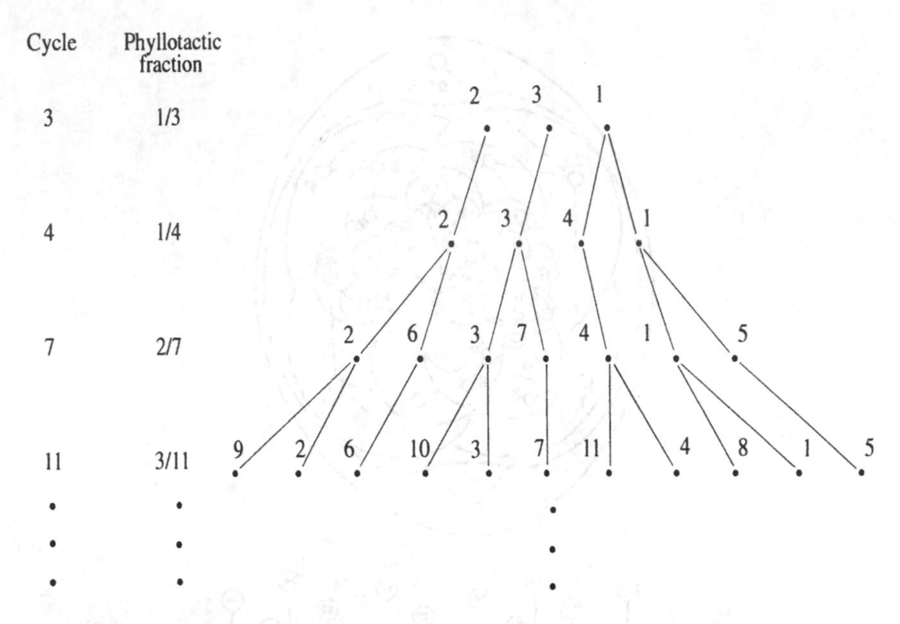

Figure 3.4. Another illustration of the fundamental theorem of Chapter 2, and of Bolle's phyllotactic law: the hierarchy $\langle 3, 4 \rangle$ of simple and bifurcated induction lines. The consecutive cycles generate the first accessory sequence and the corresponding phyllotactic fractions are the end points of the intervals in the theorem.

can be applied to the decussate, the distichous, and the multijugate patterns $J(m, n)$ where J leaves arise at the same time on the same level of the stem to constitute J spiral patterns and J "genetic" spirals. Bolle said that his approach "allows us to explain with a simple and uniform principle, patterns considered until now as particular or aberrant cases." He applied it to the spermatophytes only. For Lestiboudois (1848, p. 80) the law of duplication "explains in an absolute manner all the rules of phyllotaxis."

3.2. Zimmermann telome theory

3.2.1. Ancestral land plants

According to the Zimmermann theory (1953), vascular plants evolved from an initial dichotomized structure by means of a few elementary morphogenetic processes. Dichotomous branching is probably the most primitive pattern of growth in plants. Palms, which are often considered to be among the oldest of flowering plants and to provide the most dramatic examples of phyllotactic spiral patterns, have descended from a dichotomous ancestor. Today contrary to most palms, *Hyphaene thebaica* shows a trunk ramifying dichotomously. But the branching process called "overtopping," a variation on the theme of dichotomous branching, is probably at the origin of the spirality in the leaf pattern on the trunk of palm trees and in higher plants in general. An **overtopping** pattern arises when the doubled branch grows unequally, with the growth of one member exceeding that of the other so that an apparent main axis takes shape. "Overtopping might have been the means by which morphological differentiation and division of functions was introduced into a system of telomes (dichotomic branchlets) originally all exactly alike" (von Denffer et al., 1971). This type of branching is among the elementary processes in the Zimmermann telome theory [Fig. 3.5(1)], and has led to the structure of plants as we see them today [Fig. 3.5(2)].

The primordial **telomes** are the branchlets of the Psilophyta. In the division of Pteridophyta, the Psilophytales are the oldest known land plants. Among them, the most primitive members are *Steganotheca striata,* which lived 400 million years ago, and its descendants within the Rhyniophytes, *Rhynia gwynne-vaughanii* and *Horneophyton* [see Fig. 3.6(2)]. They had dichotomously forked stems with a single vascular strand, but entirely lacked leaves and roots. The Zimmermann telome theory asserts that these plants represent the kind of primitive vascular plants from which all others have evolved. In the same division, the Primofilices show forms that range almost without discontinuity between early land plants and the morphologically complex plants of today [see Fig. 3.6(3)]. The process of so-called fusion in the axis in particular [Fig. 3.5(1)] is responsible for the generation of the vascular system.

The telome theory, influenced to a considerable degree by paleobotanical discoveries, has led to fundamental changes in the concepts of plant morphology. It has permeated the thinking of morphologists and paleontologists, and it is often said to be the only coherent explanation of the evolution of plants (see Cusset, 1982 and Hagemann, 1984 for alternatives). The concept encompasses a consideration of all vascular plants, and supports the hypothesis that the algal ancestor of land plants was a green dichotomously ramified alga of the type depicted in Figure 3.6(1).

3.2.2. Algal ancestors

In the Zimmermann telome theory it is held that the dichotomous branching of early land plants was derived from forked marine algae, which

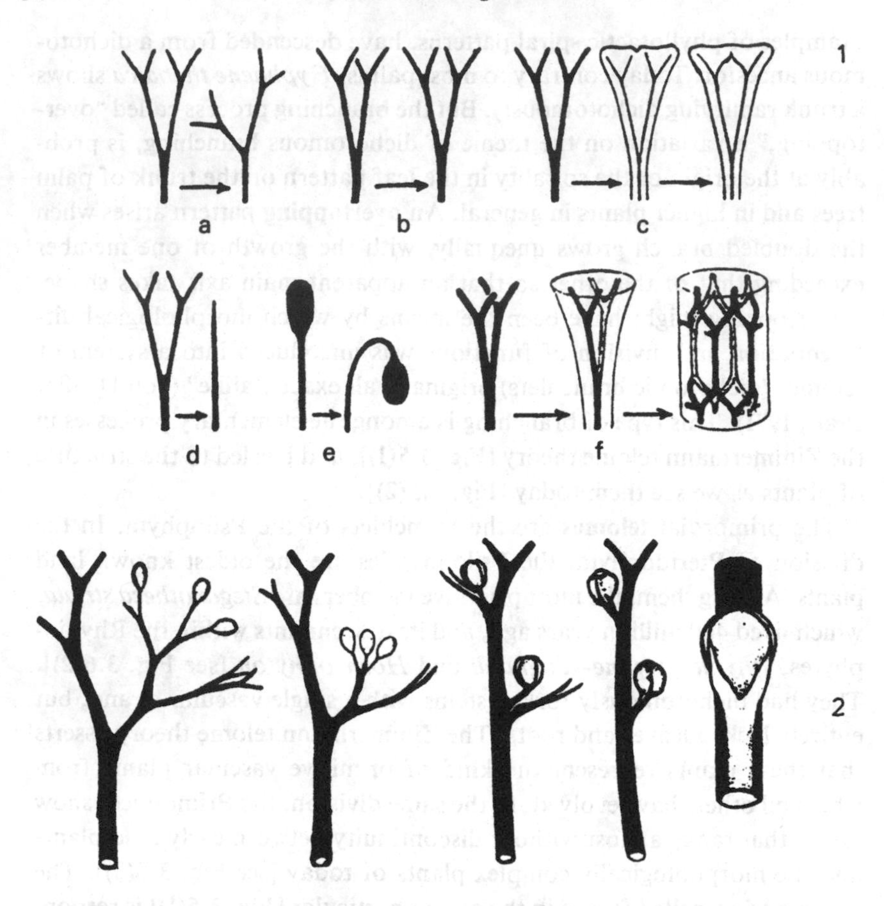

Figure 3.5. (1) Elementary processes of the telome theory leading to vascular plants: (a) overtopping, (b) planation, (c) fusion in the leaf, (d) reduction, (e) recurvation, and (f) fusion in the axis to form vascular systems (from Zimmermann, 1953). (2) Evolutionary stages leading to sporangial position on a *Psilotum*. (From Stewart, 1964.)

exhibited a mechanism of branching by apical bifurcation, and that both the stem and leaf in the higher plants were derived by an overtopping process in particular. At the beginning of the century the botanist Church (1920a,b) proposed (later followed by others such as Corner, 1981) that we must look into the sea, at marine brown algae, to understand the remarkable phyllotactic patterns found in higher plants, even though brown algae are not ancestral to higher plants.

For Church (1920a, p. 49) "the modern equipment of a land-plant, though often apparently admirably suited for the necessities of the present

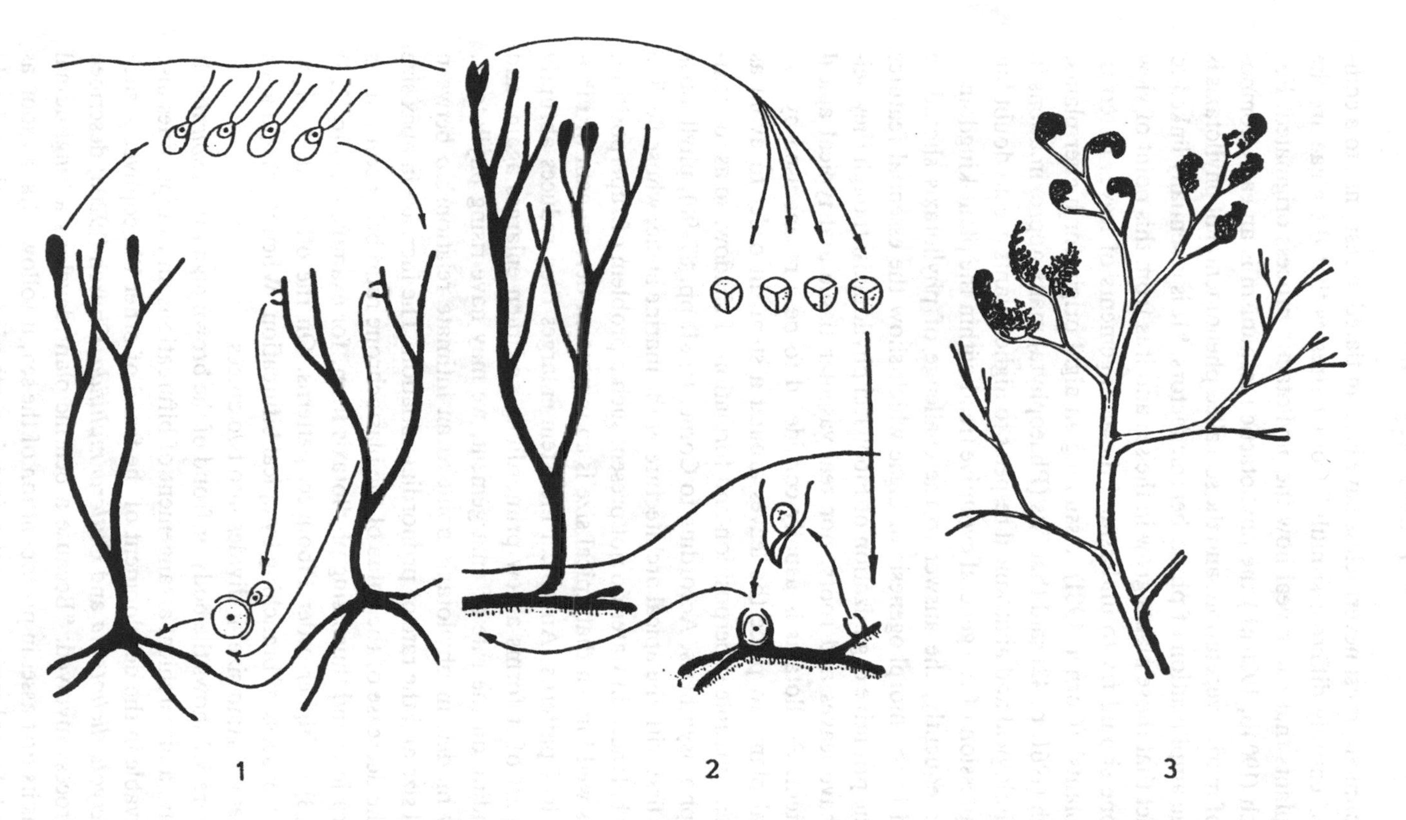

Figure 3.6. (1) Hypothetical algal ancestor of early vascular plants. (2) *Rhynia major*, the ancestral land plant. (3) A primitive frond of *Protopteridium hostimense*, from the Devonian. [(1) and (2) from Stewart, 1964; (3) from von Denffer et al., 1971.]

environment, must nevertheless have been initiated in response to a complex of entirely different stimuli." Corner stresses that the algae are the only plants that can reveal how the phyllotactic process originated. For Church (1904b, 1920a,b) the phyllotactic mechanism is an early component of marine vegetation, and the so-called phenomenon of phyllotaxis is but the amplification of phyto-benthic factors. "It is essentially linked to the fact that those plants grow in the sea and it is from this point of view that one should try to understand the arrangements of leaves in terrestrial plants" (Corner, 1981). Assuming an algal origin for higher plants, Fritsch (1961, p. 42) said that "its (Phaeophyta) many approximations in morphological and anatomical respects to higher plants are no doubt but an expression of the general evolutive trend within the plant-kingdom."

Consequently, the answer to the challenge of phyllotaxis should be found in the morphogenesis of algae which show the essential features and the primitive organization of Fibonacci phyllotaxis, though they neither have leaves and roots, nor real vascular tissue and typical apical meristem. Phyllotaxis is usually considered to be a means used by terrestrial plants to place the leaves around a stem, in order to avoid as much as possible superposition and formation of shadow, so as to maximize photosynthesis. According to Corner (1981, pp. 87–90), phyllotaxis starts first with the apical architecture of the marine algae, whose flexible stems balanced in water do not present such a problem of superposition.

It is well known that apical size is an important determinant of Fibonacci leaf patterns. An apical meristem enlarges then reduces abruptly as a part of it forms a new primordium, and then enlarges again, and depending on the rate of enlargement, we may have rising phyllotaxis. Many models in phyllotaxis point to an intimate relationship between apical size and the rate of primordial initiation. The increase in apex size and the decrease of the value of the plastochrone ratio bring continuous bifurcations and thus rising phyllotaxis from, for example (2, 1) to (2, 3) to (5, 3) or higher-order Fibonacci patterns. On the other hand, algae show a process of branching by apical bifurcation. When the frond apex reaches a certain size, it divides into two apices.

Figure 3.7 shows the body or frond of the brown alga *Fucus spiralis.* It displays a hierarchical arrangement of bifurcations and simple lines also observable in the development of the frond of other Phaeophyceae such as *Dictyota dichotoma* and *Cutleria multifida.* Church (1920a) described the process involved: "Because a benthic plant develops a longitudinal axis as its first asset in protobenthon of the sea, it follows that as soon as it builds lateral ramuli, one at a time, in rhythmic sequence, these should

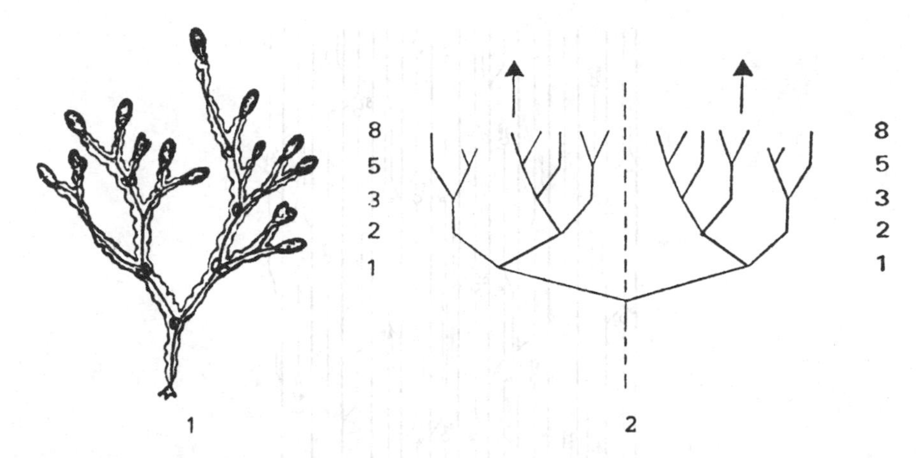

Figure 3.7. (1) Body (frond) of a brown alga. (2) Schematic representation of the frond, a multilevel system obtained by the Horton-Strahler method for ordering ramified patterns, showing the rhythmic production of bifurcations.

follow the Fibonacci rule. Observation of the somatic organization of Phaeophyceae and Bryophyta shows that this has been the case . . . . If in such a plastic and plasmic organism, outgrowth in one direction is followed by a compensating growth movement in the next direction for balanced equilibrium, the attainment of the latter implies the presentation of a divergence angle of the Fibonacci series, as an inevitable consequence . . . . Fibonacci symmetry is seen to be the expression of an oscillatory balancing effect in two dimensions."

3.2.3. Vascular phyllotaxis

The study of vascularization reveals the underlying branching process and the hierarchical control or organization of phyllotactic patterns. Sterling (1945) and Esau (1954) working on *Sequoia sempervirens*; and Gunckel and Wetmore (1946), Girolami (1953), and Jensen (1968) working respectively on *Ginkgo biloba, Linum usitatissimum,* and *Kalanchoe tubiflora,* clearly illustrate, with hierarchies similar to the one in Figure 3.8, the close relationship between phyllotaxis and vascularization. Without getting into the polemical theory of acropetal induction of leaves by the foliar traces, phyllotactic transitions can be described after anatomical analysis of bundle connections.

For Larson (1977, 1983), vascular phyllotaxis carries the real structural meaning of current concepts such as orthostichy, Fibonacci sequence,

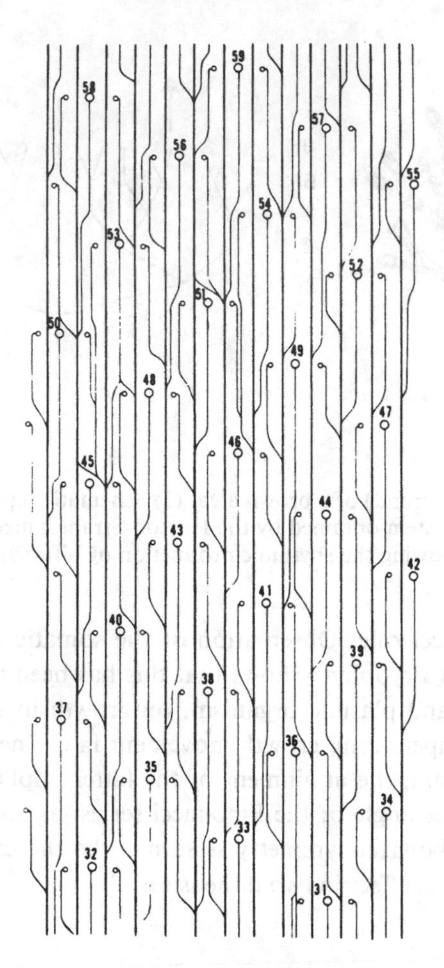

Figure 3.8. Portion of a stem of *Kalanchoe tubiflora,* showing 13 vascular bundles from which leaf traces arise. Bundles (indicated by broad lines and located between orthostichies) supply median traces to leaves situated 13 primordia apart. The divergence is seen to be equal to 5/13, revealing, as it is the case in many plants, the equality between the number of bundles and the denominator of the phyllotactic fraction. The pairs (2, 3) and (5, 3) are conspicuous. (From Jensen, 1968.)

and ontogenetic helix (genetic spiral), which are not abstractions imposed on the plants by investigators. In terms of strict botanical usage and when applied to the external morphology of leaf arrangement, the orthostichies define leaf primordia joined by their central traces; they may be either vertical or strongly inclined. The Fibonacci sequence "is a real phenomenon which provides a convenient way to describe the vascular array." The genetic spiral "is essential to understanding vascular organization." Lar-

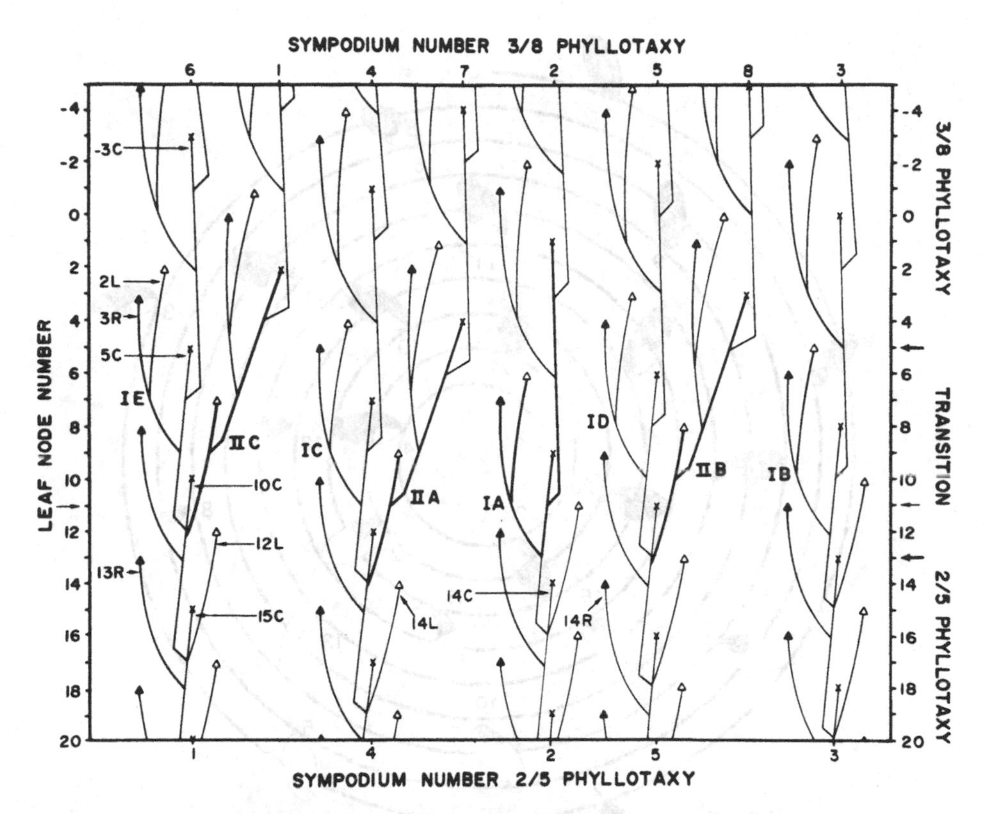

Figure 3.9. Vascular cylinder of *Populus grandidentata* or of *Populus deltoides* is displayed as if unrolled and laid flat. The three traces leading to each primordium are represented by an × (the central trace), a black triangle (the right trace), and a white triangle (the left trace). The transition from 2/5 to 3/8 divergence occurs in a precise and systematic pattern. (From Larson, 1977, p. 244.)

son's research on *Populus* is based on anatomical examination and analysis of the vascular system and on carbon-14 translocation profiles. He explained the transitions from 2/5 to 3/8 to 5/13 divergence angle, having identified the position of these transitions within the vascular system. His diagram (see Fig. 3.9) shows what is happening between two levels. Notice that the orders of the leaves at the bottom and at the top of this figure are the same as in Figures 3.1 and 3.2 (even though the links between the two levels are different).

3.3. Translocation of substances in plants

The ramified structure of phyllotactic patterns can be underlined furthermore by considering the phenomenon of sectorial translocation of

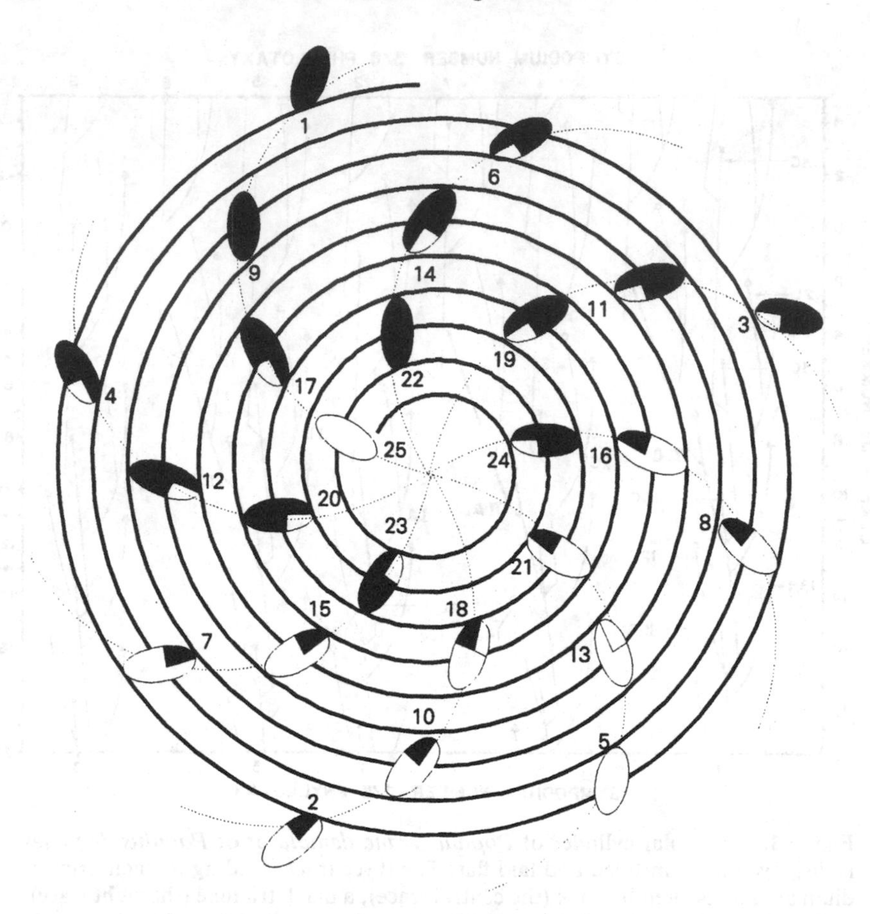

Figure 3.10. Fibonacci phyllotaxis of leaves around a stem of apple (adapted from Barlow, 1979). The source leaf is marked 1 and successive younger leaves are numbered 2 to 25. The source leaf was exposed to $^{14}CO_2$ for 1–2 hours. The pattern of transport may be followed on a leaf-by-leaf basis. Whole-leaf autoradiography for one to several days after treatment showed the distribution pattern of ^{14}C. Leaves that lie superimposed in vertical ranks and that are located closest to the donor leaf take up label over their entire surface. Leaves on the opposite side of the stem from the donor leaf received no labeled assimilate. Leaves in adjacent ranks acquire label in a small sector on one side of the midrib or another, depending on their positions with respect to the donor leaf. Leaf 25 was not examined.

assimilates in plant shoots, that is, the transfer of substances in sections of the plants in contrast to uniform distribution. This phenomenon is clearly seen in Figure 3.10, where the leaves have an average divergence angle of 136.7°. This figure illustrates the restricted movement of labeled carbon applied to a donor leaf into specific sectors of the growing shoot. The three 8-parastichies 4-12-20, 1-9-17, and 6-14-22 contain the leaves that are assumed to have the most direct vascular connections with the donor leaf 0 ((5, 8) vascular phyllotaxis). Only the dominant pattern found in each leaf is shown.

Notice that leaves 2, 5, 7, 8, 10, 13, 15, 16, 18, 21 have been only partially reached by $^{14}CO_2$. This can be correlated with the right-hand part of the hierarchy in Figure 3.1, developing under point 2 (linked to points 5, 7, 10, 13, 15, 18) and representing a sector of the whole hierarchy that has been only partially touched by the assimilate. Moreover, the primordia having the most direct vascular connections with the donor leaf 1, that is 4, 6, 9, 12, 14, 17, 20, 22, correspond to the sector of the hierarchy in Figure 3.1 developing under point 1. The distribution patterns displayed by carbon-14 assimilates are thus predictable on the basis of the ramified structure of phyllotaxis.

Ho and Peel (1969) demonstrated a marked correlation between the phyllotactic configuration of leaves in shoots of *Salix viminalis* and the transport of tracers between these leaves. They present a figure (their Fig. 1) almost identical to Figure 3.10, but with ten leaf primordia only and numbered differently. It can be verified by marking their leaf a_1 as leaf 1 and the other leaves as leaves 2, 3, 4,... along the genetic spiral, that when carbon-14 is applied to leaf 6, leaves 2, 5, 7, and 10 show almost no carbon-14 activity. These four leaves correspond again to the right-hand part of the hierarchy in Figure 3.1, representing the (2, 3) phyllotaxis of the plant. The assimilate was transported from leaf 6 to leaves 3, 4, 8, and 9 via leaf 1, on the left-hand side of the hierarchy. Thus, the authors found two main channels in the transport of tracers.

As another example, in the sunflower shoot the pattern of assimilate movement is also highly sectorial. In the domestic sunflower, "deriving from branched ancestors relatively recently, perhaps the pronounced sectorialization reflects an architectural constraint that is remnant of this branched ancestry" (Watson and Casper, 1984). The Zimmermann theory would not deny that point of view. The algorithmic production of sunflowers can also generate branched structures, hierarchies with simple and double nodes, and reveal sectorialization.

3.4. Hierarchies arising from modeling

3.4.1. Van der Linden model

Van der Linden (1990) proposed a simple model for generating the capitulum of a sunflower. This model supposes an initial placement of a few floret primordia. Figure 3.11 illustrates the mechanism. It starts with the three circles on the left, which grow in size according to the sigmoid

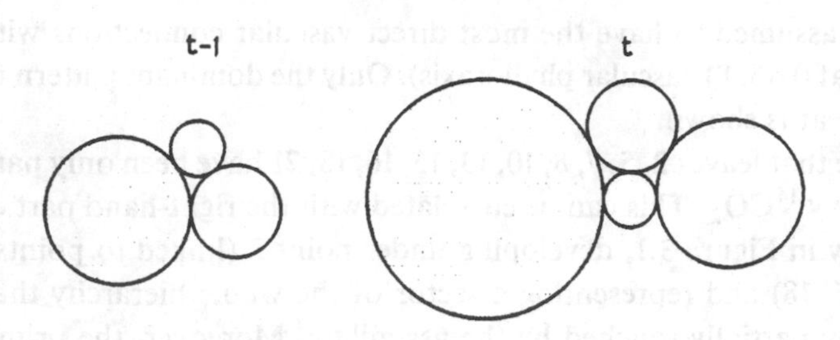

Figure 3.11. An initial placement of four flower primordia in a model of sunflower capitulum. The largest circle on the left has grown to become the largest circle on the right, and a primordium has been added by growth (the smallest one on the right). The drawing of the structure at time t is such that if it is turned in some way, it fits onto the drawing of the structure at time $t-1$. The model thus allows us to construct the structure at t from that at $t-1$. (From van der Linden, 1990; reprinted by permission of Elsevier Science Publg. Co., Inc., copyright, 1990.)

growth law $r_t = R(1+e^{c(t'-t)})^{-1}$. Here r_t is the radius of the circular primordium t, R is the maximal radius a disk can have, c is a growth constant, t' is the ordinal number of the unit with radius $R/2$. Then a floret primordium is added so that the structure at plastochrone t (the plastochrone is the time elapsed between the initiation of two successive primordia) reproduces the structure at plastochrone $t-1$. That is, the relative arrangement of primordia $t-1$ and its two nearest neighbors p and q must be repeated with primordia t (to be placed), $p+1$ and $q+1$. But the sizes of other primordia between $p+1$ and $q+1$ may not allow the contact of both of them with t. That is why the model gives supplementary rules allowing us to choose between possible places for primordium t between existing primordia. The mechanism takes in charge the developing process primordium by primordium, in a dynamical way, till the capitulum is fully reproduced. In that way phyllotaxis can rise for example to any pair of consecutive Fibonacci numbers. It follows that the divergence angle between consecutively born primordia can settle around the limit divergence angle τ^{-2}, with any degree of accuracy, for reasons explained in Chapter 2.

A variant of this model is based on the similarity between the soap-bubble-like unit coalescences and aggregates of primordia. This model can be used to generate phyllotactic patterns with apparently any kind of primordial form and on any surface. The model does not deal with the most important problem of phyllotaxis, namely the reasons for the initial placements of primordia. Initial configurations are simply chosen that will initiate one pattern or another. Figure 3.12, which is made with 5,000 such units, illustrates the result of the application of the mechanism: a

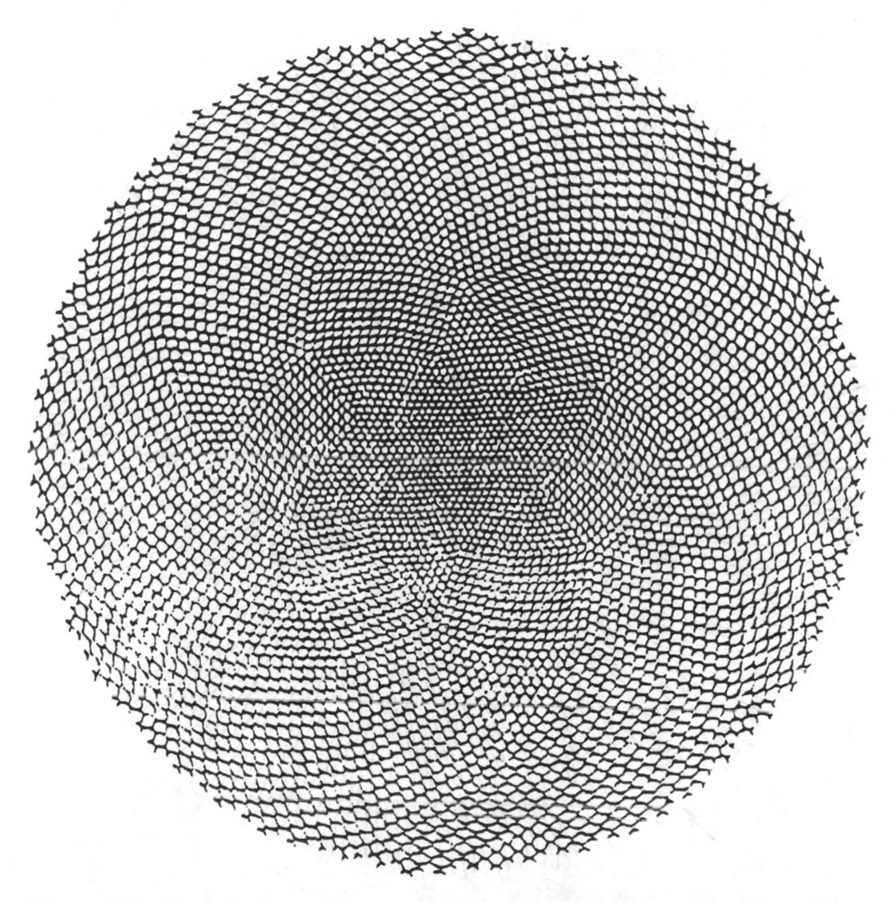

Figure 3.12. Model of a sunflower capitulum (89, 144) constructed without using divergence angles and plastochrone ratios. (From van der Linden, 1990; reprinted by permission of Elsevier Science Publg. Co., Inc., copyright 1990.)

capitulum (89, 144) rising to (233, 144) on the rim. Figure 3.13 reveals the underlying symmetry of Figure 3.12: lines bifurcating or not, from the center to the rim (233 lines on the rim, starting to bifurcate). These lines represent the connections between circle centers. Figure 3.13 can easily be transformed into a treelike structure, a hierarchy of simple and double nodes, simply by pulling apart the two lines on each side of the arrow. The figure gives a hint about branching and sectorialization in sunflowers.

3.4.2. The fractal nature of phyllotaxis

For Hermant (1946), an architect, phyllotaxis is a purely formal problem, and the essential architectural characteristics of the main plant-shoot

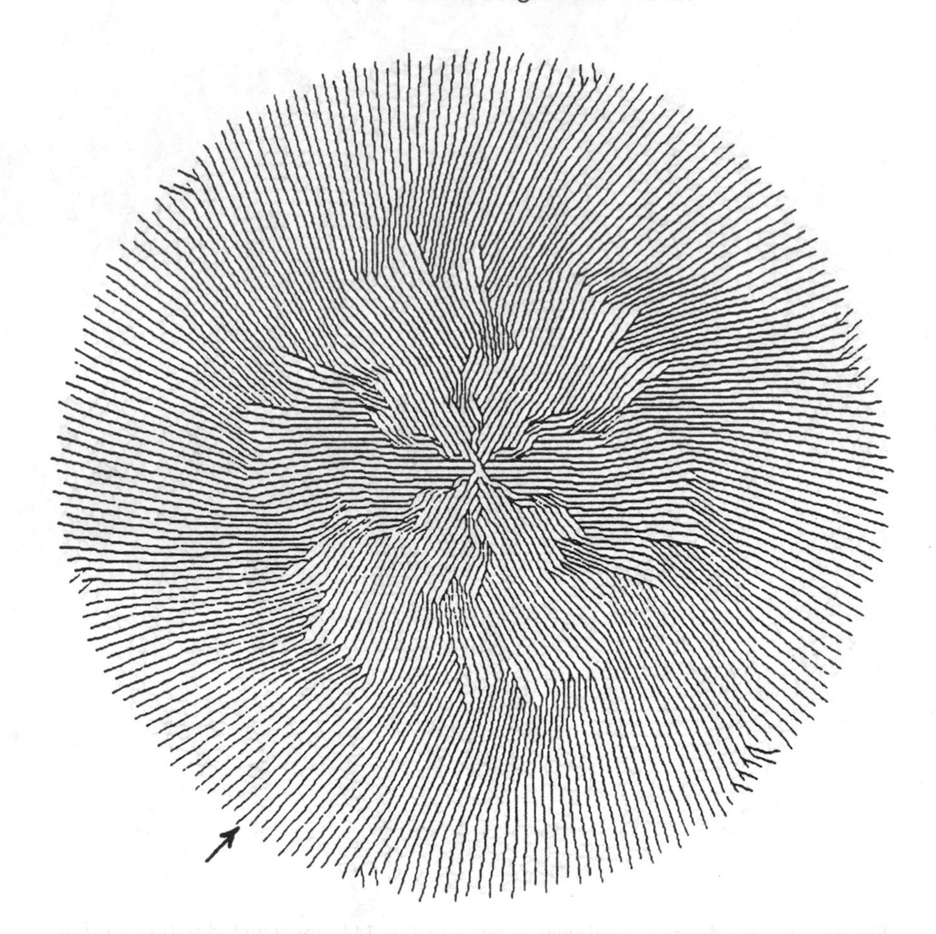

Figure 3.13. Symmetry and hidden structure of the model depicted in Figure 3.12. It reveals bifurcated and simple induction lines and the rise of phyllotaxis from the inner to the outer region. (From van der Linden, 1990; reprinted by permission of Elsevier Science Publg. Co., Inc., copyright 1990.)

structures, the main functions of phyllotaxis, can be deduced from Figure 3.14(1). The figure can be developed further with a two-step algorithm for the generation of consecutive rings. From this figure we easily obtain Figure 3.14(2), which shows binary ramification.

Let us consider now the fractal in Figure 3.15(1), also proposed by Hermant in his endeavor to understand phyllotaxis. The word "fractal" means irregular or fragmented. Fractals are intriguing objects as complicated on the small scale as in the large, where the part is identical to the whole, where the dimensions are most of the time fractions, and where a

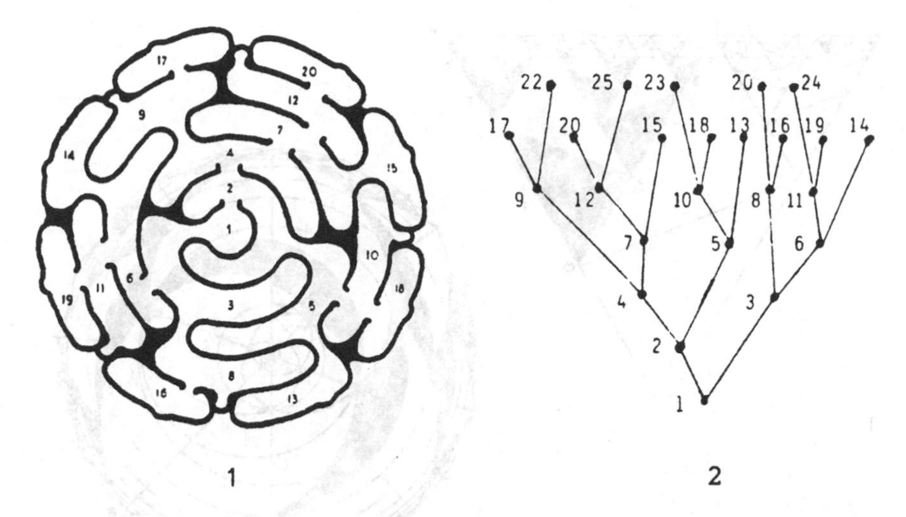

Figure 3.14. (1) Illustration of a simple mechanism used to generate phyllotactic patterns expressed by the main sequence. (From Hermant, 1946.) (2) Hierarchy of simple and double nodes that can be derived from (1).

line can totally fill a surface. The essence of fractals is that they are self-similar; that is, smaller parts are condensed versions of the whole. For example, each branch on a tree looks like a miniature tree having branches that are even smaller trees, and so on. No definition can be given really; one must take a look at Mandelbrot's book on fractals (1982) or at the illustrated literature on the subject (e.g., Barnsley, 1988).

Hermant wanted to demonstrate that a simple principle of cellular division in plants is sufficient to explain the complex patterns of parastichies in pine cones and sunflower capitula. The three sides of the largest equilateral triangle are divided into the golden ratio. The same thing is done with the three sides of the consecutive equilateral triangles thus generated in the obvious way. This is a variation on the theme of the Sierpinski triangle (see Barnsley, 1988 or Prusinkiewicz & Lindenmayer, 1990, Chapter 8 entitled "Fractal Properties of Plants"), which reminds of Thom's (1975) generalized catastrophes. From such a scheme we can apparently reconstruct buds and flowers. Indeed the hierarchy of scalene black triangles in the figure is slightly more complex than the hierarchy in Figure 3.2, but it is essentially the same. When points A and B are brought together in a circular movement, the result [Fig. 3.15(2)] resembles to some extent the familiar cross section of a vegetative bud with its leaf primordia [Fig. 3.15(4)], and the floral diagram of a flower showing the relative

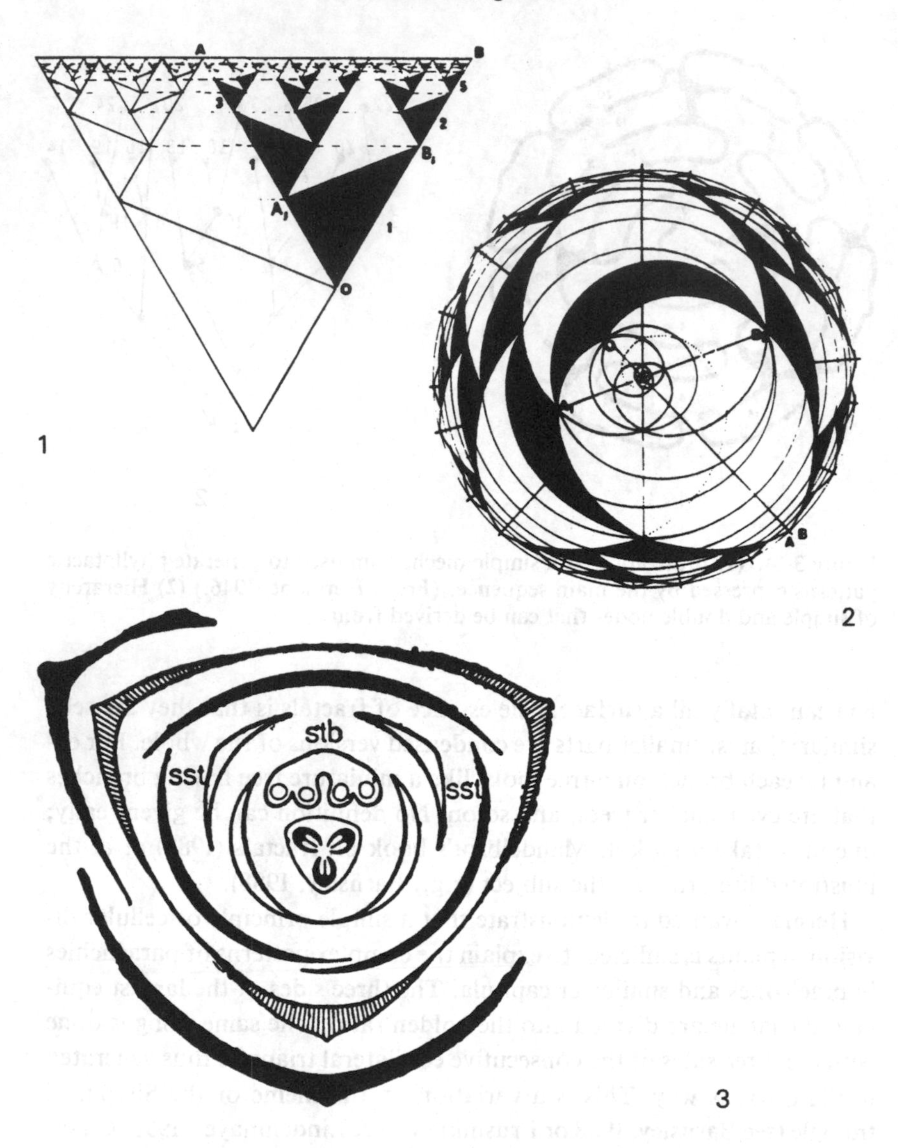

Figure 3.15. (1) Hierarchy of congruent scalene black triangles, presenting levels with consecutive Fibonacci numbers and revealing the architecture of the complexity found in the next diagrams. (2) Topological deformation of the fractal, obtained by bringing together points *A* and *B* in a circular movement around 0. (3) Floral diagram of *Kaempferia ovalifolia,* Zingiberaceae. (4) Transverse section of a foliage shoot of *Saxifraga umbrosa,* showing the phyllotactic system (3, 5). [(1) and (2) from Hermant, 1947; (3) from von Denffer et al., 1971; (4) from Church, 1904b.]

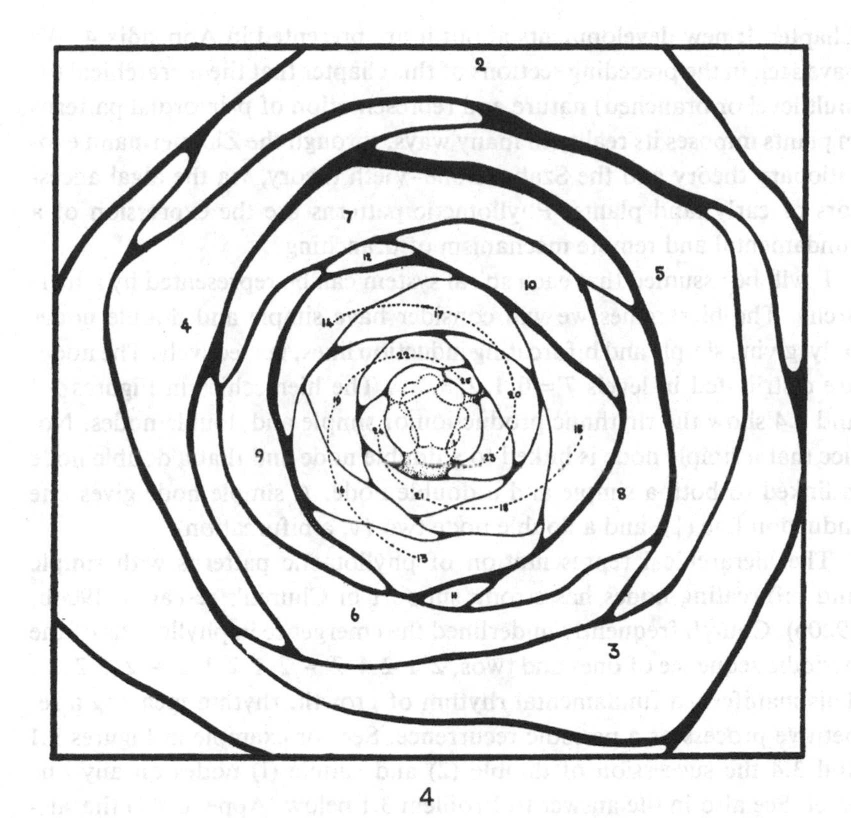

arrangement of its floral parts [Fig. 3.15(3)]. Further research in this direction will hopefully reveal more convincing examples.

When developed further on, under the same rules, a hierarchy (e.g., Figs. 3.1 & 3.4) becomes a fractal expressing the self-similarity of phyllotactic branching structures. The dependence of the number of parastichies on the size of the field of attention is another intriguing aspect of spiral phyllotaxis (see Fig. 1.4) related to fractal geometry via self-similarity (a concept explored in Section 9.2). The allometry-type model (object of Chapter 4) using the plastochrone ratio shows a fractal dimension 2 in phyllotactic patterns.

3.5. The hierarchical representation of phyllotaxis

3.5.1. Hierarchies with only simple and double nodes

Chapter 1 introduced the centric representation of phyllotaxis. We presented in Chapter 2 the cylindrical representation that will be used in

Chapter 4; new developments about it are presented in Appendix 4. We have seen in the preceding sections of this chapter that the hierarchical (or multilevel or branched) nature and representation of primordial patterns in plants imposes its reality in many ways, through the Zimmermann evolutionary theory and the Szabo–Bolle–Vieth theory, via the algal ancestors of early land plants. Phyllotactic patterns are the expression of a fundamental and remote mechanism of branching.

It will be assumed that each spiral system can be represented by a hierarchy. The hierarchies we will consider have simple and double nodes only, giving simple and bifurcating induction lines, respectively. The nodes are distributed in levels $T=0,1,2,3,\ldots$. The hierarchies in Figures 3.1 and 3.4 show the rhythmic production of simple and double nodes. Notice that a simple node is linked to a double node and that a double node is linked to both a simple and a double node. A simple node gives one induction line (|), and a double node two (V, a bifurcation).

The hierarchical representation of phyllotactic patterns with simple and bifurcating nodes has strong support in Church's research (1904b, 1920b). Church frequently underlined the emergence in phyllotaxis of the periodic sequence of ones and twos, 2 1 2 1 2 - 2 1 2 1 2 - 2 1 2.... This manifests a fundamental rhythm of growth, rhythm meaning a repetitive process or a periodic recurrence. See for example in Figures 3.1 and 3.4 the succession of double (2) and simple (1) nodes on any one level. See also in the answer to Problem 3.1 below (Appendix 2) the succession of pairs (2) of points and of single (1) points at each level. Church compared phyllotaxis with cellular division. "The law of arranging members of a higher cycle on a preceding lower one consists in the method of dividing them in sequence in the order indicated" (Church, 1904b).

This is reminiscent of the Lestiboudois law of duplication (Section 3.1), which underlines the sequence of ones and twos (Lestiboudois, 1848, p. 85). And by studying algorithms introduced to reproduce Fibonacci phyllotaxis – for example, Church's (1904b) and Mathai and Davis's (1974) algorithms to reconstruct the sunflower head; Franquin's (1974) algorithm to reproduce ramification of branches (see Jean, 1978b); Berdyshev's (1972) diagrams to explain the role of Fibonacci numbers in plants; Holland's (1972) algorithm to reproduce cellular division; and McCulloch's algorithm for constructing cones (see Kilmer, 1971) – the sequence of ones and twos can be easily retraced (see the Problems in Section 3.6 in which four of these algorithms are analyzed).

In the hierarchies the points or nodes will not even be identified, as we did in Figure 3.1. This identification is useful for reading phyllotactic fractions, but in a hierarchy the divergence angle is implicitly present via the theorem of Chapter 2 as we saw. In the hierarchical representation, the visible opposed parastichy pairs are represented by the numbers of nodes in any two consecutive levels, and the pair indicates an approximation of the divergence. Bolle states that "the phyllotactic organization of the vascular system is essentially unaltered by and independent of contact parastichies, divergence angles, and wandering of orthostichies." It is assumed that the hierarchies with simple and double nodes show the essential features, functions, and structures of phyllotactic patterns, recalling their ancestral origins. In the hierarchical representation it is considered that *phyllotactic arrangements are partially ordered systems of interrelated elements (the primordia) interacting in an aggregative fashion.*

In most studies of phyllotaxis, in the cylindrical or centric representation, the divergence angle is the main parameter. The rise or plastochrone ratio simply establishes which pair, among the visible opposed parastichy pairs of the system, is conspicuous. That is why in recent models (e.g., Thornley, 1975a; Marzec, 1987) only the divergence angle is considered, giving the visible opposed parastichy pairs. But some models do not even use the divergence angle, just as in the hierarchical representation. This is the case for van der Linden's model considered above. This is also the case for Bursill and XuDong's (1988) model of the packing of growing disks in which model the authors remove the necessity for artificially constraining the divergence angle, and in which a genetic spiral naturally arises. The rules for packing the disks are: (1) Disks are added sequentially as close to the origin as possible, touching at least two existing disks. (2) Successive disks increase monotonically in radius according to a growth law $G=f(n)$ where G is the radius of the nth disk. Figure 3.16 illustrates the procedure with the law $G=0.5n$ and two positions of the origin. The result is that when the origin is centered on the first disk, the third disk has a divergence closer to 99.5° than to 137.5°, and this effect propagates to the next few disks, thus setting a pattern which finally converges on the Lucas divergence angle. If the origin is placed midway between the first two disks, the third and successive disks have a divergence angle closer to 137.5° and thereafter the divergence converges on the Fibonacci angle. An alternative choice of origin produces a chaotic behavior.

3.5.2. Growth matrices, L-systems, and Fibonacci hierarchies

To generate those hierarchies and to give a mathematical meaning to the word rhythm, we can use growth matrices and L-systems (these systems

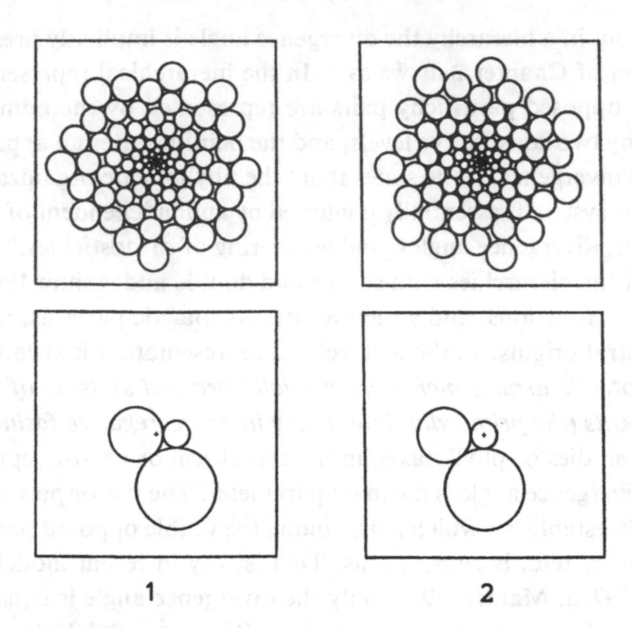

Figure 3.16. Aggregates of growing disks determined only by the position of the first three disks and by the choice of origin C. (1) First three disks at $(0.75, 0)$, $(-0.75, 0)$, and $(0.75, -2)$ generate a counterclockwise genetic spiral, $(8, 13)$ phyllotaxis, and the Fibonacci angle. (2) First three disks at $(0, 0)$, $(-1.5, 0)$, and $(0, -2)$ generate a counterclockwise genetic spiral, $(11, 18)$ phyllotaxis, and the Lucas angle. (Both figures from Bursill & XuDong, 1988.)

are studied for example in Herman and Rozenberg, 1975). For phyllotaxis a growth matrix is defined here as a square matrix $C=(c_{ij})$ of order $w=2, 3, 4, \ldots$, whose entries are ones and zeros only, where $0<\sum_i c_{ij} \leq 2$ for every j, and whose directed graph is strongly connected (irreducibility of the matrix). The next paragraphs give two examples of growth matrices but first we need to define other concepts in order to make the examples fully intelligible. A growth matrix of order w generates a language called an L-system (A, P, w_0) where A, called the alphabet, is a set of w symbols denoted by $a_1, a_2, a_3, \ldots, a_w$, and w_0 called the axiom is any string of symbols. P is a function defined on A by

$$P(a_j)=(a_1 * c_{1j})(a_2 * c_{2j})(a_3 * c_{3j}) \cdots (a_w * c_{wj}),$$

where $*$ means that a_i is in $P(a_j)$ if $c_{ij}=1$, and is not otherwise, for $j=1, 2, 3, \ldots, w$. The **strong connectedness** in the graph of the matrix means that from any of the symbols in A we can reach any other using P.

If $P(a_j) = a_i a_k$, then a_j is called a **double node**.
If $P(a_j) = a_i$, then a_j is called a **simple node**.

The definition of C means that we have only simple and double nodes, given the binary nature of phyllotaxis. The hierarchy generated by C is the sequence $w_0, w_1, w_2, \dots$ of levels $w_T = P(w_{T-1})$, $T = 1, 2, 3, \dots$, where P behaving like an homomorphism, that is $P(a_i a_j) = P(a_i)P(a_j)$, defines the links between consecutive levels.

To show how a growth matrix generates a hierarchy with simple and bifurcated induction lines, or simple and double nodes, let us consider how the hierarchy in Figure 3.1 is derived from the growth matrix

$$\begin{bmatrix} 1 & 1 \\ 1 & 0 \end{bmatrix}$$

when a convenient w_0 is chosen. This matrix is known in the literature as the **Q-matrix**. It is a growth matrix of order 2 where the sum of the entries in each column is, as requested, greater than 0 and smaller than 3. It is interesting to notice that the consecutive powers Q^n, $n = 1, 2, 3, \dots$ of Q are:

$$\begin{bmatrix} F_{n+1} & F_n \\ F_n & F_{n-1} \end{bmatrix}.$$

The Q-matrix gives an alphabet with $w = 2$ symbols denoted by a_1 and a_2. The rules of production P of the words of the L-system are, based on the entries of the matrix, given by $P(a_1) = a_1 a_2$ and $P(a_2) = a_1$. (The second rule is due to the presence of the zero in the second column of the matrix.) Notice that from a_1 we can go to a_2, and from a_2 we can go to a_1 because of P. This is the meaning of the expression "strong connectedness." Starting with the axiom or word $w_0 = a_1 a_2$ on level $T = 0$, and applying the rules P, the skeleton of the hierarchy in Figure 3.1 is generated: In level $T = 1$ we have three points and the string $w_1 = a_1 a_2 a_1$ linked to $a_1 a_2$ in the same way as in the figure; in level $T = 2$ five points are generated from those of level $T = 1$, and the two levels are linked as in the figure.

The hierarchies with simple and double nodes only (bifurcating induction lines or not) generated by the Q-matrix with at least one double node in level 0, are called **Fibonacci hierarchies**. The numbers of nodes in the consecutive levels of a Fibonacci hierarchy are the consecutive terms of the Fibonacci-type sequence $\langle n, m, m+n, 2m+n, 3m+2n, \dots \rangle$, where $n < m \leq 2n$. The hierarchy is denoted by $\langle n, m \rangle$, where n and m respectively represent the numbers of nodes in levels $T = 0$ and 1 of the hierarchy. The

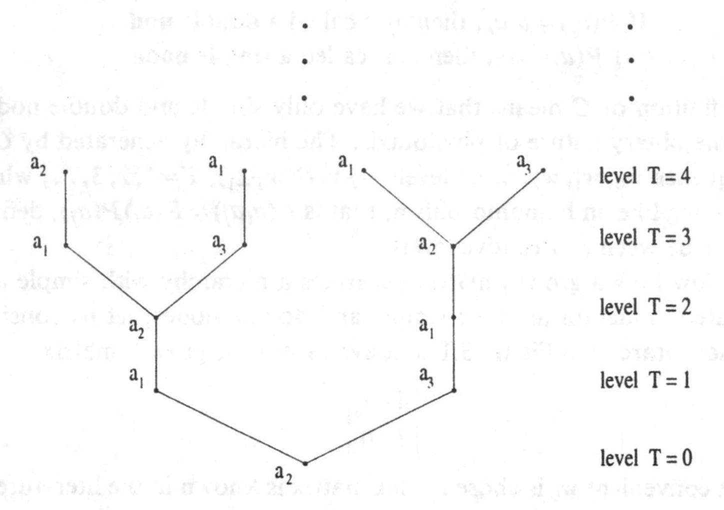

Figure 3.17. Hierarchy with simple and double nodes called the companion hierarchy of order 3, generated by the companion matrix of order 3. Rhythm of the hierarchy is $w=3$.

first inequality means that level 0 contains at least one double node. The second inequality relates to the fact that the hierarchy is made up by simple and double lines only. Church (1920b) observed that "in no case is the number in the parastichy pair (n, m) more than twice the other, ranging from (n, n) to $(n, 2n)$," so that $n \leq m \leq 2n$. The case $n=m$ concerns whorled patterns, and will be considered in Chapter 8. Figures 3.1 and 3.4 represent the hierarchies $\langle 2, 3\rangle$ and $\langle 3, 4\rangle$, respectively. They are both generated with the same alphabet and the same rules of production but from different axioms: $a_1 a_2$ in the first case and $a_2 a_2 a_1$ in the second case.

Here is another type of hierarchy with simple and double nodes. The companion hierarchy of order $w=2, 3, 4, \ldots$ is the hierarchy generated by the growth matrix $C=(c_{ij})$ where $c_{1\,w-1}=c_{1w}=c_{21}=c_{32}=c_{43}=\cdots=c_{w\,w-1}=1$, all the other entries being zero, known as a companion or **Frobenius matrix,** where a_{w-1} in the alphabet A is the only double node and where the axiom $w_0=a_{w-1}$. Figure 3.17 shows the companion hierarchy of order $w=3$, generated by the growth matrix

$$\begin{bmatrix} 0 & 1 & 1 \\ 1 & 0 & 0 \\ 0 & 1 & 0 \end{bmatrix},$$

with $P(a_1)=a_2$, $P(a_2)=a_1 a_3$, $P(a_3)=a_2$, and $w_0=a_1$. The companion hierarchy of order 2 is the Fibonacci hierarchy $\langle 1, 2\rangle$.

An L-system is a language or interface used to link the growth matrices and the hierarchies so as to be able to generate those hierarchies from the matrices. The property of strong connectedness in the graph of a growth matrix of order w does mean that the process by which a hierarchy is generated follows a rhythmic pattern coming back when w levels of the hierarchy have been generated. When this is the case, the other levels are generated according to the same rule defined by the matrix, so that the system begins to exhibit rhythm and to produce loops. A hierarchy can be generated by two or more growth matrices of different orders. The smallest value of w is called the preponderant period of the rhythm or simply the **rhythm of the hierarchy.**

In this chapter we have illustrated the theorem of Chapter 2 with hierarchies or multilevel systems and have learned to mathematically describe and generate those hierarchies that are relevant in phyllotaxis, that is, the hierarchies with simple and double nodes only. We will come back to the hierarchical representation of phyllotaxis in Chapter 6, in order to use it in a model for pattern generation. In the meantime, we will continue our investigation on pattern recognition with a fruitful model analyzed in Chapter 4 and applied in Chapter 5.

3.6. Problems

3.1 (1) Imagine a tree where each branch grows according to the following rule. The branch does not produce a new branch during the first year of its growth. The second year it generates a new branch, rests during one year, then ramifies again, and so on. The trunk and its prolongation is considered as a branch. Construct such a tree and show that for each year there is a Fibonacci number of branches (from Steinhaus, 1960).

(2) An organism has two states: young (y) and mature (m). It takes one time unit for a young organism (y) to become mature (m), and one time unit for a mature organism (m) to produce one offspring (y). The mature organisms live forever. Given that at time $t = 1$ there is one mature organism (m), generate the total numbers of organisms at successive time intervals (from Thornley and Johnson, 1990).

(3) If the mature organisms are considered as mother cells and the young organisms are daughter cells, reconstruct Berdyshev's (1972) multilevel system showing the cells linked to adjacent levels. Notice that the cells in each level are grouped in bunches of two's and one's, and that Church's sequence 2 1 2 1 2 - 2 1 2 1 2 - ... can be retraced.

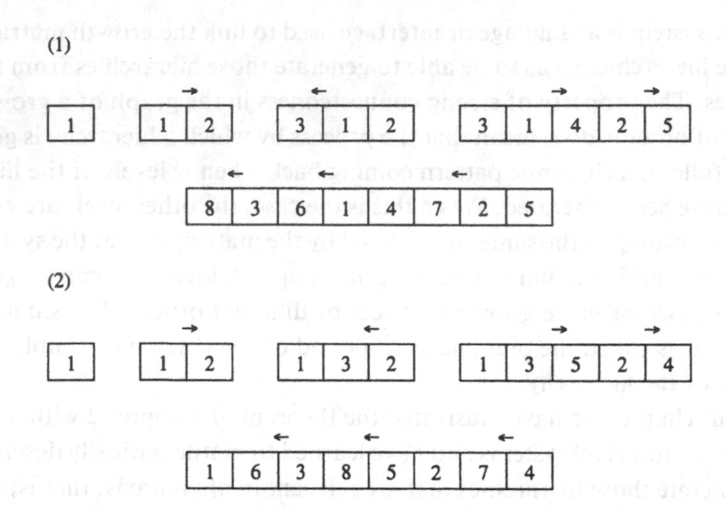

Figure 3.18. Two processes of cell division that generate Fibonacci numbers. (From Holland, 1972.)

3.2 A cell divides itself according to the diagram in Figure 3.18(1), where the direction of division alternates, starting with a right division. Each new cell skips a generation before starting to reproduce. Compare the sequence of numbers in each cycle of this process to the sequence of numbers in the consecutive levels of the hierarchy in Figure 3.1, and to the sequence of numbers obtained by projecting radially the points in Figure 1.4 on the circumference of the disk. Make the same comparison if each cell (except the first one that divides once), divides two times only while changing direction, as in Figure 3.18(2).

3.3 Kilmer (1971) reports McCulloch's algorithm for growing pine cones and other fruit displaying the Fibonacci sequence (from a posthumous publication of McCulloch). Figure 3.19 depicts a cone with phyllotaxis (8, 5). First a circle α is drawn and two scales β_1 and β_2 are placed. The scales γ_1, γ_2, and γ_3 are added at the junctions determined by the previous scales. There are junctions j^* between two scales of the same level and junctions $j^\#$ between two scales of two adjacent levels. At each level there is a Fibonacci number of scales. Retrace Church's sequence of one's and two's by working on the junctions. A fruit (13, 8) would give a sixth level, and the sequence 2 1 2 1 2 - 2 1 2 that can be extrapolated from the figure.

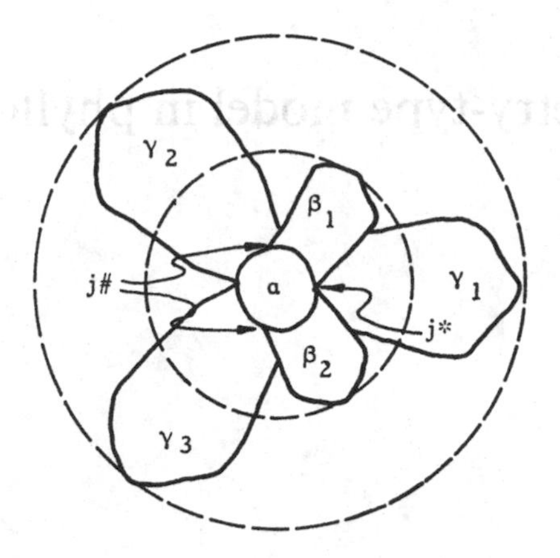

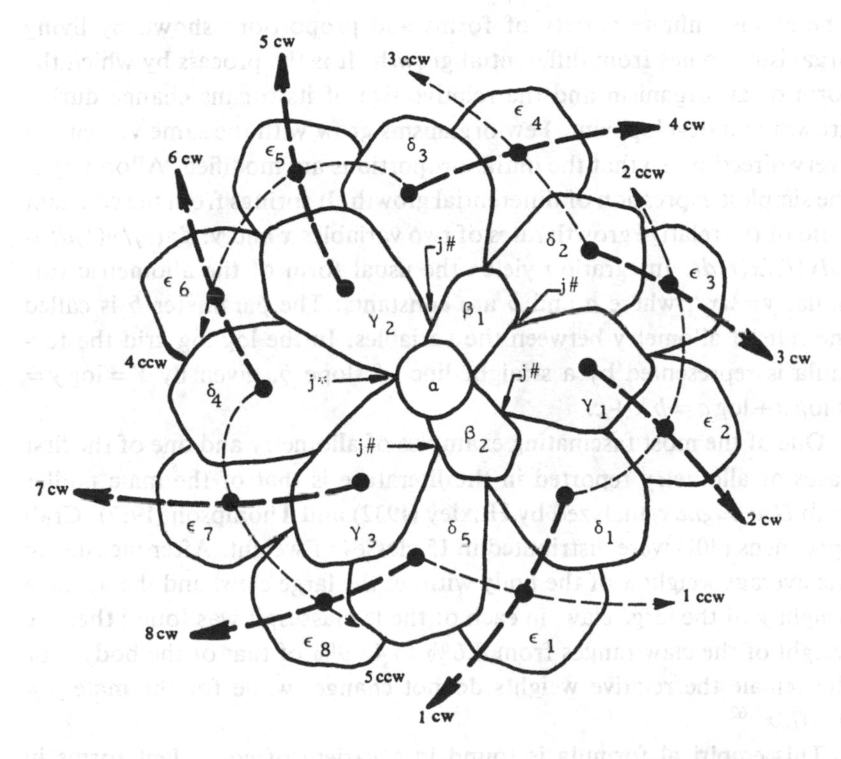

Figure 3.19. Illustration of McCulloch's algorithm by which the morphology of cones is compared to cellular division and to hierarchies of simple and bifurcated lines determined by two kind of junctions between the scales. (From Kilmer, 1971; reprinted by permission of Elsevier Science Publg. Co., Inc., copyright 1990.)

4

Allometry-type model in phyllotaxis

4.1. Differential growth in the plant apex

The almost infinite variety of forms and proportions shown by living organisms comes from differential growth. It is the process by which the form of an organism and the relative size of its organs change during growth and development. Few organisms grow with the same velocity in every direction, so that the initial proportions are modified. Allometry is the simplest expression of differential growth. It springs from the constant ratio of the relative growth rates of two variables x and y: $dy(t)/y(t)dt = bdx(t)/x(t)dt$. Integration yields the usual form of the allometric formula: $y = ax^b$, where a and b are constants. The parameter b is called the rate of allometry between the variables. In the log-log grid the formula is represented by a straight line of slope b, given by $Y = \log y = b \log x + \log a = bX + c$.

One of the most fascinating examples of allometry and one of the first cases of allometry reported in the literature is that of the male fiddler crab *Uca pugnax* analyzed by Huxley (1932) and Thompson (1917). Crab specimens (401) were distributed in 15 classes of weight. After measuring the average weight x of the body without the large claw, and the average weight y of the large claw, in each of the 15 classes, it was found that the weight of the claw ranges from 8.6% to 42.9% of that of the body. For the female the relative weights do not change, while for the male $y = 0.0073x^{1.62}$.

This empirical formula is found in a variety of equivalent forms in all sectors of biology, from embryology, taxonomy, and paleontology to ecology and zoology, between all sorts of variables. The formula is related to logistic formulae of growth and to Thompson's (1917) famous theory of transformations. Its great appeal is derived primarily from its simplicity. It is the most versatile expression for intra- and interspecies comparisons (see Jean, 1984c, for a general introduction to the subject of allometry).

It is well known that the organization and functioning of the shoot apex in plants is profoundly influenced by differential growth. According to Wardlaw (1965b), the position and size of the leaf primordia are fundamentally determined by the genetically controlled allometric growth pattern. However from the point of view of allometry, phyllotaxis has received very little if any mathematical or botanical consideration. Dormer (1965a, p. 468) wrote that

> practically nothing is known concerning the correlations of discontinuous variables with each other, and very little more of their correlations with variables of other classes. It is surprising that the subject should be so completely neglected. One should have expected at least that the extensive literature on meristematic variation, especially in flowers, would have led to some examination of the correlations involved, but nothing significant seems to have emerged. It is difficult to imagine the kind of relationships likely to arise, though presumably the Fibonacci numbers will occupy a specially prominent position.

This chapter proposes an allometry-type relation between phyllotactic parameters.

In his pioneering book Thompson (1917) complained that the progress in the domain of phyllotaxis was limited to the study of the divergence angle, representing the tangential spacing of the leaf primordia. This chapter is concerned with their radial spacing. It puts forward a mathematical model involving the plastochrone ratio R, and the angle of intersection γ of the opposed parastichy pair (m, n). This model will be used for practical assessment of phyllotactic patterns in Chapter 5, in conjunction with the theorem of Chapter 2 relating (m, n) and the divergence angle d.

4.2. The model

4.2.1. Derivation of the model

Figure 4.1 shows a n-parastichy and a m-parastichy, of a visible opposed parastichy pair (m, n), meeting at point mn in a cylindrical lattice (Section 2.5). The coordinates of points n and m, on each side of the y axis, are respectively

$$(x_n, nr) \quad \text{and} \quad (x_m, mr), \tag{4.1}$$

where r is the rise (see Section 2.5.1),

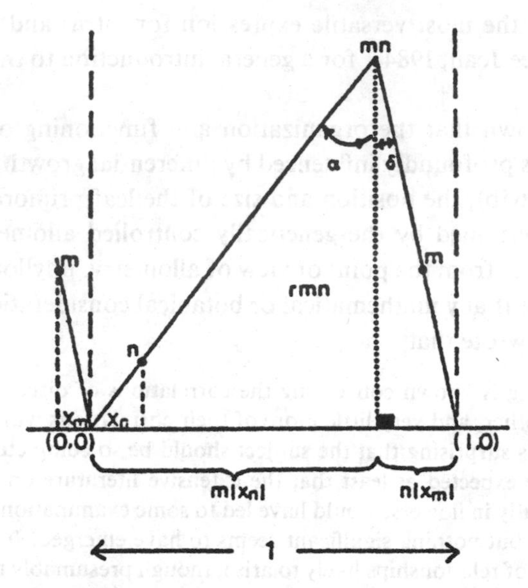

Figure 4.1. Parameters involved in the cylindrical representation of phyllotaxis. Two opposed parastichies, of the pair (m, n), going through points m and n are represented. Points (0, 0) and (1, 0) represent the same point, and x_n and x_m are the secondary divergences of points n and m.

$$x_n = v - nd \quad \text{and} \quad x_m = u - md, \tag{4.2}$$

(the secondary divergences of n and m), u and v, called the **encyclic numbers**, being the integers nearest to nd and md respectively, and

$$nu - mv = \pm 1. \tag{4.3}$$

The abscissas of m and n are of opposite signs. One can easily deduce from the figure that

$$mnr^2 - r\cot\gamma + x_n x_m = 0, \tag{4.4}$$

where $\gamma = \alpha + \delta$ is the angle of intersection of the parastichy pair.

Compare these relations with Equations (1.5), and (1.6). Notice that Equation (4.4) is similar to Equation (1.10). In fact with the usual relation

$$r = \ln(R)/2\pi \tag{4.5}$$

the cylindrical lattice becomes the spiral or centric lattice. Putting $\gamma = 90°$ in (4.4) or (1.10) yields the Richards formula [Eq. (1.11)], that is $\ln R = (-\Delta_n \Delta_m / mn)^{1/2}$ or $r = (-x_n x_m / mn)^{1/2}$.

Working from Equation (4.4) [or Eq. (1.10)] we easily obtain that

$$r = [\cot\gamma + (\cot^2\gamma - 4mnx_n x_m)^{1/2}]/2mm. \tag{4.6}$$

In the case of normal phyllotaxis [see sequence (2.1)] where

$$m = J(F_k t + F_{k-1}) \quad \text{and} \quad n = J(F_{k-1} t + F_{k-2}),$$

$t = 2, 3, 4, \ldots$, and $J = 1, 2, 3, \ldots$, it can be calculated using Binet's formula ($F_k = [\tau^k - (-\tau)^{-k}]/\sqrt{5}$, giving $F_k \approx \tau^k/\sqrt{5}$) that for large k

$$x_m \approx (-1)^k/\tau^{k-1} J(\tau t + 1), \qquad x_n \approx (-1)^{k-1}/\tau^{k-2} J(\tau t + 1), \tag{4.7}$$

$$m \approx J\tau^{k-1}(\tau t + 1)/\sqrt{5}, \qquad n \approx J\tau^{k-2}(\tau t + 1)/\sqrt{5}. \tag{4.8}$$

In the case of anomalous phyllotaxis [see sequence (2.2)], for

$$m = 2F_{k-1} t + F_k + F_{k-2} \quad \text{and} \quad n = 2F_{k-2} t + F_{k-1} + F_{k-3},$$

we have

$$x_m \approx (-1)^{k-2}/\tau^{k-2}(2\tau t + \tau + 2), \qquad x_n \approx (-1)^{k-3}/\tau^{k-3}(2\tau t + \tau + 2), \tag{4.9}$$

$$m \approx \tau^{k-2}(2\tau t + \tau + 2)/\sqrt{5}, \qquad n \approx \tau^{k-3}(2\tau t + \tau + 2)/\sqrt{5}. \tag{4.10}$$

Putting these values in Equation (4.6) gives (Jean, 1983a)

$$r = p(\gamma)(m+n)^{-2}, \tag{4.11}$$

or using Equation (4.5),

$$\ln R = 2\pi p(\gamma)(m+n)^{-2} \tag{4.12}$$

where

$$p(\gamma) = \tau^3[\sqrt{5} \cot \gamma + \sqrt{(5 \cot^2 \gamma + 4)}]/2\sqrt{5}. \tag{4.13}$$

The model desired is given by Equation (4.11) or (4.12), depending on the representation used, and is true for any Fibonacci-type sequence. The model shows an explicit relation among γ, r (or R), and $m+n$, and an implicit relation between $m+n$ and the divergence angle d (governed by the theorem of Chapter 2). In contrast to the mechanistic models in phyllotaxis trying to control the behavior of the parameters d and r, the model expressed by Equation (4.11) proposes a description of the patterns on the basis of r and (m, n), because of the close relationship between (m, n) and d.

The value of $p(\gamma)$ decreases monotonically from ∞ to 0 as γ increases from 0 to 180°. Then for r fixed, $m+n$ decreases monotonically. For any pair (m, n) and two symmetrical values of γ around 90°, that is $90 \pm x$, where x is in degrees, we have:

$$p(90-x)\,p(90+x) = \tau^6/5 = p(90)^2.$$

4.2.2. Interpretation of the model

As emphasized by Wardlaw (1965b) "for a given divergence angle d, the contact parastichy pair (m, n) depends on the plastochrone ratio R and the primordium shape" (that is, the angle γ). Equation (4.12) is the general algebraic expression of this observation; it exhibits the relationship between these parameters. In order to establish this relationship we need to replace the ideas of the shape of the primordia and of contact parastichy pairs by more general concepts. This is a model for pattern recognition, which expresses the relation between the main parameters involved at the shoot apex. Equations (4.11) and (4.13) also control the dynamics of the cylindrical lattice experiencing hyperbolic transformations (see Appendix 9) under the variation of (m, n), d, r, γ, and g (the girth of the cylinder).

Equations (4.11) and (4.12) are approximations of the exact relations existing in the cylindrical and centric representations respectively, such as Equations (4.4) and (1.10). This approximation has been derived mathematically, that is, without using computerized approximation procedures. The derivation comes from the very simple fact that for large values of k, $F_k \approx \tau^k/\sqrt{5}$. It can be calculated that the percentage of deviation $s(\gamma) = 100|1 - r/r^*|$ between the exact value r^* of the rise in the cylindrical representation, and the value r of the rise in the model, for γ in the interval $(0°, 180°)$, has an absolute minimum around 90°. This is convenient given that the conspicuous parastichy pairs are those we have a practical interest in.

The best correlation available is given by the conspicuous pair. When $\gamma = 90°$, $p(\gamma) = \tau^3/\sqrt{5}$, and we have, in the centric and cylindrical lattices respectively

$$y = 5.16942x^{-2}, \tag{4.14}$$

$$y = 1.89443x^{-2}, \tag{4.15}$$

where $y = \ln R$ or r respectively, and $x = m + n$. If R and (m, n) are such that Equation (4.14) is verified, then it can be concluded that the opposed parastichies in the pair are orthogonal. The allometric line passes through the origin of coordinates for $\gamma = 121.43°$; then we have $r = (m+n)^{-2}$. The accuracy and reliability of the model even for patterns expressed by low secondary numbers [the numbers at the beginning of the phyllotactic sequences (2.1) and (2.2)] can be estimated by considering that the values of R given by Equation (4.14) and those given by the Richards formula $\ln R = (-\Delta_n \Delta_m / mn)^{1/2}$ are about the same (see Problem 4.2).

Table 4.1. *Value of $s(90°)$ (in %) and value of γ minimizing $s(\gamma)$ for common patterns $(m(k), n(k))$*

(m, n)	$s(90°)$	γ min
$(1,2)$ $(k=1)$	0.868	98°
$(2,3)$ $(k=2)$	0.016	91°
$(3,5)$ $(k=3)$	0.049	94°
$(5,8)$ $(k=4)$	0.011	93°
$(3,4)$ $(k=2)$	0.271	86°
$(4,7)$ $(k=3)$	0.182	95°
$(7,11)$ $(k=4)$	0.023	92°
$(4,5)$ $(k=2)$	0.588	84°
$(5,9)$ $(k=3)$	0.288	96°
$(9,14)$ $(k=4)$	0.027	92°
$(2,5)$ $(k=1)$	3.02	103°
$(5,7)$ $(k=2)$	0.106	88°
$(7,12)$ $(k=3)$	0.117	95°

Moreover for γ fixed, $s(\gamma)$ tends towards zero as k increases, so that r is asymptotically equal to r^*. For (1, 1) patterns $s = 2.524\%$ and $s(m, n) = s(Jm, Jn)$. For fixed k the value of s increases as t increases: at the limit, in the case of normal phyllotaxis, the value of s is, for $k = 1, 2, 3, 4, \ldots$, 100%, 2.52%, 0.87%, 0.016%, ..., respectively. It follows that the only significant deviations of r from r^* occur when $k = 1$: for the patterns (1, 3), (1, 4), (1, 5), (1, 6), ..., we have respectively $s(90°) = 5.61$, 10.97, 15.91, 20.28, ... The deviation is generally smaller than 1% for orthogonal pairs. [See Table 4.1, established for normal conspicuous parastichy pairs $(F_k t + F_{k-1}, F_{k-1} t + F_{k-2})$, $t = 2, 3, 4$, and anomalous conspicuous parastichy pairs $(4F_k + F_{k+1} + F_{k-1}, 4F_{k-1} + F_k + F_{k-2})$.]

In the allometric formula $y = ax^b$, putting $x = m + n$, $b = -2$, and $y = r$ or $\ln R$, gives the Equation (4.11) or (4.12), respectively. It can be concluded that for any given value of γ, the allometric line of slope -2 in the log-log grid, given by Equation (4.11), contains all types of spiral phyllotaxis, including the visible opposed parastichy pairs (m, n) expressed by low secondary numbers. In most of the published figures on allometry, the number of plotted points is generally small, and the general standard of graphical work from which the relation is established is often poor. Here we have, for γ fixed, a definite line free from subjective bias and involving an infinite number of points.

If one observes a plant shoot with a visible opposed parastichy pair (m, n), plastochrone ratio R, and angle of intersection γ of the parastichies in the pair, and if the observed divergence angle between the primordia is around the angle corresponding to the Fibonacci-type sequence $\langle \dots, n, m, m+n, 2m+n, 3m+2n, \dots \rangle$, then

the point $(m+n, \log R)$ *is on the straight line of slope* -2

and intercept equal to $2\pi p(\gamma) \log e$ in the log-log grid. Also, for any given γ, whatever be the pair (m, n), the values of r and $\log R$ can be read on log-log paper from the line of slope -2 and intercept $p(\gamma)$ and $2\pi p(\gamma) \log e$, respectively.

The Williams–Brittain model (1984, see Appendix 5), with which it is possible to reproduce patterns of the types $(t, t+1)$ and (t, t) for various values of γ, is in agreement with the allometry-type model. Equation (4.12) shows that for fixed $m+n$, if γ increases, R decreases. And for γ fixed, if $m+n$ increases, R decreases. The results in the Williams–Brittain model exhibit similar behavior. From the values of two of the parameters R, γ and $m+n$, the third one, deduced from any of the two models, has practically the same value when R or γ is deduced, and has always the same value when (m, n) is deduced (using the correspondences in Table 2.1). The deduced values of R or γ are equal to many decimal places when the divergence angle produced by the Williams–Brittain model is close to the limit-divergence angle corresponding to (m, n), even for low systems such as $(2, 1)$ and $(3, 2)$.

Equation (4.12) is a remarkable symmetry property. The exponent 2 represents the fractal dimension of the set of phyllotactic patterns. This dimension is the slope of the linear relation between the logarithm of the surface of a disk to the logarithm of its radius. This means, for example, that by plotting on log–log paper the area covered by the capitulum of a sunflower versus the radius of the capitulum, the packing has a fractal dimensionality of exactly 2 (see also Bursill & XuDong, 1989, and Rigaut, 1987, for remarks on this subject).

4.3. Generalized Coxeter formula

The model in Equation (4.11) gives an approximation of the value of r as a function of the visible opposed pair (m, n) with angle of intersection γ. It can be used to generalize and approximate what I call (Jean, 1984a) the **Coxeter formula** which arises from the study of pineapples. This formula is given by

$$r = (F_k F_{k+1})^{-1/2} \tau^{-k-1/2}. \tag{4.16}$$

It was established for the Fibonacci sequence $\langle 1, 1, 2, 3, 5, 8, \ldots \rangle$ (for which $d = \tau^{-2}$) and orthogonal intersection of the pair (F_k, F_{k+1}) such that F_k and F_{k+1} are the neighbors of the origin of the cylindrical lattice (thus giving a conspicuous pair). Coxeter (1972) looked for a value of r such that the three points F_{k-1}, F_k, and F_{k+1} are the closest to 0, given that with this value the scales of the pineapple are hexagonal and the whole fruit shows three families of contact parastichies and thus two contact parastichy pairs. He calculated the value of r for which the pair (F_{k-1}, F_k) is also orthogonal, and then chose the "good value" $r = 1/F_k \tau^k$, intermediate between the former two values of r.

Putting $\gamma = 90°$, $m = F_{k+1}$, and $n = F_k$ in Equation (4.11), $r = \sqrt{5}/\tau^{2k+1}$ by using the approximation $F_k \approx \tau^k/\sqrt{5}$. Putting this approximation of F_k in the Coxeter formula [Eq. (4.16)] gives the same value of r. For the orthogonal pair (F_{k-1}, F_k) the two formulae give $r = \sqrt{5}/\tau^{2k-1}$. The "good value" of r is thus

$$r = \sqrt{5}/\tau^{2k}, \tag{4.17}$$

and it approximates the "good value" obtained by Coxeter.

The same process of finding a "good value" of r can be performed by means of Equation (4.11), whatever be the two contact parastichy pairs. Consider the Fibonacci-type sequence

$$\langle a, b, a+b, 2b+a, \ldots, n = F_{k-1}b + F_{k-2}a, m = F_k b + F_{k-1}a, \ldots \rangle,$$

and the value of d corresponding to that sequence (see Table 2.1). When (m, n) is orthogonal,

$$r = \tau^3/\sqrt{5}(m+n)^2 = \sqrt{5}/\tau^{2k-3}(\tau b + a)^2; \tag{4.18}$$

and when $a = 1$ and $b = t$,

$$r = (mn)^{-1/2} \tau^{-k+3/2}/(\tau t + 1), \tag{4.19}$$

is a formula that generalizes Equation (4.16) (put $t = 1$). For three neighbors $m-n, n, m$ of the origin the value of r is approximately given by

$$r = \sqrt{5}/(\tau a + b)^2 \tau^{2k-4}. \tag{4.20}$$

Putting $a = b = 1$ in Equation (4.20) yields Equation (4.17).

4.4. Derivation of the Richards phyllotaxis index

Particular implications of Equation (4.12) include Richards's (1951) phyllotaxis index. This index is the strange expression

$$\text{P.I.} = 0.379 - 2.3925 \log\log R, \tag{4.21}$$

implicitly centered on the Fibonacci sequence. Even with this restriction, this index, meant for practical pattern assessment, is still in use today. The index is generally considered to be one of the articles of the botanists' faith. One of the reasons for this is Richards's intuitively precise but abstruse presentation of it. Considering the historical importance of the Richards index and his pioneering work in general, this section shows that the index can be deduced from Equation (4.12) which is able to take care not only of the Fibonacci sequence but of any Fibonacci-type pattern and even of whorled patterns.

Consider the sequence of normal phyllotaxis $\langle 1, t, t+1, 2t+1, 3t+2, 5t+3, \ldots \rangle$, with general terms $m = F_k t + F_{k-1}$, and $n = F_{k-1} t + F_{k-2}$, and the orthogonal opposed parastichy pairs (m, n) and $(m, m+n)$, with corresponding plastochrone ratios $R(m)$ and $R(m+n)$. If $m(t)$ and $n(t)$ are high secondary numbers (k is large), then it is easily deduced from Equation (4.12) with $\gamma = 90°$ that

$$[\log\log R(m) - \log\log R(m+n)]/\log \tau^2 \approx 1.$$

Define $x(m)$ by

$$x(m) = 1 - [2.3925 \log(m^2/5.169)],$$

where $[\cdot]$ means the fractional part of, where 2.3925 represents $1/\log \tau^2$, and 5.169 is the approximate value of $2\pi\tau^3 \log e/\sqrt{5}$ given by Equation (4.14) when the parastichy pair is orthogonal. Then the value of y in the following expression is always (almost) an integer:

$$y = x(m(t)) - 2.3925 \log\log R(m(t)). \tag{4.22}$$

For the Fibonacci sequence $(t = 2, F_k)$, and the first accessory sequence or the Lucas sequence $(t = 3, L_k)$ we have, respectively

$$y_1 = 0.379176 - 2.3925 \log\log R(F_k),$$

$$y_2 = 0.706900 - 2.3925 \log\log R(L_k).$$

We recognize the P.I. in y_1. This index has other forms, known to Richards, which can also be deduced from the model giving

$$y = 4.785 \log(m+n) + x(m) - 1.707$$

where 4.785 represents $1/\log \tau$. For the Fibonacci and Lucas sequences it follows that

$$y_1 = 4.785 \log F_k - 1.327734221,$$

$$y_2 = 4.785 \log L_k - 1,$$

representing two parallel straight lines giving the integer y as a function of $\log F_k$ and $\log L_k$ respectively.

From Equation (4.19) we have

$$\log R = 2\pi(mn)^{-1/2}\tau^{-k+3/2}/(\tau t+1)\log e. \qquad (4.23)$$

And for the Fibonacci sequence it follows that (m and n are functions of k)

$$\log R = 2\pi(F_k+2F_{k+1})^{-1/2}\tau^{-k-1/2}\log e,$$

that is $\log R = 1.68646192(0.618033989^{2k-1}/F_{k+2}F_{k+1})^{1/2}$, a formula introduced by Richards (1951, with that many decimal figures) working with the Fibonacci angle and orthogonal Fibonacci parastichy pairs in the centric representation. [Equation (4.12) has been derived from the cylindrical representation.] From Equation (4.23) we have for the Lucas sequence

$$\log R = 2\pi[(1/\tau^{2k-1})/L_{k+1}L_k]^{1/2}\log e/(\tau^2+1),$$

that is $\log R = 0.7542086988(0.6180339887^{2k-1}/L_{k+1}L_k)^{1/2}$, a formula introduced, with all those decimals, by Richards (1951) for the orthogonal intersection of the parastichies in a distribution of leaves with divergence angle $(3+1/\tau)^{-1}$.

From Equations (4.22) and (4.23), it can be deduced that

$$y = x(m) - (\log\tau^2)^{-1}\log[2\pi(mn)^{-1/2}\tau^{-k+3/2}\log e/(\tau t+1)].$$

In particular, we can express y_1 and y_2 by

$$y_1 = k/2 + 1.19625\log(F_{k+1}F_k) - 0.414,$$
$$y_2 = k/2 + 1.19625\log(L_kL_{k-1}) + 0.75.$$

(Complementary developments on this index can be found in Jean, 1979b.)

Equation (4.21) indicates that if measurements taken on a shoot apex conforming to the Fibonacci divergence angle lead to a phyllotaxis index of 4, it may be inferred that the fourth Fibonacci system (8, 5) is exactly orthogonal. If the index obtained is 4.5, no parastichy pair intersects orthogonally, but the systems (8, 5) and (8, 13) are equally removed from orthogonality. The P.I. is obviously an integer but only when the orthogonal parastichy pair is expressed by the Fibonacci sequence.

4.5. The Pattern Determination Table

The allometry-type model has a very practical value. It allows us to organize phyllotactic data in a more coherent manner than is permitted by earlier models, and it considers the diversity of relative sizes and shapes of leaf or reproductive primordia in relation to the sizes of shoot apices. By examining phyllotactic patterns in plant shoots, we can determine some parameters, but others may be more difficult to determine, such as the value of the plastochrone ratio R. We thus need a reliable and general practical tool that will allow us to deduce easily certain parameters from other given parameters, whatever be the phyllotactic sequence involved.

The aim of this section is to learn how to characterize a phyllotactic pattern, by using Equation (4.12). All the possible situations involving the parameters (m, n), d, γ and r or R can be dealt automatically with in a Pattern Determination Table. On a theoretical level, given some of the parameters, the others can be deduced. Table 4.2 gives the possibilities available from the model.

The Pattern Determination Table (Table 4.3, from Jean, 1987a) illustrates the relation between R and (m, n) given by Equation (4.12) for various values of γ. Take the natural logarithm of the figures in any column to see a line of slope -2. This table is a very easy and accurate tool to use; it represents an improvement over earlier methods of phyllotactic pattern characterization as we shall see in Chapter 5. It provides a useful way to approximate the value of R in particular. The table is built for values of γ between 62° and 118° and can be extended to show any angle between 0° and 180°. Then we would notice that the numbers in each row, representing the possible values of R to three decimal places, decrease monotonically to 1 from left to right.

Let us use the table to decide some of the cases reported in Table 4.2:

Case 2: The fundamental theorem of phyllotaxis (Chapter 2) gives an interval for d, while the Pattern Determination Table gives the value of γ. For example if $(m, n) = (5, 6)$, $m + n = 11$, and if $R = 1.12$, then the table gives $\gamma \approx 82°$.

Case 3: Settled in much the same way as case 2, except that this time R is deduced. For example, if the pair $(m, n) = (3, 5)$ shows $\gamma \approx 80°$, then the table gives, in the row where $m + n = 8$, $R \approx 1.255$. This case is considered in detail in Section 5.3.

Case 4: Consider, for example, a pattern for which d is near 99.5°, and for which $R = 1.09$. We know from Table 2.1 that the visible opposed pairs are (1, 3), (4, 3), (4, 7), ..., for which $m + n = 4, 7, 11, \ldots$. More generally, given the value of d we can use Table 2.1 for the sequence involved. The problem is now to determine the conspicuous pair. It is the plastochrone

Table 4.2. *Deduction of phyllotactic parameters from known data, using the proposed method*

Case	Given	Can be deduced
1	(m,n) and d	Insufficient data for γ and R
2	(m,n) and R	γ, interval for d
3	(m,n) and γ	R, interval for d
4	d and R	Visible opposed pairs (m,n), the conspicuous pair for which γ is closest to 90°
5	d and γ	Visible opposed pairs (m,n), insufficient data for R
6	R and γ	$m+n$ is known, insufficient data for d
7	d, γ, and R	(m,n)

Note: (m,n) is the visible opposed parastichy pair with angle of intersection γ, R is the plastochrone ratio, and d is the divergence angle.
Source: From Jean, 1987a.

ratio that determines among the numerous visible opposed parastichy pairs, the one that is conspicuous. In column $m+n$ of Table 4.3, look for the presumed values of $m+n$, and in each of the corresponding rows look for the given value of R (using interpolation if necessary). In the particular example here, the value $m+n=11$ gives $\gamma \approx 94°$, while the other values of $m+n$ give angles farther from 90°. We can thus conclude that the pair (4, 7) is the conspicuous pair. In cases of normal phyllotaxis with $J \geq 2$, m and n will have a common factor.

The Pattern Determination Table is a sensitive numerical tool whose utility and versatility will be abundantly illustrated in Chapter 5 with data from various authors. The Richards phyllotaxis index and the Maksymowych–Erickson method (1977) for evaluating phyllotactic patterns will also be discussed in this chapter.

4.6. Apical size and phyllotaxis of a system

4.6.1. Church bulk ratio

Equation (4.12) provides a basis for the analysis of the relations between apical size, primordium size, the **plastochrone** P (the time elapsed between the emergence of consecutively initiated primordia) and the phyllotactic pattern (m,n). Equation (4.12) indicates that the higher the value of $m+n$, the smaller R becomes. It is known that when R decreases, the size of the apical meristem increases relatively to the size of the primordium. It is

Table 4.3 *Pattern Determination Table. Possible values of R for patterns (m, n) with angle of intersection γ between 62° and 118°*

$m+n$	γ														
	62	66	70	74	78	82	86	90	94	98	102	106	110	114	118
3	10.225	8.463	7.142	6.128	5.332	4.695	4.179	3.753	3.398	3.099	2.844	2.624	2.434	2.268	2.122
4	3.698	3.325	3.022	2.772	2.564	2.387	2.235	2.104	1.990	1.889	1.800	1.721	1.649	1.585	1.527
5	2.309	2.157	2.029	1.921	1.827	1.745	1.673	1.610	1.553	1.503	1.457	1.415	1.378	1.343	1.311
6	1.788	1.706	1.635	1.573	1.520	1.472	1.430	1.392	1.358	1.327	1.299	1.273	1.249	1.227	1.207
7	1.533	1.480	1.435	1.396	1.360	1.329	1.300	1.275	1.252	1.231	1.212	1.194	1.178	1.162	1.148
8	1.387	1.350	1.318	1.290	1.265	1.243	1.223	1.204	1.188	1.172	1.158	1.145	1.133	1.122	1.112
9	1.295	1.268	1.244	1.223	1.204	1.187	1.172	1.158	1.146	1.134	1.123	1.113	1.104	1.095	1.087
10	1.233	1.212	1.194	1.177	1.163	1.149	1.137	1.126	1.116	1.107	1.099	1.091	1.083	1.076	1.070
11	1.189	1.172	1.157	1.144	1.133	1.122	1.112	1.103	1.095	1.088	1.081	1.074	1.068	1.063	1.058
12	1.156	1.143	1.131	1.120	1.110	1.101	1.093	1.086	1.079	1.073	1.067	1.062	1.057	1.053	1.048
13	1.132	1.120	1.110	1.101	1.093	1.086	1.079	1.073	1.067	1.062	1.057	1.053	1.049	1.045	1.041
14	1.113	1.103	1.094	1.087	1.080	1.074	1.068	1.063	1.058	1.053	1.049	1.045	1.042	1.038	1.035
16	1.085	1.078	1.072	1.066	1.061	1.056	1.052	1.048	1.044	1.041	1.037	1.035	1.032	1.029	1.027
21	1.049	1.045	1.041	1.038	1.035	1.032	1.030	1.027	1.025	1.023	1.022	1.020	1.018	1.017	1.015

Source: From Jean, 1987a.

thus expected that the relative apical radius L, the relative apical area A, and the relative apical volume V will vary in the same direction as the increasing order of phyllotaxis: relations would apply such as $L=c_1(m+n)$, $A=c_2(m+n)^2$, and $V=c_3(m+n)^3$, where c_1, c_2, c_3 are constants. We will see in this section that this is the case, with $L=1/B$ where B is the Church bulk ratio and A is the Richards area ratio. These relations deal with the range of relative apical and primordial sizes observed in the variety of phyllotactic patterns. The values of B and P will be shown to decrease proportionally to $m+n$ and $(m+n)^2$ respectively (Propositions 2 and 5 below). The relative apical size expressed, for example, by A or B has never before been related to the parastichy pair (m, n) by mathematical formulas, although it is intuitively widely stated, in agreement with the proposed formulas, that apical enlargement produces a shift in the contact pair from (m, n) to $(m, m+n)$, given that d remains unchanged.

Church (1904b) defined the bulk ratio for orthogonal systems expressed by consecutive terms of the sequence $\langle \ldots, n, m, m+n, 2m+n, 3m+2n, \ldots \rangle$ as the ratio of the radius of the primordium, assumed to be circular, to the apical radius. It is the sine of half the angle, at the center of the apex, made by the primordium. Church's system of opposed parastichies in the (s, μ) plane of polar coordinates is defined by $\ln s = \ln c + m\mu/n + (2k-1)\pi/n$, $k=1,2,3,\ldots,m-1$, $\ln s = \ln c - n\mu/m + (2j-1)\pi/m$, $j=1,2,3,\ldots,n-1$. These opposed families of spirals are orthogonal given that the product of the coefficients of μ is -1.

Proposition 1 (Church, 1904b). *The bulk ratio for the system of orthogonal opposed parastichies* (m, n) *with circular primordia is given by*

$$B = \sin(\pi/(m^2+n^2)^{1/2}). \tag{4.24}$$

Proposition 1 is easily proved by transforming the spirals into straight lines with the relations $x=\ln s$ and $y=\mu$. These lines produce a tessellation of squares whose sides are equal to $2\pi/(m^2+n^2)^{1/2}$. The center of the square made by the lines $j=0$, $j=1$, $k=0$ and $k=1$ is $(\ln c, 0)$, and the equation of the inscribed circle is $(\ln(s/c))^2+\mu^2=\pi^2/(m^2+n^2)$.

Proposition 2. *For the orthogonal system* (m, n),

$$B^2 \approx (\pi \ln R)/2, \tag{4.25}$$

and the relative apical radius $L=1/B$ *is given by*

$$L \approx (m+n)/(\pi(\tau^3/\sqrt{5})^{1/2}). \tag{4.26}$$

Putting $W = \pi \ln R/2$, the following expressions are better approximations for B than the one given in Proposition 2: $(W - W^2/3)^{1/2}$, and $(W - W^2/3 + 2W^3/45)^{1/2}$. Equation (4.26) will be applied to data in Chapter 5.

4.6.2. Richards area ratio

Richards (1951) wrote that the plastochrone ratio R is rigidly related to the ratio of the transverse components of two areas – that of the central apex and that of the newly initiated primordium. Richards assumed that any change in size of the bare apex from one plastochrone to another is sufficiently small to be negligible. The apex expands uniformly during the plastochrone to its maximal area C, and then surrenders a certain constant amount of its area to initiate a primordium. The area ratio, or relative mean apical area, denoted here by A, is defined by Richards by

$$A = \frac{\text{mean apical area}}{\text{primordium area on initiation}}.$$

Considering that the system expands exponentially from area c to area C, the mean apical area is $(C-c)/(\ln C - \ln c)$, and the primordium area is $C-c$, so that $A = 1/\ln(C/c)$. The maximal and minimal shapes of the apex being geometrically similar, their areas are similarly proportional to the square of corresponding linear dimensions. Considering thus that the apex expands from radius 1 to radius R it follows that $C/c = R^2$. The following proposition is proved.

Proposition 3 (Richards, 1951). *The area ratio is given by*

$$A = 1/(2 \ln R). \tag{4.27}$$

Assuming that p is the radius of the primordium on initiation and q the radius of the apex at its mean transverse area, we have $A = (q/p)^2$, $p^2 = 2q^2 \ln R$, and, from Equation (4.25), $p/q = 2B/\pi^{1/2}$ showing the relation between two similar parameters, p/q and B, in two different systems, B being obviously smaller than p/q. It is easily seen from Equations (4.25) and (4.27) that AB^2 is approximately equal to $\pi/4$ for all orthogonal systems (m, n), that is

$$A \approx \pi L^2/4,$$

a new relation involving both the Richards area ratio A and the Church bulk ratio $1/L$.

Assuming that the angle at the apex of the cone tangent to the apical surface at the level of initiation of primordia is 2μ, $p^2 = 2q^2 \ln R/\sin\mu$. This relation is valid for the dome-shaped apex (Richards, 1951) (where p is now the primordium area on the apical surface), and p/q becomes still larger than the previous value in the plane.

A similar parameter, r_p/r_c, the constant ratio of the radius of the primordium to its distance from the center of the system, is determined by trial and error by Williams (1975). He uses it in a simple geometrical procedure that allows him to construct the spiral pattern (m, n) of tangent circular primordia in the plane by distributing the centers of the consecutive primordia with the limit divergence angle corresponding to (m, n), and with the value of R given by Richards for (m, n). For the cases he illustrated, this value is slightly smaller than B, but closer to B than B is to p/q.

Proposition 4 (Jean, 1987a). *For the orthogonal system* (m, n), *where m and n are consecutive terms of* $\langle \ldots, n, m, m+n, 2m+n, 3m+2n, \ldots \rangle$, *the area ratio is given by*

$$A = (\sqrt{5}/4\pi\tau^3)(m+n)^2. \tag{4.28}$$

The phyllotaxis (m, n) of a system may be regarded as the resultant of the relative velocities of two growth processes: the rate of expansion of the apex and the rate of production of the primordia. Equation (4.28) is a mathematical confirmation of the observation that when the size of the apical meristem increases in proportion to the size of the primordia (that is when A increases or R decreases), the phyllotaxis (m, n) of the system rises. It is surprising that Hackett, Cordero, and Srinivasan (1987), working on *Hedera*, arrived at the opposite conclusion. They noticed an increase in apical area and a decrease in phyllotaxis. But these authors never mention the area of the primordium: it is the area ratio that must be observed to increase, not the apical area only. In agreement with Equation (4.28), Richards (1951) stated that "the absolute sizes of apex and primordium are immaterial, and provided their ratio is constant, phyllotaxis is fixed." Working on *Silene*, Lyndon (1977) reached experimentally the same conclusion that it is the size of the apical dome relative to the size of the primordia which is important, rather than the absolute sizes, so that "if the size of the primordia remains constant then an enlargement of the apical dome will lead to a change to a higher order of phyllotaxis, which is what occurs in *Silene*" (Lyndon provided the apical volumes, but not the phyllotaxes of the systems).

4.6.3. The plastochrone P

Besides the relative mean apical area A given by $1/(2\ln R)$, we can consider the relative mean apical radius given by $1/\ln R$ and the relative mean apical volume shown to be given by $1/(3\ln R)$.

Richards (1951) emphasized that R may be expressed entirely in terms of rates of change within the apex. The expression $\ln(C/c)$ discussed earlier is the relative growth rate $d(a(t))/a(t)\,dt = r_A$ of the apical area per plastochrone, that is,

$$\ln(C/c) = \int_{t=0}^{t=1} d(a(t))/a(t)\,dt = r_A,$$

where $a(1) = C$ and $a(0) = c$. Then $2\ln R(P) = r_A P$ where P is the plastochrone.

Richards considered the radial relative growth rate of the apex, denoted here by r_L. When the factors affecting phyllotaxis are constant, he concludes that $\ln R(P) = r_L P$. We can also consider the relative growth rate of the apical volume, denoted by r_V, which takes the form $3\ln R(P) = r_V P$. Lyndon (1977) considered the relative growth rate r_V, and a parameter V. This is $\ln(C/c)$ when C and c are not measured one plastochrone apart (as in r_V) but at the same plastochrone. Then the relation $r_V > V$ expresses an enlargement of the apical dome and $r_V < V$ a decrease (his measurements show a slight difference between these two parameters). Charles-Edwards et al. (1979) consider the relative growth rate r_W of the mass W_a of the apical dome: $dW_a/W_a dt = r_W$.

Lyndon (1977) asks what controls the magnitudes of the relative growth rate of the plastochrone and of the proportion of the meristem that is used to form the new apical dome? Thornley and Cockshull (1980) propose that $P = \ln(1 - W_p/W_a)/(r_W - b)$ where W_p is the dry weight of the primordium just after initiation, W_a is the dry weight of the apical dome just before primordial initiation, and b is a constant. From the relations for r_L, r_A, and r_V, and from Equation (4.12) we can state a formal relation controlling the parameters.

Proposition 5 (Jean, 1987a). *The plastochrone P is related to the phyllotaxis (m, n) of a system by*

$$P = (2\pi p(\gamma)/r_G)(m+n)^{-2}, \tag{4.29}$$

where $r_G = r_L$ or $r_A/2$ or $r_V/3$, r_L, r_A, and r_V being respectively the radial relative growth rate of the apex, the relative growth rate of its surface, *and of its volume, given by* $\ln R(P)/P = r_G$, *and where* $p(\gamma)$ *is given by Equation* (4.13).

Equation (4.29) is in agreement with Richards's observations according to which (1) the phyllotactic pattern depends on the radial relative growth rate of the apical surface, the rate of leaf initiation (represented by the value of P), and the size of a leaf primordium on initiation; and (2) when the phyllotaxis of a system rises, P normally decreases given that r_L (and thus $2\pi p(\gamma)/r_L$, γ fixed from one pattern to the next) does not generally experience any permanent change. Using gibberellic acid, (GA), Maksymowych and Erickson (1977) report a change in P from 3.3 days for the control plants to 1.9 days for GA-treated apices of *Xanthium*. Their analysis shows that r_L is not significantly altered by the treatment. However, since the rate of leaf initiation is altered (P decreases), the system moves from (2, 3) to (3, 5), a situation that will be analyzed in Chapter 5 with respect to the corresponding decrease of R from 1.35 to 1.19.

4.7. Problems

4.1 Verify Binet's formula $F_k = [\tau^k - (-\tau)^{-k}]/\sqrt{5}$ for $k = 1, 2, 3, 4$, and show that the formula delivers the approximation $F_k \approx \tau^k/\sqrt{5}$. Hint: Prove first the relations $\tau^2 = \tau + 1$, $\tau + 2 = \tau\sqrt{5}$, and use these relations to replace any power of τ by an expression of the type $a\tau + b$, where a and b are constants.

4.2 (1) Prove that when $\gamma = 90°$, we have

$$\ln R = (-\Delta_n \Delta_m / mn)^{1/2} \quad \text{(Richards formula, 1948)} \qquad (4.30)$$

[Δ_n in the spiral lattice corresponds to x_n in the cylindrical lattice. See Chapter 1; compare with Equation (4.4) with $\gamma = 90°$],

$$\ln R = (2\pi\tau^3/\sqrt{5})/(m+n)^2, \quad \text{and} \qquad (4.31)$$

$$\ln R = 2\pi/(m^2+n^2) \quad \text{(Church's formula, 1904b)} \qquad (4.32)$$

[The divergence angles corresponding to the first two formulas are those given by Table 2.1, while the divergence angle in Church's system (orthogonal by definition) is given by $d = [(m-n)^2 + n^2]/(m^2+n^2)$, a value close to those in the table for large k.]

(2) With various Fibonacci visible opposed pairs and $d = \tau^{-2}$, compare the respective values of R given by the expressions in part (1).

(3) Do the same as in part (2) with Lucas pairs and $d = (3 + 1/\tau)^{-1}$.

4.3 (1) For $d = \tau^{-2}$ and $R = 1.14$, what is the conspicuous pair?

(2) Apply the same question for $d = (3 + 1/\tau)^{-1}$ and $R = 1.14$.

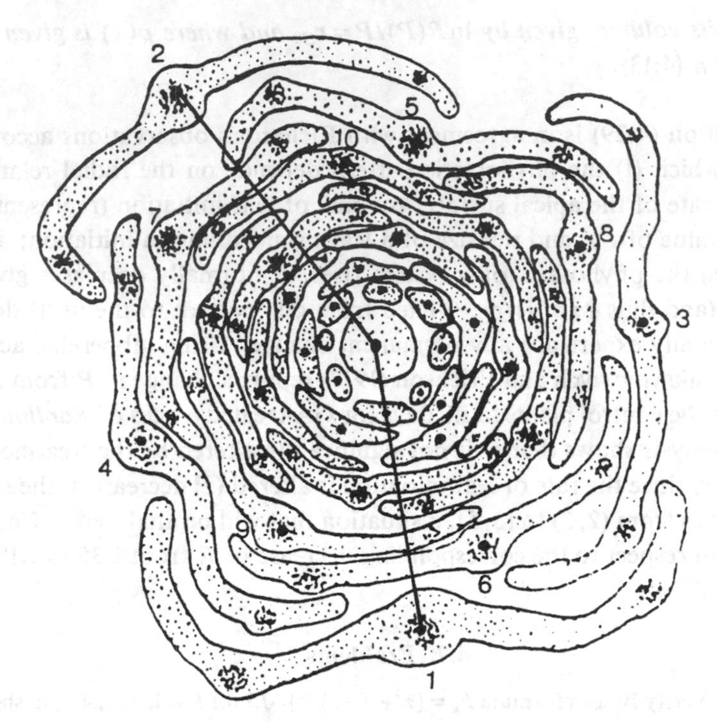

Figure 4.2. Illustration of a more complex system observed from a microscopic cross section through a growing apical bud of flax (*Linum usitatissimum*). Center of system and midpoints of primordia have been marked by black dots to help determine the divergence angle and the plastochrone ratio. Also, the outer ten primordia have been numbered according to their decreasing ages. The pair $(6-1, 9-1) = (5, 8)$ is (apparently) conspicuous (as Problem 4.5.1. confirms). The pairs (2, 3) and (5, 3) are contact parastichy pairs. (Modified from a figure in Richards & Schwabe, 1969.)

(3) Apply the same question for the systems showing visible opposed parastichy pairs given by consecutive terms of the sequence $\langle 1, 4, 5, 9, 14, \ldots \rangle$ corresponding to the divergence angle $360(4+1/\tau)^{-1}$.

4.4 For $d=\tau^{-2}$ and $\gamma=90°$, use the calculator and the Pattern Determination Table to verify that the phyllotaxis index is almost an integer, whatever be the visible opposed spiral pair made with consecutive Fibonacci numbers. Establish a table showing the values of P.I. for the consecutive visible opposed pairs. Conclude that whenever a value of R gives a value of P.I. close to an integer we have a specific Fibonacci system if the divergence angle is close to τ^{-2}. This P.I. is sometimes used to analyze the patterns of growth at the plant apex (see Problem 4.5) when the divergence angle is close to the Fibonacci angle. However, in Figure 1.3(1) or in systems expressed by other sequences than the Fibonacci sequence, this P.I. is useless. On a conical apical surface, the equivalent phyllotaxis index is used, incorporating the angle 2α at the vertex of the cone. It is obtained by adding to P.I. the term $2.395 \log \sin \alpha$.

4.5 Using Figure 4.2, W. W. Schwabe (Jean & Schwabe, 1990) arrived at the following mean values for the divergence angle and for the plastochrone ratio: 138.77° and 1.0792, respectively, with standard deviations of 11.57 (8%) and of 0.0869 (8%), respectively.

(1) Using the Pattern Determination Table, determine, among the visible opposed parastichy pairs of the system, the conspicuous parastichy pair and its angle of intersection γ.

(2) How would you use the P.I. defined in Problem 4.4 to obtain the conspicuous pair?

(3) Verify by taking measurements in Figure 4.2 that for the pairs (2, 3) and (3, 5) the respective values of γ are larger than 100°, as predicted by the Pattern Determination Table.

4.6 (1) Considering the mean values of R and d obtained in Problem 1.3, determine if the conspicuous parastichy pair obtained from the Pattern Determination Table is the contact parastichy pair observed in Figure 1.3(1) or 1.3(2).

(2) Taking the mean value of d from Problem 1.3 and the contact parastichy pair (m, n) and the value of γ from Figure 1.3(1) or 1.3(2), deduce the value of R.

(3) Compare the measured and deduced values of γ and R obtained in parts (1) and (2).

5
Practical pattern recognition

5.1. The necessity of theoretical frameworks

According to Erickson (1983),

> One of the difficulties in empirical studies of leaf initiation is in the small size and relative inaccessibility of the shoot apex, enclosed as it usually is by the young leaves of the bud. Usually rather extensive dissection is required to expose it to view, and it is dubious if its further growth is normal. Much of our knowledge of its structure is therefore indirect, based on inferences from sectioned material, or dissections viewed by stereo light microscopy or scanning electron microscopy.

Erickson believes that the classical geometrical models allow one to present results in more objective and general terms than would be possible without a model. In Chapter 4 I proposed a theoretical tool to organize the data so as to achieve that goal more easily and to assess phyllotactic patterns accurately and quickly, as we will see in Chapter 5. The patterns are characterized by the parameters (m, n), d, r or R, and γ. These parameters can be measured on plant shoots. The last parameter has been neglected in previous practical analyses. Its theoretical importance is however acknowledged by van Iterson (1907) and by Erickson (1983) who explains formulae involving γ in various representations of phyllotaxis.

The method of evaluation put forward here will be illustrated with data from various authors, and the conclusions drawn will be compared to theirs. The model allows us to interpret the data objectively, in more general terms than possible with the classical models. A number of the approaches that other authors have taken will be summarized and reevaluated. The model can have great utility in empirical studies, as it can form a guide for the analysis of descriptive data and experimental results. It can also help in making inferences about the activity of the shoot apex.

It has been shown in Chapter 4 that when a theoretical assessment of R is needed, the allometry-type model given by Equation (4.12) is reliable. This model is defined by the equations

$$\ln R = 2\pi p(\gamma)(m+n)^{-2}, \tag{4.12}$$

$$p(\gamma) = \tau^3[\sqrt{5}\cot\gamma + \sqrt{(5\cot^2\gamma+4)}]/2\sqrt{5}, \tag{4.13}$$

with

$$\ln R = (2\pi\tau^3/\sqrt{5})/(m+n)^2 \approx 11.903(m+n)^{-2}. \tag{4.31}$$

in the case of orthogonal intersection ($\gamma = 90°$). The values of R obtainable from the model are very close to those provided by the foremost classical models, and almost identical to those given by the Richards formula

$$\ln R = (-\Delta_n \Delta_m / mn)^{1/2} \tag{4.30}$$

(see in Appendix 2, the table in the answer to Problem 4.2). This is surely a good quality for a system to have, if one considers, as Williams (1975) does, that "the quantitative system devised by Richards for the objective description of phyllotaxis in plants is strongly recommended."

For most analyses, including the one here, *phyllotaxis is fundamentally a two-dimensional phenomenon* (Church, 1920b). The transverse section of a shoot apex with young leaf primordia presents the picture of an array of points or figures in a plane. As Williams (1975) in particular made it abundantly clear, shoot apices are usually domed. The dome is relatively high for lower-order patterns [such as (1, 2) and (2, 3)] and relatively flat for larger parastichy numbers [such as (21, 34) and (34, 55)]. One should thus consider the positioning of primordia on domed surfaces. Van Iterson (1907) proposed that in practical pattern assessment, the apical angle of a cone tangent to such a dome be considered. This gave rise to his folioids, used as we will see by Erickson and co-workers to assess phyllotactic patterns. However, it will be shown that van Iterson's sophisticated concepts do not produce a better assessment than the one proposed here which is, moreover, easier to obtain. It is true that no one has proposed an adequate analysis for a domed surface. It is far from being certain however that such an analysis would shed more light on the mechanism of leaf initiation or will bring more precision to the evaluation of the patterns.

Data collecting may not be a simple routine listing of observations. Interpreting the observations can prove to be a challenging task as we will see. Many difficulties and biases await investigators who venture into the task of phyllotactic pattern recognition. They are revealed in discrepancies

Table 5.1. *Data from the computer simulation of an expanding apex, according to Mitchison (1977)*

Stem radius	Divergence angle	Contacts
1.0	138.9°	(1, 2)
1.3	139.7°	(2, 3)
2.0	138.3°	(3, 5)
2.6	137.1°	(5, 8)
4.0	137.8°	(8, 13)
6.4	137.6°	(13, 21)

between independent evaluations. As a preventive measure, the following practices are not recommended. They may be error-inducing, erroneous, or dependent on too particular assumptions: (1) the use of a single phyllotactic fraction to represent a pattern, (2) the use of contact parastichy pairs instead of conspicuous parastichy pairs, (3) the use of R only, as is sometimes the case for those working with Richards's phyllotaxis index, and (4) the dependence on models based on a presumed orthogonality of the opposed parastichy pair. It is this very dependence that weakened Church's equipotential theory. Richards (1951) advocates approaches independent of any particular assumption, but uses the Fibonacci sequence in his phyllotaxis index given by Equation (4.21)! Methodological suggestions will be made in Section 5.6 regarding the necessity to improve the relevance and reliability of the data.

5.2. Applications of the allometry-type model

5.2.1. Linear relations in an expanding apex

We can find in the literature an expression of a linear relation between the apical radius and the phyllotaxis (m, n) of a system. The straight line, from King (1983), is based on values found in Table 5.1, from Mitchison (1977). In King's graph the value of the stem radius is reported in the usual way on the vertical axis, but King did not give any reason why he distributed the systems (m, n) as he did on the horizontal axis. Using a ruler it can be seen in his graph that the pattern (5, 8) is at 8 units from the origin, that (13, 21) is at 13 units from it, and so on. Thus it can be checked that the line makes an angle of 30° with the horizontal.

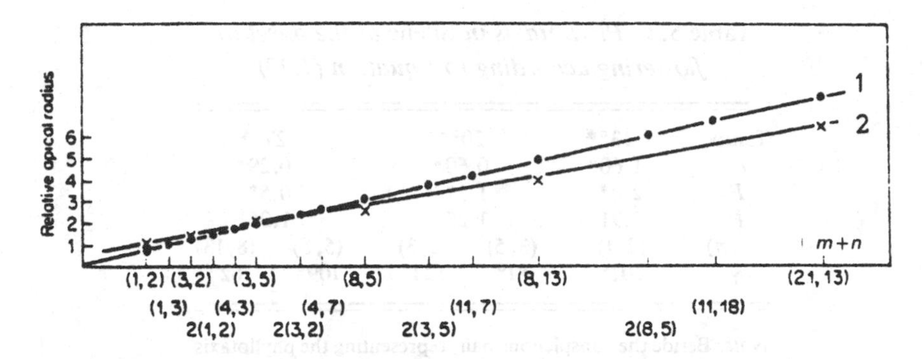

Figure 5.1. Line 1: Theoretical line expressing the relation [cf. Eq. (4.26)] $L=(m+n)/(\pi(\tau^3/\sqrt{5})^{1/2})$ for orthogonal intersection of the opposed parastichy pair (m,n) for values of m and n corresponding to three of the sequences in Table 2.1; the line makes an angle of about 13° with the horizontal axis. Line 2: Arrangement of the data in Table 5.1 according to the presupposed relation $L=(m+n)/\pi p(\gamma)^{1/2}$, assuming an angle of intersection $\gamma\approx 60°$ between the opposed parastichies in the pair (m,n).

The model given in Equation (4.12) allows us to define mathematically the relation expressing the behavior of Mitchison's data, and thus to bring precisions to King's representation of it. Figure 5.1 is a correction of King's drawing. In the figure the pattern (m,n) is represented on the horizontal axis as $m+n$. An angle of about 13° is obtainable from

$$1/B=L\approx(m+n)/(\pi(\tau^3/\sqrt{5})^{1/2}). \quad (4.26)$$

(deduced from (4.31), that is (4.12) for orthogonal intersection of the parastichies), where B is the Church bulk ratio.

Mitchison's contact parastichies do not seem to correspond, unlike Equation (4.31), to the orthogonal intersection of opposed parastichies. Mitchison's data seem to be the ones at the transitions, that is, when the contact parastichy pair $(a+b,b)$, showing an angle of intersection of the opposed parastichies of about 60°, has just replaced the contact parastichy pair (a,b) showing an angle of intersection of 120° for various values of a and b. Church's system involving the bulk ratio B is essentially orthogonal, but assuming nevertheless that

$$B^2\approx(\pi\ln R)/2, \quad (4.25)$$

still holds when the angle of intersection γ of the opposed parastichy pair is different from 90°, we get from Equation (4.12), $1/B=(m+n)/\pi p(\gamma)^{1/2}$.

Table 5.2. *Phyllotaxis in* Silene *at the onset of flowering according to Equation (4.29)*

Temp.	13°*	20°*		27°*	
r_V	1.60*	0.60*		0.29*	
P	2.0*	1.1*		0.5*	
R	2.91	1.25		1.05	
(m,n)	(2, 1)	(3, 5)	(2, 3)	(5, 8)	(8, 13)
γ	101°	81°	124°	109°	62°

Note: Beside the conspicuous pair, representing the phyllotaxis of the system, the table also indicates the visible pair where γ is still relatively close to 90°.
Source: The data obtained by Lyndon (1977, tables 1 and 5) is marked by an asterisk. Jean, 1987a.

Assuming an angle of about 10° for the slope of the line $1/B$ as a function of $m+n$, we get $\gamma \approx 63°$. Mitchison's computer simulation is thus in a good agreement with our model.

5.2.2. *Phyllotaxis in* Silene, *a function of temperature*

The phyllotaxis (m,n) of leaves in *Silene* can be deduced, together with the angle of intersection of the opposed parastichy pair, from data provided by Lyndon (1977). In the flower "the sepals, stamens and carpels are initiated singly in positions which are predictable on the basis of Fibonacci phyllotaxis." This means that the angle of divergence of the system is around 137.5°. From Table 2.1 we can infer that $m+n=3$, or 5, or 8,..., and that $(m,n)=(2,1)$, or (2, 3), or (3, 5),... . The problem is to determine which pair is conspicuous, presenting an angle of intersection γ of the opposed parastichies closest to 90°. Considering that Lyndon's data deal with volumes, we must therefore use formula

$$P=(2\pi p(\gamma)/r_G)(m+n)^{-2}, \tag{4.29}$$

with $r_G=r_V/3$, r_V being the relative growth rate of the volume, and P the plastochrone. Table 5.2 summarizes the data and the deductions. Unfortunately, Lyndon neither provided the information on the specific Fibonacci patterns involved, nor drawings corresponding to the data, that would have allowed us to check the adequacy of the deductions. Table 5.2 shows the rise in the phyllotactic pattern from (2, 1) to (5, 8) conspicuous pairs, induced by an increase in temperature. Indeed, from the data one can see that the increase in air temperature during growth (13° in one batch of plants, 20° in another, and 27° in the third), brought a decrease in the plastochrone P and in the relative growth rate r_V as well. Theoretically, this results in a significant decrease of the plastochrone ratio R because of the relation $3\ln R = r_V P$. And it is well known that a decrease in R brings a rise in the phyllotactic pattern. More precisely, the rise in the pattern must be interpreted dynamically: If the conditions existing at the onset of flowering continue to prevail, the phyllotaxis of the system will be higher at higher temperatures. From another more complete collection of such data, it would be possible to check these predictions, and perform a more elaborate analysis.

5.3. On the theoretical determination of the plastochrone ratio

5.3.1. Various models to use

It is known from Richards (1951) and Williams (1975) that the determination of a reliable value for R, directly on an apex, is not an easy task. It involves measurements on various transverse sections and the determination of the centers of the primordia and of the apex. If however the other parameters are more easily available, values of R can be deduced from geometrical models. The problem then is which model should be used. This section concerns case 3 of Table 4.2: given (m, n) and γ, R will be deduced.

Erickson (1983) suggested that a phyllotactic pattern might be specified by citing the opposed parastichy pair that is most evident visually, and that this pair may be the one that intersects most nearly at right angles. This would mean the contact parastichy pair, but it is known that such a pair can depart considerably from orthogonality. It is proposed here that the most evident parastichy pair is the conspicuous parastichy pair. In the technical meaning of the term, this is the pair intersecting most nearly at right angles. Of course the divergence angle implied by the pair, obtainable from Table 2.1, must be the one we can actually measure on the apex.

It is suggested that the method for objectively estimating the plastochrone ratio of a system showing the conspicuous and approximately orthogonal parastichy pair (m, n) be based on Equation (4.31), whatever be the measured values of the divergence angle, even for low patterns such as (2, 1) and (3, 2) where d can depart considerably from the corresponding limit-divergence angle given in Table 2.1. The case where the conspicuous pair (m, n) is not orthogonal will be analyzed in the next sections.

One of the reasons to support the preceding suggestion is that the value of R obtained from Equation (4.31) is reliable. It is, as we said, very significantly close to those obtained from the classical geometrical models elaborated by Richards and Church [Eqs. (4.30) & (4.32)]. The differences for R obtained for the various models become negligible for higher-order patterns. For lower-order patterns the value of R given by Equation (4.31) is particularly close to the one given by the Richards formula [Eq. (4.30)]. Moreover, the value of R is most easily obtained from Equation (4.31), while van Iterson's model (next subsection), for example, requires numerical analysis and gives a value of R which does not deviate significantly from the former value.

Another advantage of Equation (4.31) is that, compared to the Church formula

$$\ln R = 2\pi/(m^2+n^2) \tag{4.32}$$

for system of logarithmic spirals in the plane, it gives the value of R for patterns (m, n), irrespective of the limit-divergence angle. If, for example, one wants to calculate R for the patterns $(1,6)$, $(2,5)$ and $(3,4)$, corresponding to divergence angles of about 54.4°, 151°, and 99.5°, respectively, the same value of R will be obtained from Equation (4.31), given that $m+n=7$ in each case. In the Church model the value of R is a function of m^2+n^2. On the other hand, in the log-log plot for $(\log R, m+n)$, the above three particular patterns are represented by the same point on the line with slope -2.

The value of d, and the value of R obtained from Equation (4.32), is identical with the values for the patterns of orthogonal opposed contact parastichies through the centers of contiguous circles on the surface of a cylinder, each circle being tangent to four neighbors, obtained from the variant for the cylindrical lattice ($r=((-x_n x_m)/mn)^{1/2}$ in Section 4.2) of the Richards formula [Eq. (4.30)] [see for example Table 3.1 in Erickson (1983), putting $h=\ln R$].

5.3.2. *Advantages of the allometry-type model*

Equation (4.31) is derived from the cylindrical lattice transposed in the plane of logarithmic spirals by the correspondence $R=e^{2\pi r}$. In the cylindrical lattice, where the observed spirals are represented by straight lines, the relations between the parameters are more manageable. They are expressed by functions that are simpler than the following one from van

Iterson for a plane lattice in which the circular primordia are in contact along m and n opposed logarithmic spirals:

$$\cos(m\alpha/2)/\cos(n\alpha/2) = R^{(n-m)/2}(1+R^m)/(1+R^n), \qquad (5.1)$$

where α is the divergence in radians. Erickson (1983) used the equivalent formula

$$\cos(m\alpha/2)/\cos(n\alpha/2) = \cosh(m \ln R/2)/\cosh(n \ln R/2). \qquad (5.2)$$

The values of R and d for orthogonal intersection of the contact parastichies are those in which Equations (4.30) and (5.1) hold at the same time. This gives values of d slightly different from the limit divergence angles, and R is close to the value given by Equation (4.30) for the limit divergencc angles. [For example, Erickson (1983, Fig. 3.15) gives superposed graphs of Equations (4.30) and (5.1) for $\ln R$ as a function of d].

The value of R obtained from Equation (4.31) is immediate, while Equation (5.1) does not yield a simple algebraic solution. In the latter case, a computer interpolation has to be used, incorporating for example the Newton–Raphson or interval-halving approximation method. If tangent circles on a conical surface are considered, the question is further complicated by the angle at the apex of the cone. However Erickson and co-workers successfully used this apparatus as we will see in Section 5.4. Their idea was to compare the curves of Equations (4.30) and (5.1), giving $\ln R$ as a function of d, with their plotted curves of the observed relations between the measured values of R and d, so as to determine the pattern (m,n). This also allowed them to say whether the opposed pair was close to or far from orthogonality. But they did not deduce the angle of intersection γ of the opposed parastichy pairs. Equation (4.12) allows us to deduce that angle at once, as we will see. The point is that in the cylindrical lattice the approximation procedure can be done mathematically instead of numerically, and it leads to Equations (4.12) and (4.31) which deliver immediate values for the angle of intersection, while Equation (5.1) requires elaborate methods to obtain that angle.

The next sections will show that, for the best results, the theoretical assessment of R should be made using Equation (4.12), which generalizes Equation (4.31). Putting the values of the observed parameters γ and (m,n) in the Pattern Determination Table (Table 4.3), R will be estimated with reliability, thus settling case 3 of Table 4.2. Section 5.4 will also show how to determine γ from R and the conspicuous pair (m,n) (cases 2 and

4 in Table 4.2). Section 5.5 will show in particular how to obtain (m, n) from the measured value of R and the value of the angle of intersection of the conspicuous parastichy pair by using the estimated value of d (cases 5, 6, and 7 of Table 4.2).

5.4. Assessing phyllotactic patterns

5.4.1. Maksymowych–Erickson method using **Xanthium**

Maksymowych and Erickson (1977) proposed a method for assessing phyllotactic patterns based on the estimation of the plastochrone ratio R and the divergence angle d. In their approach they considered the two parameters as constant. The section of the shoot apical meristem, being irregular in shape and changing at each plastochrone, has a center that cannot be located objectively. Consequently, their derivation of R and d is based on the measurement of the chord lengths between the estimated centers of three consecutive primordia, these centers being represented in older primordia by the earliest formed protoxylem vessels. Calling d_1 the length of the chord between the centers of primordia 1 and 2, and d_2 that between the centers of primordia 2 and 3, then the larger angle between those chords becomes d, and $R = d_1/d_2$.

In the case of *Xanthium* treated with gibberellic acid, Maksymowych and Erickson plotted the mean values of $\ln R$ as a function of the mean values of d. Then they superimposed on the experimental curve the ideal curves for various patterns (m, n) as expressed by Equations (4.30) and (5.1), representing patterns of orthogonal parastichies and patterns of tangent circles in the plane, respectively. They concluded that the system underwent an apical enlargement and shifted from $(2, 3)$ to $(5, 3)$ under the influence of the growth regulator. The experimental points being close to the point of intersection of the ideal lines for the $(5, 3)$ pattern, they concluded that the opposed parastichies intersected almost orthogonally. Then, they introduced contiguous folioids, that is, the plane projections of the contiguous circles on a cone, resembling the shape of the primordia seen in transverse sections [expressed by a variation of Equation (5.1) deduced by van Iterson, and involving the apical angle of the conical surface]. It appeared that the data points were "somewhat closer" to the new theoretical lines. These authors registered a change in R from a mean value of 1.35 to a mean value of 1.19, but no significant change in the divergence angle, which remained around 139°. Let us thus take those two values of R for granted.

From a consideration of their Figure 6, showing the transverse section that provided the value $R=1.35$ for the control apices, it is clear that a genetic spiral can be traced, linking the consecutively born primordia according to their increasing size. Then it is easily seen, and it will be theoretically confirmed in a next paragraph, that both the patterns (2, 3) and (5, 3) are conspicuous, while the authors report (2, 3) only. Instead of using the Maksymowych–Erickson method to obtain the pattern (m, n), we can use the experimentally obtainable value of (m, n) to deduce another phyllotactic parameter, the angle of intersection of the opposed parastichy pair. Maksymowych and Erickson simply mention that "the control apices do not agree closely with any orthogonal pattern." Their method does not provide a value for γ. The angle can be estimated from Table 4.3: For $R=1.35$, one can read that γ is approximately equal to 114° for $(m, n)=(2, 3)$, and to 66° for $(m, n)=(5, 3)$. This means that the visible opposed pairs are equally removed from orthogonality (an equal difference of 24°).

For the treated apices, giving $R=1.19$, the pattern (m, n) cannot apparently be estimated from the authors' drawings. From Table 4.3 with $R=1.19$, it can be seen that $m+n$ can take any integral value. Indeed, each column and each row to the right tends to 1, and each row to the left goes to infinity. But the values measured by the two authors for the divergence angle will allow us to determine the pattern anyhow. Knowing indeed that for the treated apices the mean value of the divergence angle is close to 137.5°, it can be seen from Table 2.1 that the conspicuous pair is expressed by two consecutive terms of the main sequence. In Table 4.3 we then look for the value $R=1.19$ in the rows where $m+n=3, 5, 8, 13, \ldots$, and such that γ is the closest to 90°. We obtain that $\gamma=93°$ and $(m, n)=(3, 5)$, thus confirming and adding precision to Maksymowych and Erickson's findings. But the previous discussion shows that the change in phyllotaxis is less striking than suggested by these authors, given that in the control apices the (3, 5) pattern is already quite visible, as visible as the (2, 3) pattern that they reported.

5.4.2. A first method using the Pattern Determination Table

It is proposed here to use Table 4.3, based on the mathematical model expressed by Equation (4.12), for the purpose of assessing phyllotactic patterns. This is an automatic pattern determination table, a tool that is flexible, accurate, and very easy to use. It is not based on orthogonality, and it provides the angles of intersection of the opposed parastichy pairs. Given that the application of Table 4.3 is based on Table 2.1, let us make additional remarks on the latter. The limit-divergence angles corresponding to normal phyllotaxis, that is $360(t+\tau^{-1})^{-1}$, become very close to one another as t increases beyond the values 6 or 7. In these cases it may be difficult to decide to which sequence the data correspond. But for $t=2$, 3, or 4 in particular, if the measurements of d cluster significantly around the values 137.5°, 99.5°, or 78°, it will be easy to identify the sequence from Table 2.1.

In other words, the proximity of the measured divergence angles to one of these limit divergence angles conveys the knowledge that a particular sequence is under consideration. The visible opposed parastichy pairs are determined by consecutive terms of the sequence. Richards (1951) put it as follows: the relation between the plastochrone ratio and the parastichy intersection angle is not seriously affected by such deviations of the divergence from the ideal angles, as usually found in apices. This question is discussed further in Section 5.5. Many authors devised ways to try to produce an exact limit divergence angle. However, the same phyllotactic system expressed by consecutive terms of the corresponding sequence given in Table 2.1 can be found within a more-or-less wide interval around such an angle.

Let us reconsider the two cases we have just analyzed ($R=1.35$ and 1.19). The divergence angle is found to vary between approximately 132° and 145°. It follows that the corresponding sequence is $\langle 1, 2, 3, 5, 8, 13, \ldots \rangle$, and that the visible opposed parastichy pairs are expressed as consecutive terms of this sequence. For the case where $R=1.35$, the conspicuous pair is to be sought for among the visible pairs $(1,2)$, $(3,2)$, $(3,5)$, $(8,5)$, ... Next, in the Pattern Determination Table, the value of $m+n$ that delivers the value of γ closest to 90° is $m+n=5$, or 8, for which $(m,n)=(2,3)$ and $(5,3)$, as we already noticed in Maksymowych and Erickson's drawings. Finally, we will have a definite $(5,3)$ pattern only if R decreases. The application of gibberellic acid brought a lower value of R and a $(5,3)$ orthogonal system.

5.4.3. *Evaluating the phyllotactic patterns for* **Proserpinaca** *and* **Xanthium**

Maksymowych and Erickson (1977) analyzed Figures 30–33 in Schmidt and Millington (1968), depicting phyllotaxis in *Proserpinaca*. Their conclusions are that long-day apices with $R=1.285$ and $d=138.3°$ show a

(2, 3) pattern, and short-day apices with $R = 1.104$ and $d = 139.5°$ show a (3, 5) pattern. Table 4.3 gives a (3, 5) conspicuous pair with an angle of intersection of 75° [(2, 3) gives 122°; that pair is more removed from orthogonality], and a (5, 8) conspicuous pair with an angle of 74° [(3, 5) gives 120°]. Maksymowych and Erickson are obviously basing their assessment on the contact parastichy pairs, but their drawings show that the contact parastichy pairs deviate much more from 90° than the pairs recognized here.

Erickson and Meicenheimer (1977) applied the same method used by Maksymowych and Erickson (1977) to analyze the photoperiodically induced change in the phyllotaxis of *Xanthium*. The angles of divergence in the cases of vegetative and of photoperiodically induced plants gravitate meaningfully around 137.5°, so that Table 2.1 shows again the main sequence. Then we look in the rows 3, 5, 8, and 13 of Table 4.3 for the values of R for each of the six days reported by the authors for vegetative and induced plants. Our conclusion, in agreement with that of the authors, is that the control plants exhibit a (2, 3) pattern throughout. But the margin of error reported by the authors for the control plants gives the possibility of a (5, 3) pattern on day 6.

The same analysis for the photoperiodically induced plants confirms the findings of the authors for days 2, 3, 4, and 5 - a (2, 3) pattern for days 2 and 3, and a (5, 3) pattern for days 4 and 5. More precision can easily be obtained. Indeed, day 3 manifests a (5, 3) pattern as well, according to the Pattern Determination Table. The angles of intersection of the conspicuous parastichy pairs can be easily produced. The mean values of R expressed day after day for the induced plants (1.556, 1.479, 1.353, 1.221, 1.150, and 1.142, respectively) clearly show a (2, 3) pattern becoming a (5, 3) pattern. However, Erickson and Meicenheimer say that on day 6 the pattern is "best described as (5, 8)." The value $R = 1.142$ on day 6 rather gives a (5, 3) pattern with an angle of intersection of approximately 106°; the pattern (5, 8) is visible but it is a bit more removed from orthogonality with an angle of intersection of approximately 58°. Even their photomicrograph of the photoperiodically induced plants on day 6 shows a (5, 3) pattern, and is not explicit enough to allow us to see a (5, 8) pattern. However, taking into consideration the standard errors reported by the authors, days 5 and 6 show the possibility that (5, 8) may be more conspicuous than (5, 3). Finally, the overall mean value $R = 1.317$ for the induced plants gives a (2, 3) pattern and a (5, 3) pattern equally removed from orthogonality ($\gamma = 118°$ and 70°, respectively).

5.5. Other applications of the Pattern Determination Table

5.5.1. The practical limits of the Richards phyllotaxis index

A tool such as the one found in Table 4.3 is much more versatile than the Richards phyllotaxis index (Section 4.4). This index, as we already know, is meant to give integral values for the orthogonal patterns expressed by consecutive terms of the main sequence (see Problem 4.4). But the values of the index for all the other patterns are as different from integers as are the values of R itself. Of course we can write, as Richards did, an index that will give integral values for orthogonal patterns corresponding to another sequence, but then for other sequences this index will register decimal places. The drawback in the Richards index is not exactly, as Maksymowych and Erickson (1977) put it, that it assumes the limit divergence angle of 137.5°. It is that, though allowing a range of approximations for that angle, it leaves no place for other such ranges around other limit divergence angles.

It is clear that a value of R, or of the phyllotaxis index, does not specify a pattern. Knowing however that the divergence angles are close to 137.5°, the value of R or of the index will allow us to specify the conspicuous pair. It will allow us to reach the preceding conclusions, such as when $R = 1.35$ (P.I. = 2.5), (2, 3) and (5, 3) are equally removed from orthogonality, and when $R = 1.19$ (P.I. = 3), (5, 3) is approximately orthogonal. But again the angles of intersection of the parastichy pairs are not given. Authors such as Williams (1975) and Schwabe (1971) made good use of this index.

What happens, however, when the system presents divergence angles clustering around another limit divergence angle? Richards's P.I. is not applicable, and the systems are not anymore "fully defined mathematically by a single parameter" (Williams, 1975). This index is an incomplete tool, a somewhat artificial function. Table 4.3, based on the model in Equation (4.12), provides an appropriate tool to analyze all patterns of the type $J(m, n)$, $J = 1, 2, 3, \ldots$ called the multijugate patterns (m and n relatively prime). Finally, Equation (4.12) supersedes the forms of the Richards index, given that all the latter can be deduced from the former as we have seen in Chapter 4.

5.5.2. A second method using the Pattern Determination Table

The divergence angle of a system and the visible opposed parastichy pairs are closely related, as we already know (Chapter 2). The method proposed above for assessing the patterns is based on the geometrical relation between the former two concepts, on which Table 2.1 is built. However, sequences such as ⟨3, 8, 11, 19, 30, ...⟩, and ⟨3, 7, 10, 17, 27, ...⟩, give limit divergence angles of 132.2°, and 106°, respectively. These angles are not far from 137.5° and 99.5°, corresponding to two well-known sequences. Given that measured divergence angles may vary a lot around an angle such as 137.5°, thus bringing doubts as to the sequence concerned, it is proposed to double check the assessment of the pattern with an alternative method in the following reverse way.

Suppose again that $R = 1.35$ (see Section 5.4.1). The idea is first to look for the row in Table 4.3 that will give a γ nearest to 90°. It is found that $m+n=6$ and $\gamma = 94°$. This means that $(m, n) = (1, 5)$ or $2(1, 2)$ or $3(1, 1)$. But the divergence angles corresponding to these patterns (around 64, 68, and 60°; see Table 2.1) are far away from the divergence angle of 139° measured by Maksymowych and Erickson. So we must look for the next closer value of γ to 90°. The value $\gamma = 78°$ gives $m+n=7$, meaning that we may have the patterns (1, 6), (2, 5), or (3, 4). But again the corresponding divergence angles (54, 151, and 100°) are far away from the measured angle. The next values of γ are precisely those we have obtained in the first place, giving (2, 3) and (3, 5) patterns equally removed from orthogonality.

5.5.3. *Evaluating phyllotactic patterns in* **Chrysanthemum** *and* **Linum**

Schwabe (1971) registered a striking change in both d and R by applying triiodobenzoic acid (TIBA) to *Chrysanthemum*. R increases from 1.47 to 1.65, d increases from 136° to 158° (mean values), and the angle in the latter case appears to approach 180°. Maksymowych and Erickson (1977) compared the data with theoretical curves (see Section 5.4.1) and concluded that Schwabe's system changed from a (2, 3) pattern to a (1, 1) or distichy.

From Table 4.3 the (2, 3) pattern is conspicuous given that $\gamma = 100°$, but $d = 158°$ for the TIBA-treated apices is still far from 180° giving distichy. Rather, this suggests a (2, 7) anomalous pattern (see Table 2.1 at 158°) with $\gamma = 36°$. Schwabe's Figures 1a and 1b contain only five and four primordia, respectively. In the former case a (2, 3) pattern is seen very likely to emerge, and in the latter case two clockwise spirals can be drawn but the (2, 7) pattern is not yet apparent.

As a final illustration of the method we analyze data from Williams (1975), who provided a detailed reconstruction of the phyllotactic system of *Linum*. According to Maksymowych and Erickson (1977) it shows a change from (1, 1) to (5, 8). We have, certainly, a (5, 8) conspicuous pattern rising gradually to a (8, 13) conspicuous pattern. This occurs from day 8 to day 50 of growth, where the divergence angles are significantly close to 137.5°. During this period we also have a gradual increase of the angle of intersection of the parastichy pair, as R decreases from 1.101 to 1.034: from 74° to 114° (on day 25, $R=1.045$) for the pair (5, 8), and from 66° (on day 25) to 86° and then to 79° for the pair (8, 13). Williams's evaluation is based on the contact parastichy pairs. He proposed a single contact parastichy pair, while other contact pairs are sometimes evident from his drawings. On days 11 and 18 we can see from them a (5, 8) conspicuous pair (in agreement with Table 4.3) with the (3, 5) contact pair. On day 22 the contact pair coincides with the conspicuous pair [(5, 8), $R=1.049$, $\gamma=110°$]. But on day 29 ($R=1.041$) there is an amazing drop to (3, 5) contact, while Table 4.3 reports an (8, 13) pattern with $\gamma=70°$ [$R=1.041$ also gives a (8, 5) pattern but more removed from orthogonality with $\gamma=118°$].

Williams's assessment offers interesting common points with the assessment based on the Pattern Determination Table. On day 15 he indeed reached the conclusion that "we have an orthogonal (5, 8) system; and this must not be confused with the fact that the same apex is a definite (3, 5) in terms of contact parastichies." Table 4.3 gives indeed a (5, 8) pattern ($R=1.073$) with an angle of intersection of 90°. Also, on day 50, Table 4.3 provides an angle of intersection of the conspicuous pair (8, 13) of approximately 79° ($R=1.034$), corresponding to the interpretation by Williams that the system is "climbing towards" an (8, 13) orthogonal system. However, on day 39 the system (8, 13) was closer to orthogonality ($R=1.030$) with an angle of intersection of approximately 86°, while on day 43 Williams reports a (3, 5) contact!

Before day 8 in Williams's data, the values of the divergence angle "fluctuate wildly between 180° and 90°." Applying the alternative method shown in Section 5.5.2 it can be seen that many patterns are possible, such as the normal patterns (2, 3) and (3, 4), the anomalous patterns (2, 5) and (2, 7), and the whorled pattern 3(1, 1). When there are very few primordia in a growing system it is premature to decide what the system is to be, though day 1, with two primordia, shows something that is or may become a distichous pattern, and day 4 (with six primordia) suggests decussation. Williams's drawing at day 4 shows (1, 2), (2, 3), and (2, 5) contact patterns as well....

In brief, a glance at Table 4.2 reveals the possibilities of the new method proposed here, by which phyllotactic parameters can be deduced from observations. A guide for the analysis of data about the activity of the shoot apex has been presented. The illustrations of the method given here have shown its versatility and reliability. A number of other assessments made possible by the allometry-type model have been reconsidered. Chapter 8 will stress the fact that whorled patterns behave like multijugate systems, thus allowing us to use the Pattern Determination Table for either spiral or whorled patterns.

5.6. Difficulties involved in pattern recognition

5.6.1. Data collection and Fujita's normal curves

When a theory is being constructed, the observations and relevant facts available are of primary importance. From these data, theories and models emerge, and because of them theories and models are confirmed or invalidated. It is common in science for theoretical concepts to be overthrown by experimental or observational data. On the other hand, when observations are gathered, bias can be easily introduced by observers insufficiently aware of the inherent difficulties. Also it makes no sense to build a theory on, or to confirm it from, ill-established facts, when there are shortcomings in the methdology for collecting data, or when there is a weak understanding of the fundamental concepts involved. The first step for a theorist is thus to question the existing data in order to try to grasp their real meaning. Adler (1987) argues that past failures to solve the riddle of phyllotaxis were due in part to faulty methods employed in the construction of theories or in the assembling of observational data. He underlines hidden errors of reasoning in Tait's (1872), Schwendener's (1878), and Thompson's (1917) reports, and methodological flaws in the organization of data in Davies (1939). With all this in mind, let us examine Fujita's plots of frequency distribution of divergence angles.

Fujita's (1939) frequency diagrams in phyllotaxis happen to be puzzling when compared to results and observations in the field. Fujita measured to the nearest integer thousands of divergence angles for the systems (1, 2), (2, 3), (3, 5), (5, 8), (8, 13), (13, 21), (2, 4), (4, 6), (6, 10), and (4, 7). For each system and for each species, Fujita drew the graphs of the frequency of the divergence angles as a function of these angles. He obtained approximately normal curves with peaks at the corresponding limit-divergence angles: 137.5° for the first six systems, 68.75° for the next three, and 99.5° for the last one. For all the species displaying the system (2, 1), Roberts (1984a) drew the same function; he did the same for all the species where Fujita observed a (3, 2) system. Obviously, the resulting graphs are also approximately normal curves centered at 137.5° and as in Fujita's graphs, two small peaks are visible: for both systems at 132°, at 140° for (2, 1) and at 142° for (3, 2). For the (2, 1) systems, 266 angles were measured by Fujita on three species, and for the (3, 2) systems, 2,385 angles were measured on thirteen species. The same work was done for other systems, such as 2,201 measurements on nine species for the (5, 3) systems, and so on.

The theorem in Chapter 2 says that for the systems (2, 1) and (3, 2), the divergence angles should be found in the ranges [0°, 180°] and [120°, 180°], respectively. The theorem does not give any priority to any angle in these ranges. Why should the frequency diagrams be approximately normal curves centered at the Fibonacci angle? How can we rationalize Fujita's diagrams? Given that we are dealing with static observations that can be taken at any time during the life of the plant, and given that we must be cautious with such ideas as finality or intention on the part of the plant toward the Fibonacci angle, we are led to expect the frequency diagrams of these systems to be horizontal lines expressing a uniform probability distribution. Also thinking in terms of distribution more highly weighted around the Fibonacci angle as for the systems (8, 5) or (13, 8) is tacitly assuming a finality in the lower systems due to eventual rising of the phyllotaxis.

It is suggested that Fujita's report is not simply a commonplace listing of the frequency of observations. Before compiling the measurements and making the graphs, Fujita had to determine what the systems were in every case he analyzed and to measure divergence angles. Therefore, a protocol had to be put forward, which included the determination of the centers of the primordia of various shapes and of the apex. In the scientific papers in which biologists describe observations and experiments, a section called "Materials and Methods" is usually found. Fujita's method for collecting and interpreting the data is not described. That does not mean of course that it does not exist. In Fujita's 1938 paper we simply learn that he examined many species, that he cut sections through shoot apical buds, that he observed the parastichies in the buds often different from those seen in mature organs (where the numbers are usually lower), and that he observed in dicotyledons a transition from decussation to spirality. All this is not exactly a protocol.

Working on the (3, 2) system of *Xanthium*, Maksymowych and Erickson (1977) arrived at a mean value of 139.4°, though it would have been reasonable to expect a mean value of 137.5°. They say that "Fujita does not give detailed directions for measuring divergence angles, and therefore one is perhaps entitled to question the universality of the angle" of 137.5°, doubting that this angle produces a peak as in Fujita's diagrams.

The subject is discussed further in Appendix 6, where possible flaws in Fujita's methodology are identified for the reader to consider. In this appendix a proposal for an interpretation of Fujita's normal curves is given. A discussion of Roberts's (1984a) unsuccessful attempt to rationalize them and to harmonize them with the contact pressure theory of phyllotaxis is also presented. The contact pressure theory predicts the existence of intervals for the divergence, included in the intervals given by the theorem of Chapter 2, with no particular emphasis on any angle in the intervals.

The value of Fujita's research lies at least in the conclusions we can make, which transcend the difficulties it raises, about the relative frequencies of occurrence of types of patterns. From it we can put forward a fact, statistically ascertained from compilation of data from many sources, that Fibonacci phyllotaxis is more frequent than bijugate phyllotaxis which is more frequent than phyllotaxis based on the Lucas sequence (also called Schoute's sequence and the first accessory sequence). This subject of relative frequencies is developed in Chapter 7.

5.6.2. Interpretation of particular lattices

Let us consider the lattice in Figure 5.2. Since the lattice is regular, we can analyze the parastichy pairs in the figure by choosing an arbitrary point in it, say 15. Clearly, points 12 and 13 are closer to 15 than is any other point of the lattice. Moreover, points 12 and 13 are on opposite sides of the axis I issued from 15 (the thickest line). This means that $(15-12, 15-13) = (3, 2)$ is a visible opposed parastichy pair. It can be seen that it is also a conspicuous parastichy pair and that the phyllotaxis of the system is (3, 2). The next point closer to 15 is 10, which is not on the same side of the axis as 12. This means that $(15-12, 15-10) = (3, 5)$ is a visible opposed parastichy pair. In this setting the pair (8, 5) is also obviously visible and opposed. So the pattern represented in this figure corresponds to the sequence $\langle 1, 2, 3, 5, 8, 13, \ldots \rangle$, and (3, 2) phyllotaxis is to become (3, 5) phyllotaxis and eventually (8, 5) phyllotaxis by a reduction of the rise, or internode distance.

The pattern in Figure 5.2 is an exact replica of one of Zagorska-Marek's (1985) lattices (found in her Fig. 1, p. 1846). She proposes a different interpretation of it; that is, she theorizes on the sequence $\langle 3, 8, 11, 19, 30, \ldots \rangle$ and the pairs (3, 8) and (11, 8). The divergence angle corresponding to this sequence is 132.18°, as it can be proved by using the fundamental theorem of Chapter 2 (more precisely the diophantine algorithm), a value close to the Fibonacci angle. Zagorska-Marek considers an axis passing between

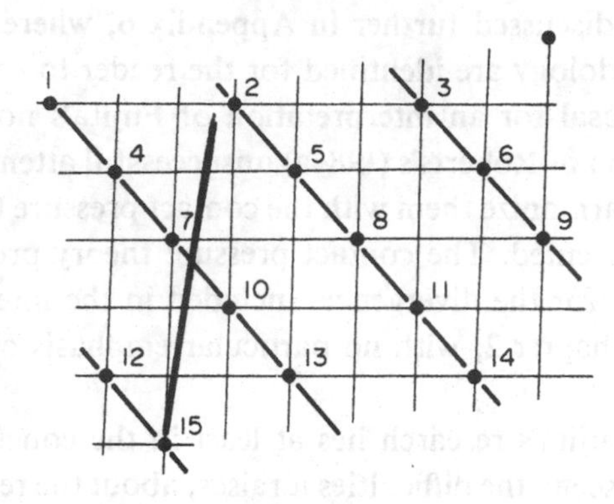

Figure 5.2. A regular lattice of points representing the centers of the primordia on a stem surface, and showing the genetic spiral going through the consecutive primordia, a family of 3 parastichies joining the primordia by 3 (3, 6, 9, 12, 15, ...; 2, 5, 8, 11, 14, ...; 1, 4, 7, 10, 13, ...) and a family of 8 parastichies joining them by 8. The line I issued from 15 and passing on the right of 7 and the left of 2 is presumably parallel to the axis of the plant. The pairs (3, 2), and (3, 5) (not illustrated) are visible and opposed. The pairs (3, 8) and (11, 8) are neither opposed nor conspicuous.

the points 15, 12, 7, 4, ... But if this axis issued from 15 passes between 4 and 7, the pairs (2, 3) and (5, 3) are still the conspicuous pairs. The points 7 and 4 are much farther to 15. We also notice that the angle made by the points 12-15-13 is obtuse, while the angle made by the points 12-15-10 is acute; this means that phyllotaxis is proceeding from (2, 3) to (5, 3). In Zagorska-Marek's diagrams (found in her Fig. 1) for the sequences ⟨1, 2, 3, 5, 8, ...⟩ and ⟨3, 8, 11, 19, ...⟩, it can be seen that behind two different networks of lines where no axis is imposed on the reader, the relative arrangements of the points are essentially the same.

In both interpretations the pairs (2, 3) and (5, 3) are conspicuous. What makes the main difference between the two interpretations, if both of them are considered to be valuable, is the presumed axis of the plant shoot. This points to a careful drawing of this axis in natural specimens. This task, however, may be difficult if one works on mature plants, given the distortions of the initial relations on the shoot apex, coming from subsequent growth, and from an increase in the internode distance. Church (1904b) frequently emphasizes that modern botany has little to do with the effects that appeal to the eye in a fully grown plant shoot, meaning that phyllotactic patterns must be primarily studied where they are being initiated (in the bud) and not as they appear in developed shoots.

Let us now consider Figure 5.3 from Zagorska-Marek (1985), representing a surface view of the terminal shoot of a specimen of *Abies balsamea*.

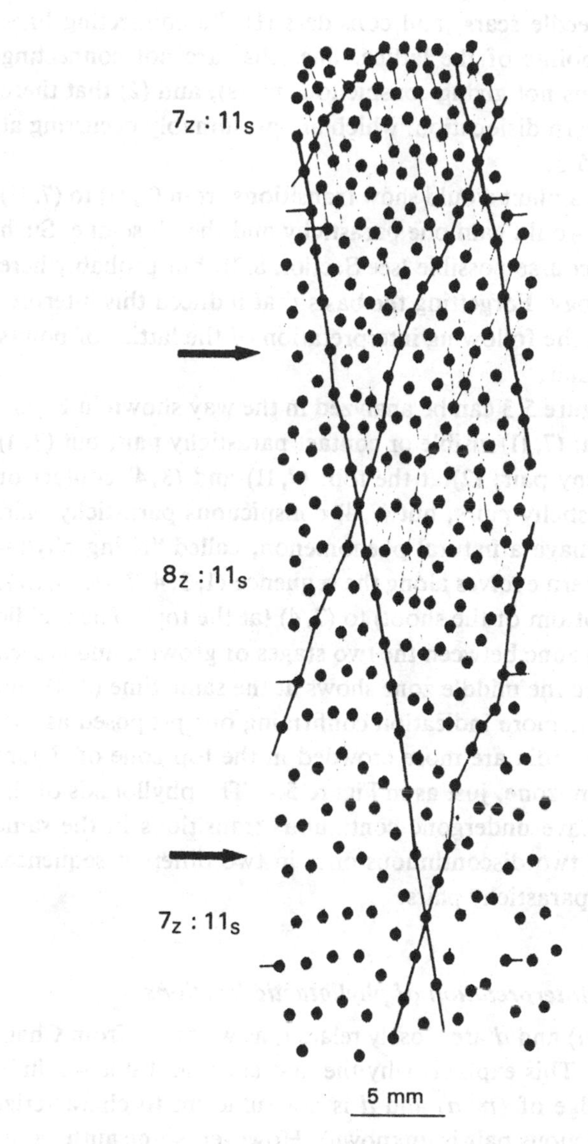

Figure 5.3. Plane representation of the needle scars on the surface of a cylindrical terminal shoot from a 12-year-old tree of *Abies balsamea*. Horizontal segments from two points on the same level mean that they represent the same point. (From Zagorska-Marek, 1985.)

The drawing has been interpreted as representing a (7, 11) pattern for the top and bottom zones, belonging to the sequence ⟨1, 3, 4, 7, 11, ...⟩, and a (8, 11) pattern belonging to the sequence ⟨3, 8, 11, 19, 30, ...⟩ for the middle zone. This interpretation is based on the use of the contact parastichy pairs made by the needle scars, and considers (1) the connecting lines drawn between the points of the lattice, lines that are not connecting nearest neighbors (thus not giving conspicuous pairs), and (2) that there are two zones of pattern dislocation, which are presumably occurring at the arrows in Figure 5.3.

It is surprising that a plant would show transitions from (7, 11) to (7, 11) via (8, 11). The plant would gain one parastichy and then lose one. Such a phenomenon is of course possible (see Section 8.3), but probably here it is a case of teratology. Forgetting the basis that induced this interpretation, we can derive the following interpretation of the lattice of points that looks rather regular.

The diagram in Figure 5.3 can be analyzed in the way shown in Figure 5.4: (1) at the bottom: (7, 11) visible or contact parastichy pair, but (3, 4) conspicuous parastichy pair; (2) at the top: (7, 11) and (3, 4) contact or visible opposed parastichy pairs, but (7, 4) conspicuous parastichy pair. This means that we have a natural phenomenon, called "rising phyllotaxis," where the pattern evolves along the sequence $\langle 1, 3, 4, 7, 11, 18, \ldots \rangle$, from (3, 4) (at the bottom of the shoot) to (7, 4) (at the top). The middle zone is the transition zone between the two stages of growth, and is seen as in Figure 5.4 where the middle zone shows at the same time (3, 4) and (7, 4) phyllotaxis. One more indication confirming our proposed assessment is that the primordia are more crowded in the top zone of Figure 5.3 than in the bottom zone, just as in Figure 5.4. The phyllotaxis of the system would thus have undergone continuous transitions in the same sequence, instead of two discontinuous ones in two different sequences with unconspicuous parastichy pairs.

5.6.3. Interpretation of phyllotactic fractions

The parameters (m, n) and d are closely related, as we know from Chapter 2 (see Table 2.1). This explains why the first entry in Table 4.2 indicates that a knowledge of (m, n) and d is not sufficient to characterize a pattern (the conspicuous pair is unknown). However, some authors assess a pattern with a single parameter, e.g., the value of a phyllotactic fraction, such as 2/5 representing 144°. This value is equally distant from the Fibonacci angle and from the angle of 151° for the main case of anomalous phyllotaxis, so that two types of spiral patterns are likely to be present. Moreover, the computational algorithm of Chapter 2 gives two sequences of visible opposed parastichy pairs for every rational divergence angle. Let us examine the question further.

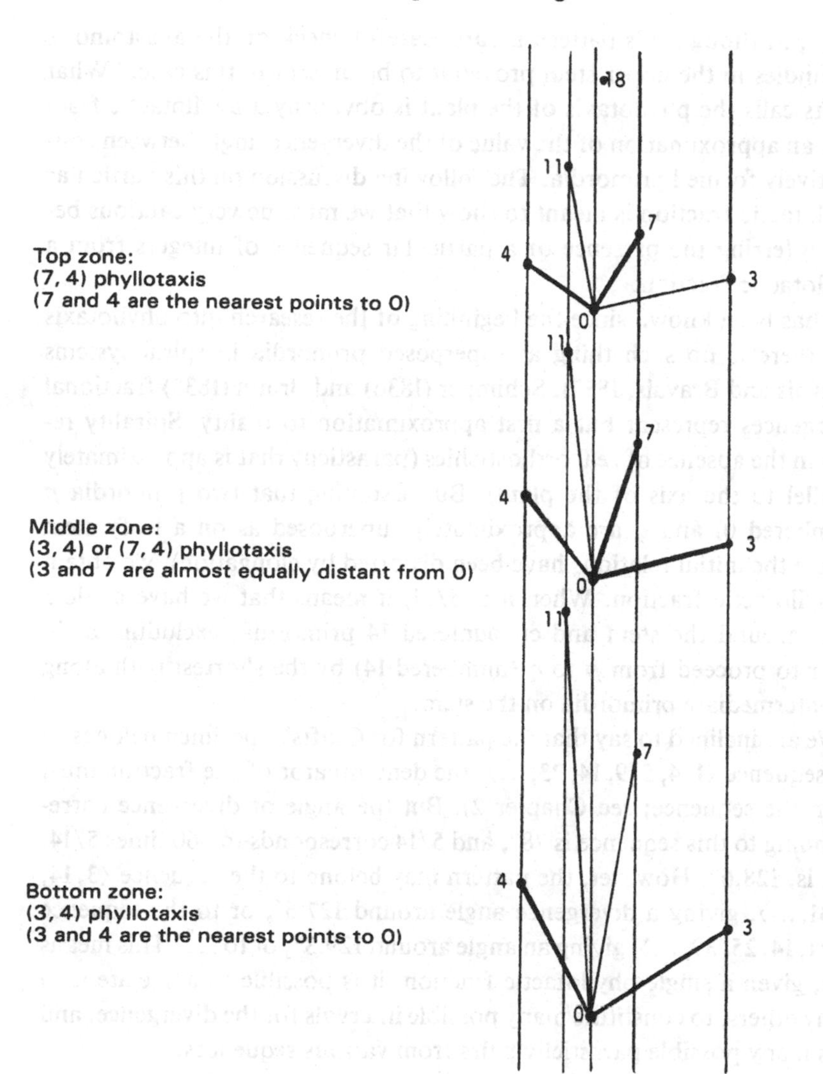

Figure 5.4. An interpretation of Figure 5.3, showing rising phyllotaxis along the first accessory sequence. Points represent the centers of the needle primordia gravitating around 0. Primordium 7 is seen gradually replacing primordium 3 as a nearest neighbor to 0 as we proceed upward in the figure, while the internode distances decrease. Eventually 11 will replace 4, and we will have (7, 11) phyllotaxis.

The patterns are sometimes very difficult to determine, even nonidentifiable (e.g., the perturbed patterns studied in Section 8.6.4). The case of *Pinus halepensis* is apparently a problematic one. Crafts (1943b) says that "after much study the phyllotaxis of *Pinus halepensis* proved to be

5/14, . . . though this pattern is rare, careful check on the anastomosis of bundles in the upper stem proved it to be present in this case." What Crafts calls the phyllotaxis of the plant is obviously a phyllotactic fraction, an approximation of the value of the divergence angle between consecutively formed primordia. The following discussion on this particular phyllotactic fraction is meant to show that we must be very cautious before inferring the presence of a particular sequence of integers from a phyllotactic fraction.

It has been known since the beginning of the research into phyllotaxis that there is no such thing as superposed primordia in spiral systems (Bravais and Bravais, 1837). Schimper (1836) and Braun (1835) fractional divergences represent but a first approximation to reality. Spirality results in the absence of real orthostichies (parastichy that is approximately parallel to the axis of the plant). But assuming that two primordia p (numbered 0) and q are approximately superposed as on a leafy stem (where the initial relations have been distorted by elongation), we obtain a phyllotactic fraction. When it is 5/14, it means that we have made 5 turns around the stem and encountered 14 primordia, excluding p, in order to proceed from p to q (numbered 14) by the shortest path along the intermediate primordia on the stem.

We are inclined to say that the pattern for Crafts's specimen belongs to the sequence $\langle 1, 4, 5, 9, 14, 23, \ldots \rangle$ (the denominator of the fraction must be in the sequence; see Chapter 2). But the angle of divergence corresponding to this sequence is 78°, and 5/14 corresponds to 360 times 5/14, that is, 128.6°. However, the pattern may belong to the sequence $\langle 3, 14, 17, 31, \ldots \rangle$, giving a divergence angle around 127.6°, or to the sequence $\langle 3, 11, 14, 25, 39, \ldots \rangle$, giving an angle around 129.3°, or to. . . . This means that, given a single phyllotactic fraction, it is possible to associate it to many others, to constitute many possible intervals for the divergence, and thus many possible parastichy pairs from various sequences.

If a single fraction can determine visible opposed parastichy pairs along various sequences in spiral systems, two such fractions can determine visible opposed pairs along the same Fibonacci-type sequence. If phyllotaxis rises, to obtain another relevant fraction, continue the movement up around the stem to another primordium q', which looks approximately superposed to p and q as mentioned earlier. Then, summing up the numbers of turns and the number of primordia encountered, respectively, another fraction is obtained. The denominators m and n of the fractions will determine a visible opposed parastichy pair (m, n). And if, for example, we are to have a (3, 14) and a (14, 17), or a (11, 14) and a (14, 25) visible opposed parastichy pairs, corresponding to the sequences $\langle 3, 14, 17, 31, \ldots \rangle$ and $\langle 3, 11, 14, 25, \ldots \rangle$ respectively, then the other fractions should be 1/3 and 6/17 (q' is then numbered 17), or 4/11 and 9/25 (q' is then numbered 25), respectively.

Crafts also reports 5/14 for *Taxodium distichum*. But Fujita (1938) found for this plant a (5, 3) phyllotaxis, corresponding to the Fibonacci sequence. Adler's theorem says that the divergence angle is then to be found between the bounds 1/3 and 2/5, that is, in the interval [120°, 144°], as it is the case for 128.6° (5/14) and 137.5°. Using the computational algorithm of Chapter 2 with $d = 5/14$ indicates that (1, 2), (3, 2) and (3, 5) are visible opposed parastichy pairs. The fraction 5/14 is close to 5/13, corresponding more likely to the main sequence.

Epilogue

Part I deals with the descriptive aspects of phyllotaxis - in its centric, cylindrical, and hierarchical representations. The parameters defining phyllotactic patterns have been introduced, and what is called here the fundamental theorem of phyllotaxis has been presented. This theorem should be in the very core of every teaching curriculum on the rudiments of phyllotaxis, whether it be in a botany or in a mathematical classroom. That is why it has been also given an intuitive presentation and applications. This phyllotactic theorem has been and will be frequently encountered.

The proofs of Equations (1.5) to (1.9), and of Richards's formula (4.30) for the centric representation have been made in the context of the cylindrical representation in Chapter 4. The treatment of the cylindrical lattice has been given in the interval $[-0.5, +0.5]$. This interval makes it easier to identify the visible opposed parastichy pairs. From past experiences the usual interval $[0, 1]$ proved to be sometimes confusing with respect to this identification. A definition of visible pair of spirals has been proposed, which generalizes the notion of visible point introduced by Hardy and Wright, and which brought more general results (e.g., the general form of the fundamental theorem of phyllotaxis in Section 2.3.1 and Proposition A4.1 in Appendix 4).

That phyllotactic patterns are fundamentally branched structures, control hierarchies, has been illustrated. From many biological approaches it can be said that there is a hierarchical control of the primordia under a law of duplication such as that of Lestiboudois–Bolle. A formal definition of the hierarchies with simple and double nodes only has been given using L-systems and square matrices. Despite the treelike appearance of the phyllotactic hierarchies, their laws of formation show that they have loops (rhythm), meaning that the same process-structure is repeated over and over again in the manner of fractals. Sattler (1986) considers that hierarchies in general are simplified versions of nets in which some relations are ignored, and that the notion of hierarchy works because these relations represent weaker interactions so that they may be negligible without a great loss of information. This consideration can be applied here. Phyllotactic hierarchies constitute the formal framework in which the model analyzed in Part II will take place.

For practical pattern recognition, a model has been put forward in Chapter 4, which has been shown in Chapter 5 to allow us to assess the patterns more accurately, rapidly, and easily than ever, with the help of the Pattern Determination Table (Table 4.3). The model expresses the dynamics between the various parameters involved in the shoot apex, that is, the divergence d, the parastichy pairs (m, n), the rise r or plastochrone ratio R, the angle of intersection γ of the parastichies, and the plastochrone P. Since phyllotaxis is an interplay between apical size and primordial size, the model of Chapter 4 has brought interesting new relations among the plastochrone P, Church's bulk ratio B, and Richards's area ratio A. Finally Chapter 5 stressed some of the difficulties involved in pattern recognition, a question that will be reconsidered in Chapter 7.

Part II
Pattern generation: a key to the puzzles

Introduction

In addition to the descriptive aspects of phyllotaxis (Part I), there are more fundamental matters concerned with the causes inducing phyllotactic patterns. How do plants achieve such pattern formation, especially when sometimes apparently quite extraordinary precision is attained, as we have seen? Where does the control come from that establishes the beautiful natural patterns described here? There are several hypotheses and mathematical models designed to explain the dynamical and causal aspects of pattern formation in phyllotaxis. The main hypotheses and/or models can be roughly categorized as:

1. Physical. Snow and Snow's first available space hypothesis (1962) that predicts the formation of a new primordium, whenever by growth a certain minimal amount of space becomes available at the apex. Esau's (1965b) and Larson's (1977) procambial strands hypothesis insisting on the acropetal influence of the vascular traces in the determination of leaves and in pattern formation. Plantefol's (1948, 1950) hypothesis (see Section 2.6) which calls upon Snow and Snow's in order to explain some observations. Schwendener's (1878) hypothesis and Adler's (1977a) model that base pattern formation on mutual physical contact pressures between primordia. Green's reinforcement patterns of cells (1985) and bulging phenomenology of the tunica (1992), for which patterns are the results of surface stresses between cells.

2. Chemical. A hypothesis that interprets the origin of the patterns as due to diffusion fields of chemical inhibitors or activating nutrients emanating from the existing primordia and/or the shoot apex (e.g., Schoute, 1913; Richards, 1951; Turing, 1952; Thornley, 1975a; Mitchison, 1977; Veen and Lindenmayer, 1977; Young, 1978; Schwabe and Clewer, 1984; Meinhardt, 1984; Marzec, 1987; Chapman and Perry, 1987). A primordium arises at that level of the apical dome where the concentration of the inhibitor or activator is below or above a critical threshold. In the case of nutrients, the primordia is assumed to grow in size toward the regions of high nutrient density (competition for nutrients), which has been called "chemical contact pressure" (Roberts, 1984a, 1987).

3. Global. The author's general theory of phyllotaxis, which considers that pattern formation is a matter of a hierarchical control of the primordia, conditioned by a principle of optimal design. It includes a model called the interpretative model of phyllotaxis, described in Chapters 6 and 8.

The first two kinds of hypotheses and the vast majority of approaches derive from an ontogenetic perspective, where a mechanism inside the apical dome is postulated that is assumed to be responsible for the generation of patterns. Their aim is to try to reproduce as faithfully as possible, from a set of assumptions, or from purely formal algorithms (for computer graphics consumption) what is observed on the apical dome or on mature sunflowers for example. We are induced to think that this cannot help much at the levels of the main problems that concern the origins and the functions of phyllotactic patterns. These hypotheses are used in attempts to answer the question: "How do the patterns arise?" In contrast, Part II of this book is mainly concerned with an interpretative approach, devoted to such questions as: "Why or how come such patterns arise?" "What are their functions?" "What purpose do they serve in the environment?" In such a global approach one does not try to imagine the subtle ontogenetic mechanism at work, thus avoiding an oversimplification of the tremendous complexities in the distal zone of the apical meristem inducing the primordia in the organogenic region. Behind the apical dome a forest is hidden which the local models cannot see. Part of this forest is concerned with the phenomena of homology (similar patterns are found in other areas of nature) and branching, and with the phyletic origins of phyllotactic patterns. The approach presented in Part II is global, inductive, axiomatic, and then deductive, predictive and explicative.

The **general theory of phyllotaxis** put forward here summarizes a body of knowledge that is encompassed by the following statements:

- Phyllotaxis has mathematical as well as botanical aspects; both are concerned with pattern recognition and with pattern generation.
- The phenomenon of phyllotaxis (Chapters 1, 2, 3) must be described with a dynamical model (the author's model elaborated in Chapter 4)

controlling the behavior of the parameters involved and making of practical pattern recognition an easy matter (the Pattern Determination Table in Chapter 5).

- Primordial patterns are multilevel systems, which can be represented by control hierarchies (the hierarchical representation of phyllotaxis introduced in Chapter 3).
- Phyllotaxis has evolutionary (Chapters 3, 8), systemic (Chapters 10, 11), and functional aspects (Chapters 9, 12). The phenomenon of phyllotaxis arises in an environment and is part of a compositional hierarchy (including cells, primordia, patterns of primordia, and plants) characterized by the phenomenon of **emergence** (new properties emerge when going from lower levels to upper levels of the compositional hierarchy; Chapter 11).
- A model of phyllotaxis should make predictions (Chapter 6) that can be compared with observations (Chapter 7), and should include all types of patterns, spiral or not (Chapter 8).
- The model should suggest experimental tests and coherent solutions to phyllotactic problems; it should have explanatory values and possible applications (Chapters 6 and 8).
- Phyllotactic patterns must be modeled with respect to an evolutionary classification of the patterns (Chapter 8), given that additional mechanisms or controls are likely to have arisen during evolution which hid elementary rules of growth.
- The dynamics of phyllotaxis must be linked to universal and simple laws of growth, such as **branching** (Chapter 3), **allometry** (Chapter 4), and **gnomonic growth** (Chapter 11).
- Pattern generation may not depend on only one chemical-physical process equally important for all plants, and in the modeling work on phyllotaxis, the ontogenetic (local) assumptions should appear as special cases of more general systemic assumptions (Chapter 9).
- The preponderance of Fibonacci phyllotaxis must be explained by a systemic (global) model taking into consideration all aspects of the large scale-optimization problem phyllotaxis represents (Chapters 6, 8, 12).
- General comparative morphology is of special importance for understanding the origins of phyllotactic patterns, because of the phenomenon of **homology** (Chapters 10, 11, 12).

The formal model or hypothetical-deductive construct of Chapter 6, offers an understanding in general terms which can be given particular meanings, and a physical basis called the hierarchical control of phyllotaxis (Chapter 3). Spiral systems are assumed to be represented by hierarchies.

These hierarchies can be generated algorithmically (with growth matrices and L-systems), and ordered according to the increasing values of a parameter E_b, which can be called the bulk entropy. The parameter is a function of three characteristics of phyllotactic patterns, and can be quantified for each control hierarchy. They are the complexity $X(T)$ of the hierarchy up to its level T, the stability $S(T)$ at T, and the rhythm of growth of period w which has completed a cycle at plastochrone P_r representing the number of nodes (primordia) in levels $T=0$ to $w-1$. The well-known principle of optimal design, operationally formulated by Rosen (1967), takes here the form of a principle of minimal entropy production. It is assumed that the larger the entropy cost, the smaller is the probability of existence of the pattern. It will be shown that when the principle is used to discriminate among the various control hierarchies, then the observed angles and sequences of normal and anomalous phyllotaxis are found.

The predictive value of the ontogenetic models is rather slender. They do not offer the kind of predictions we will see in Section 6.4, or the kind of explanatory possibilities we will present in Section 8.6 for the interpretative model. The contact pressure model seems to offer a means for verifications because of the mathematical relation it predicts between the divergence angle d and the rise r, deduced for primordia under contact. According to this relation, the point (d, r) would move on a well-defined arc of a circle. On the other hand, the predictions of the interpretative model can be directly compared with the data available. This is done in Chapter 7.

Chapter 8 shows that the interpretative model is concerned not only with spiral patterns, but also with whorled patterns (e.g., distichy and decussation), which are shown to be special cases of spiral systems. The model is used to predict the existence of patterns such as monostichy and superposed dimery, and it can be used to settle difficult problems of pattern recognition, such as the patterns on *Dipsacus* and *Costus*.

In Chapter 9 an algorithm called the τ-model (Jean, 1990) is proposed, and a comparison is made with Marzec's entropy model (1987), the interpretative model (Jean, 1980b, 1993b), Adler's model (1974, 1977a), and the model of Xudong et al. (1989) for packing of disks. In particular, the priority order of the various spiral patterns that can be predicted by these models are compared. It is shown that more general frameworks of thoughts based on more general principles are grounds for more comprehensive models, by which it becomes possible to generalize and unify the existing developments. According to Rashevsky (1965), general principles can give to models, transient by their own nature, a greater permanence.

6
An interpretative model

6.1. The necessity of defining entropy measures

6.1.1. An a-disciplinary concept

The word entropy comes from a Greek word meaning evolution. The physical meaning of the concept of entropy is much disputed; it is still considered to be not very clear. According to Poincaré, it is a "prodigiously abstract concept." There does not exist a completely rigorous mathematical formulation of thermodynamics. Wiener (1948) indicated the need to extend the notion of physical entropy when he stated that information is negative entropy. For Brillouin (1959) information and physical entropy are of the same nature, the increase in entropy corresponding to a loss of information.

The literature - in physics; in statistical theory of communications and information theory; in social sciences and the life sciences; in probability theory, graph theory, and Lebesgue's integration theory - contains many concepts and formulas for entropy and for quantity of information. Among them we find the topological and structural information content of Rashevsky (1955) and Trucco (1956a,b) defined on graphs as a measure of their complexity; the more well-known Shannon–Weaver entropy of a set of probabilities given by $I = -\sum p_i \log p_i$; the Hartley–Nyquist formula $I = -\log p$ where p is the probability of drawing an n letters message in an urn containing all messages; the chromatic information content of Mowshowitz (1968); the entropy of measurable functions and the epsilon-entropy; and the absolute S-entropy and the weighted entropy. Haynes, Phillips, and Mohrfald (1980) cite a long list of authors who have developed their own concepts of entropy by specifying a function and applying it to a particular context. Most of the time, unfortunately, the form of the mathematical function is not specified. Auger (1990) developed the notion of entropy in compositional hierarchies (the meaning of this concept of hierarchy is considered in Section 11.4; the hierarchies of Chapter 3 are control hierarchies). The word "entropy," corresponding to an a-disciplinary concept, has multiple meanings that may, however, show common denominators, such as, for example, the idea of the entropy of a partition.

Mowshowitz (1967) concludes that his investigations point up the desirability of defining and examining other entropylike measures. For Brillouin (1959) history suggests that laws have very often to be reviewed, corrected, and completed to allow their extensions to wider domains. Life looks like a struggle against physical laws and the second principle of thermodynamics, and the concepts and formulas for entropy mentioned above do not appear to be adapted to biological phenomena.

Schrödinger (1962) initiated the use of the concept of entropy in the life sciences by developing a literal interpretation of the well-known formula $E = k \log D$, where D (a large number) is a measure of the disorder and k is Boltzmann's constant ($1.38/10^{16}$ ergs by degree centigrade). Rashevsky (1968) suggested the satisfaction function $P = A \log X$ which looks like a factor of negative entropy, each organism tending to optimize P. But it is currently recognized that entropy in living organisms, what can be called bio-entropy, must be a sum of two factors, one of which represents negentropy.

6.1.2. Entropy in phyllotaxis

Jean (1976b, 1978a, 1980a,b, 1988b) was the first to introduce an entropy-like function in phyllotaxis. The idea of nutrient in Roberts's chemical contact pressure hypothesis (1984a) implicitly refers to energy brought into the systems, that is, to negative entropy. Marzec (1987) developed a model of phyllotaxis based on an entropy concept, which is presented in Section 9.5. He found that the naturally occurring divergence angles produce local minima in both the rate of production of entropy due to morphogen diffusion and the rate at which entropy is lost due to morphogen degradation. Rivier (1990; presented in Section 10.4.4) introduced an entropy concept in the area, suggested by his crystallographic approach. Barabé (1991a) argues that the transition from a regular phyllotactic system to an irregular one represents a variation in the entropy level that induces a change in the parameters of the system. Green (1992, introduced in Section 12.2) considers that the tunica bulges to reach the configuration of minimal strain energy and of maximal entropy. Levitov (1991a,b) and Douady and Couder (1992; studied in Section 12.3.3) consider that the primordia produce a structure of minimal global interaction energy.

It appears necessary to place the phenomenon of phyllotaxis in the context of evolution in order to make it more fully intelligible. Entropy is a major concept in evolutionary theories. One way to introduce an entropy-like function is via the hierarchies of simple and double nodes described in Chapter 3, as a function of three main characteristics of phyllotactic patterns. In a phyllotactic hierarchy we can define parameters able to represent these characteristics, namely the rhythm w, the complexity $X(T)$, and the stability $S(T)$. The parameter w is the order of the smallest growth matrix that generates the hierarchy, and T $(=0, 1, 2, 3, \dots)$ represents the levels in the hierarchy (also called a multilevel system; see for example Fig. 3.17).

It is considered that stability has to do with duplications. Two wheels at the end of an axle form a more stable system than one wheel standing alone. Nature would tend to duplicate an object when it does not present sufficient structural stability. "The phenomenon of stabilization by duplication by its extreme simplicity, its efficiency, is of fundamental importance" (Bruter, 1974b). The parameter $S(T)$ will thus represent the relative frequency of duplications, that is, of bifurcations (from double nodes) in the hierarchy up to level T, with respect to the total number of simple and double nodes. For example, in the case of Figure 3.1 representing the Fibonacci hierarchy $\langle 2, 3 \rangle$, it is easily seen that we have $S(1) = 1/2$ (one bifurcation out of two nodes in level 0), $S(2) = 3/5$ (three bifurcations out of five nodes in levels $T = 0$ and $T = 1$), $S(3) = 6/10$, and $S(4) = 11/18, \dots$. In the case of Figure 3.4 representing the Fibonacci hierarchy $\langle 3, 4 \rangle$, $S(1) = 1/3$, $S(2) = 4/7$ (four bifurcations out of the seven nodes in levels $T = 0$ and $T = 1$), $S(3) = 4/7$, $S(4) = 3/5, \dots$. For the companion hierarchy of order 3 (Fig. 3.17) we have $S(1) = 1$, $S(2) = 1/3$, $S(3) = 2/5$ (two bifurcations out of five nodes in levels $T = 0, 1, 2$), $S(4) = 3/8, \dots$.

And we need a mathematically interesting formula for $X(T)$ that will increase rapidly with T. The one chosen is

$$X(T) = \prod_{k=1}^{T} k^{f(k)},$$

where $f(k)$, $k = 1, 2, 3, \dots$, is the number of nodes in level k of the hierarchy. The function $f(k)$ (examined in Appendix 7) is the growth function of the L-system generating the hierarchy. For example, in the hierarchy

⟨2, 3⟩ (Fig. 3.1), we have $X(1)=1^3$, $X(2)=1^3 2^5$, $X(3)=1^3 2^5 3^8$, etc. For the companion hierarchy of order $w=3$ (Fig. 3.17), we have $X(1)=1$, $X(2)=2^2$, $X(3)=2^2 3^3$, $X(4)=2^2 3^3 4^4$, etc. The parameters $X(T)$, $S(T)$ and $f(k)$ can easily be expressed in terms of the entries of the powers of the growth matrix generating the hierarchy. Other functions for $X(T)$ could be chosen that increase more slowly with T while giving the same results (a suggestion on how this can be done is made following Proposition 8 of this chapter).

We arrive at a first proposal for a formula linking the parameters, called the entropy E:

$$E(T) = -\log[S(T)/X(T)]. \tag{6.1}$$

The value of E increases rapidly with the complexity $X(T)$, and incorporates a factor $-\log S(T)$ representing negentropy. Other combinations of $S(T)$ and $X(T)$ can be examined. For the sake of the deductions that will now be made from this setting, it is not necessary to call E the entropy, but it is interesting to emphasize that Equation (6.1) is linked to formulas for entropy. For example, the entropy of a single event of probability p has been defined by Rényi (1961) by the formula $\log(1/p)$. With $p=1/X(T)$ we have $\log(1/p)=\log X(T)=-\log(1/X(T))$, which is one of the two factors in Equation (6.1). It also represents the entropy of a set of $X(T)$ elements and corresponds to the maximum of the more general entropy $\sum p_i \log p_i$ under minimal constraints. Following are two propositions based on these concepts.

Proposition 1.

(a) *Denoting by $S_t(T)$ the stability of the Fibonacci hierarchy ⟨t, t+1⟩ at T, we have:*

$$S_t(T) = [F_{T+1}t + F_T - t]/[F_{T+2}t + F_{T+1} - (t+1)], \quad t = 1, 2, 3, \ldots,$$

$$S_t(T) > S_{t+1}(T), \quad t = 1, 2, 3, \ldots,$$

$$S_t(T+1) > S_t(T) \quad \textit{for } t > 1,$$

$$\lim_T S_t(T) = 1/\tau \quad \textit{for each } t.$$

(b) *For $t = 1, 2, 3, \ldots$, the hierarchy ⟨t, t+1⟩ has for all T a smaller E than the hierarchy ⟨t+1, t+2⟩.*

(c) *Denoting by $S_w(T)$ the relative frequency of duplications for the companion hierarchy of order $w = 2, 3, 4, \ldots$, we have:*

$$\lim_T S_w(T) > \lim_T S_{w+1}(T),$$

$$S_2(T) > S_w(T) \quad \text{for } w > 2 \text{ and } T = 2, 3, 4, \ldots$$

(This proposition can be proved by working on matrices; see Appendix 7.)

(d) *Among the Fibonacci hierarchies and the companion hierarchies,* $\langle 1,2\rangle$ *(that is the companion hierarchy of order 2) has maximal* $S(T)$ *for all* T*, and minimal* $E(1)$ *and* $E(2)$ *(these two values are identical to the corresponding values for the companion hierarchy of order* $w=3$*).*

Proposition 2.

(a) *For any given* $T>1$*, the minimal value* $E^*(T)$ *that* $E(T)$ *can take in the set of all hierarchies with simple and double nodes is*

$$E^*(T)=\log(2T-1)+2\sum_{k=1}^{T}\log k.$$

(b) *Among the hierarchies generated by growth matrices of order* $n\leq w$*, the companion hierarchy of order* w *has minimal* $E(T)$ *for* $T=1,2,3,\ldots,w$*, with* $E(T)=E^*(T)$ *for* $T=1,2,3,\ldots,w-1$*.*

Proof: Whenever there are only two nodes in the consecutive levels starting from level $T=1$, $X(T)$ is minimized and equal to $\log(T!)^2$. In this case, if w_0 (the axiom of the L-system; see Section 3.5) contains one double node, $\log S(T)=\log(2T-1)$, so that $E(T)=E^*(T)$. If w_0 contains two simple nodes, $E(T)$ is infinite.

When there are more than two nodes in the consecutive levels starting from $T=1$, then $X(T)$ is minimized by the value $\log(T!)^3$, and assuming that $S(T)=1$ for all T, then $E(T)$ is minimized by that value of $X(T)$, a value larger than $E^*(T)$ for $T=3,4,5,\ldots$. For $T=1$ the two values are equal, and for $T=2$ it can be verified that $E^*(T)$ is the smallest possible entropy.

The companion hierarchy is precisely the one for which $E(T)=E^*(T)$ for $T=1,2,3,\ldots,w-1$. If level $T=w-1$ contained two simple nodes, then the entropy would be minimized by $E^*(w)$; but the period of the rhythm, which has to bring back a double node, would be larger than w. If level w contains two double nodes, then the entropy of the hierarchy will be larger than that of the companion hierarchy of order w. ■

6.2. Dynamic versus static approaches

6.2.1. *Principle of minimal entropy production*

Given Proposition 2, the companion hierarchy of order w can be called the *path of minimal entropy of order w*. If a pattern had no rhythm, then

w could be as large as desired, and the path of minimal entropy would be a kind of vortex where entropy is minimized for every value of T. For a Fibonacci hierarchy the entropy is larger than for the companion hierarchy, at least starting from $T=3$. But Fibonacci hierarchies have the smallest rhythm ($w=2$). This might be an advantage in difficult conditions of growth (e.g., in disturbed habitats along roadsides and mountain slopes). When the rhythm is equal to $w=2$, the hierarchy that minimizes E for each consecutive value of T, corresponding to a dynamic approach, is the hierarchy $\langle 1,2\rangle$ defining the main sequence $\langle 1,2,3,5,8,13,\ldots\rangle$. This particularity of other models as well will be discussed in Chapter 9. We thus need another functional E that is able to overcome such a limitation and that is able to give other spiral patterns as well, depending on an appropriate entropy principle.

The bulk entropy of a hierarchy or pattern is defined by

$$E_b = -\sum_{T=1}^{w} \log(S(T)/X(T)), \tag{6.2}$$

where w is the rhythm of the hierarchy that is the order of the smallest growth matrix generating the hierarchy (Section 3.5.2). We look for the hierarchy that minimizes E_b at **plastochrone P_r** (P_r primordia exist). P_r is defined as follows: It is the plastochrone at which the rhythm of growth has ended producing a loop or a cycle, and a new cycle of primordial production is about to start along the same rules. The function E_b is not as formal as it looks at first sight; we will see that this function is related to the usual parameters of phyllotaxis (Proposition 8).

Principle of minimal entropy production: A plant generates primordia, and at P_r the primordia constitute the levels of the hierarchy giving the smallest E_b. Then the P_r nodes are distributed in levels $T=0,1,2,\ldots,w-1$; the value of w is known, and every further level of the hierarchy is determined by the production rules of the L-system generating the hierarchy.

This principle is a particular case of the principle of optimal design (Rashevsky, 1960, 1961) formulated by Rosen (1967) in the following way: biological structures which are optimal in the context of natural selection, are also optimal in the sense that they minimize a certain functional cost. To find the optimal solution to a particular design problem, we have:

(a) to determine the class of all possible solutions to the problem, here the hierarchies with simple and double nodes only,
(b) to assign to each solution a real number which represents the cost involved, here E_b, and
(c) to search among the set of costs to find that which is least.

According to Rosen all the art and difficulty of the principle of optimal design lies in finding the appropriate functional.

Equation (6.2) expresses a dynamic approach up to the plastochrone P_r at which a cycle of primordial production is completed, and a static approach thereafter, the system having become rhythmic and deterministic. It will be shown that at P_r, only Fibonacci hierarchies $\langle n, m \rangle$ arise. A limit divergence angle will follow, if the rhythm is maintained, the one corresponding to the Fibonacci-type sequence $\langle n, m, m+n, 2m+n, \ldots \rangle$, by the fundamental theorem of phyllotaxis.

6.2.2. Particular notions of rhythm

The study of rhythm in biology belongs to a relatively new discipline called chronobiology (Gauquelin, 1973), which proceeds from an evolutionary rather than from a cartesian or analytical stream of research. According to Reinberg (1977) its essential ideas are that rhythmic activity is a fundamental property of living matter, it can be seen at all levels of organization, and it is part of the genetic inheritance of the species. Biological rhythm manifests itself as a regular and predictable variation, and one of the first aspects of the temporal structure of an organism is the preponderant period of the rhythm.

These general ideas echo Church's more particular ideas for phyllotaxis, where the rhythm is considered to demand the essential mechanism of phyllotactic pattern formation and to represent one of the most fundamental laws of living plasma. He considers the factor inducing the rhythm as one of the most fundamental problems of physiology (Church, 1920b). For Church (1920b) "every form of phyllotaxis construction is rhythmic and works out a pattern." We have seen in Chapter 3 that in a Fibonacci hierarchy the rhythm is expressed by the fact that each simple node gives rise to a double node, and each double node produces a simple and a double node. The rhythm has to be represented in a formula such as E_b, together with the complexity and the stability.

The interpretative model offers an understanding of the phenomenon of phyllotaxis in formal terms which can be given specific "physicochemical" meanings. In the model the notion of rhythm can take particular acceptations. For example, rhythm can express the way in which the primordia-inducing nutrients (morphogens) are provided in the system as a whole. Rhythm can be the result of the coordination of various relative growth rates dY/Ydt of substances and sizes within the shoot apex. Barlow (1989a,b) studies various aspects of rhythms in plants and suggests, for example, that the alternation of tannin and nontannin cells in nodes and internodes of *Sambucus nigra* reveals a rhythmic activity in the apex, which is integrated with the initiation of each leaf. Using correlogram analysis Kumazawa and Kumazawa (1971) show that for *Erigeron sumatrensis* the variations in internode length and in successive divergence angles of young and full-grown shoots are rhythmic, with a period of 2.62 nodes (or internodes) corresponding to one revolution of the genetic spiral.

The rhythm could be the expression of the two-dimensional principle of generative resonance used by Green (1985) at the cellular level to reproduce distichy and decussation. As Goebel (1969) did, Green refers to distichous phyllotaxis as pendulumlike and likens the pattern to a swinging pendulum. Green (1992) compares decussation to a resonating metal sheet that produces potato chip-like pairs of crests and troughs, and he compares phyllotactic pattern generation to recurrent buckling processes. In Jesuthasan and Green's (1989) biophysical approach at the cellular level for the decussate pattern of *Vinca major*, a pattern-generating cycle is formed from three components: cellulose reinforcement, primordium formation, and the resulting physical influences on the cytoskeletons of the dome. The various types of patterns sometimes found in various individuals of the same species or in even the same specimen are compared by Green (1991) to the various self-reinforcing resonating harmonics in a bugle. We will come back to these ideas in Chapter 12.

Rhythm is present in the contact-pressure model of phyllotaxis. The plastochrone P_r at which a cycle is completed (with P_r primordia) can be understood as the plastochrone P_c at which contact pressure begins in Adler's contact pressure model. Adler's maximin principle is that at P_c the centers of the primordia move by growth so as to maximize, by increasing their sizes, the minimal distance between them. Then the system shows phyllotaxis (m, n). When contact pressure begins, a rhythm of growth is induced. The divergence angle starts to oscillate between end points of nested intervals of decreasing amplitudes (rising phyllotaxis), intervals that are included in the intervals given by the fundamental theorem of Chapter 2. This process leads to a particular limit divergence angle. Growth thus becomes deterministic at P_c, as it becomes deterministic at P_r, giving the hierarchy with minimal E_b.

The contact-pressure model partially produced the main result of the interpretative model. The main result in the contact-pressure model (included in Proposition 6 below) is that Fibonacci phyllotaxis is inevitable if contact pressure begins early, that is, before P_c (respectively P_r) $= 6$. To achieve more general results we need more formal concepts able to take particular acceptations. We will come back to this idea in Chapter 9 and discuss other instances where the interpretative approach can be seen to produce a generalization of fundamental ideas of other approaches.

6.3. The optimal designs

In this section we are looking for minimal entropy costs (see part c of the principle of optimal design, p. 132) and for algorithms that will allow us to obtain easily the hierarchy giving minimal entropy at any value of P_r.

Proposition 3.

(a) *For the hierarchy* $\langle 1,2\rangle$, E_b *is smaller than for the companion hierarchy of order* $w>2$, *and for the hierarchies that have double nodes only.*

(b) *If we order the first few Fibonacci hierarchies* $\langle n,m\rangle$, $n<m$, *by (strictly) increasing* $E_b = \log[X(1)X(2)/S(1)S(2)]$, *we have* (log *means base* 10):

$\langle 1,2\rangle$ ($E_b \approx 1.08$), $2\langle 1,2\rangle$ ($E_b \approx 1.98$),

$\langle 2,3\rangle$ ($E_b \approx 2.03$), $\langle 3,5\rangle$ ($E_b \approx 2.79$),

$\langle 3,4\rangle$ ($E_b \approx 2.83$), $3\langle 1,2\rangle$ ($E_b \approx 2.89$),

$2\langle 2,3\rangle$ ($E_b \approx 3.53$), $\langle 4,5\rangle$ ($E_b \approx 3.57$),

$\langle 4,7\rangle$ ($E_b \approx 3.63$), $4\langle 1,2\rangle$ ($E_b \approx 3.79$),

$\langle 5,7\rangle$ ($E_b \approx 4.24$), $\langle 5,6\rangle$ ($E_b \approx 4.27$),

$\langle 5,8\rangle$ ($E_b \approx 4.35$), $\langle 5,9\rangle$ ($E_b \approx 4.50$),

$2\langle 3,4\rangle$ ($E_b \approx 4.93$), $\langle 6,7\rangle$ ($E_b \approx 4.96$),

where the numbers in parentheses are the costs for (n,m) *phyllotaxes and where for example* $2\langle 1,2\rangle = \langle 2,4\rangle$.

(c) *For Fibonacci hierarchies when* $P_r = 3$, *and* 5 *to* 14, E_b *is minimal for* $\langle 1,2\rangle$, $\langle 2,3\rangle$, $2\langle 1,2\rangle$, $\langle 3,4\rangle$, $\langle 3,5\rangle$, $3\langle 1,2\rangle$, $2\langle 2,3\rangle$, $\langle 4,7\rangle$, $4\langle 1,2\rangle$, $\langle 5,8\rangle$, *and* $\langle 5,9\rangle$, *respectively (the sum of the terms in* $\langle\cdot\rangle$ *is* P_r*). For* $P_r = 4$ *no phyllotactic hierarchy can exist, given that* $S(1) = 0$ *and* E_b *is infinite.*

Let us show for example how to calculate E_b for the hierarchy $\langle 4,7\rangle$. For this hierarchy there are 4 nodes in level $T=0$, and 7 nodes in level 1. We thus have the diagram VVV|, containing necessarily three double nodes in level $T=0$, giving 3 duplications, and one simple node in the same level; consequently $S(1) = 3/4$. On the basis that each double node in level $T=0$ gives a double node and a simple node in level $T=1$, and that the simple node in level $T=0$ gives one double node, we can calculate

$S(2)=7/11$. Now for the complexities we $X(1)=1^7=1$, $X(2)=1^7 2^{11}=2^{11}$, and we can calculate

$$E_b=\log[X(1)X(2)/S(1)S(2)]=\log[2^{11}\times 4\times 11/(3\times 7)]\approx 3.6325.$$

In the following propositions the value of P_r is given as a term of a Fibonacci-type sequence $\langle H(1), H(2), H(3), \ldots\rangle$.

Proposition 4. *Given $P_r=H(k+3)$, $k\geq 0$, the hierarchy*

$$\langle H(k+1)+[-H(k-1)/3], H(k+2)-[-H(k-1)/3]\rangle,$$

which has P_r nodes in levels 0 and 1 altogether and where $[-H(k-1)/3]$ means the smallest integer larger than $-H(k-1)/3$ has minimal E_b among the hierarchies with $w=2$.

Proof: Whatever P_r may be, when $w=2$ we have to look for minimal E_b among the Fibonacci hierarchies generated by the Q-matrix (p. 71). Indeed the growth matrix $C=(c_{ij})$ of order $w=2$ where $c_{11}=c_{22}=0$ and $c_{12}=c_{21}=1$ gives infinite entropy. The only other growth matrix of order $w=2$ contains ones only and generates hierarchies with double nodes only. In this case the number of nodes in levels $T=1$ and 2 is a multiple of 3, that is $3p$, and $p\langle 1,2\rangle$ has a smaller bulk entropy.

The hierarchy we are looking for, giving minimal entropy, is given by $\langle m+n, 2m+n\rangle$, where $m>0$ is the number of double nodes and $n\geq 0$ is the number of simple nodes in level 0 of the hierarchy. The sequence we are looking for starts with the terms

$$n, m, m+n, 2m+n, 3m+2n=H(k+3).$$

It starts also with the terms

$$H(k-1)+3p, H(k)-2p, H(k+1)+p, H(k+2)-p, H(k+3),$$

where p is an integer to be determined. We have that $m=H(k)-2p>0$ and $n=H(k-1)+3p\geq 0$, so that

$$-H(k-1)/3\leq p<H(k)/2.$$

It can be calculated that the entropy for the hierarchy $\langle m+n, 2m+n\rangle = \langle H(k+1)+p, H(k+2)-p\rangle$ is given by

$$\begin{aligned}E_b&=E(1)+E(2)\\&=\log[(H(k+1)+p)H(k+3)2^{H(k+3)}/(H(k)-2p)(H(k+2)-p)].\end{aligned}$$

This is an increasing function of p in the above interval for p so that the minimal E_b is given by $p=[-H(k-1)/3]$. ■

Proposition 5. *The bulk entropy for the hierarchy $\langle H(k+1), H(k+2)\rangle$ is smaller than for any other hierarchy with rhythm of period $w>2$ having $H(k+3)\geq 3$ nodes in the first w levels (including level 0).*

Proof: For the Fibonacci hierarchy, $E_b = E'$ is smaller or equal to

$$\log[H(k+3)2^{H(k+3)}/H(k)] \approx \log[\tau^3 2^{H(k+3)}].$$

Consider a hierarchy where $w>2$ having $H(k+3)$ nodes in the first w levels. Putting for this hierarchy $S(T)=1$ and $X(T)=(T!)^{H(k+3)/w}$ for $T=1,2,3,\ldots,w$, the bulk entropy will be minimized by the value $E''=\log X(2)+\log X(3)+\cdots+\log X(w)$. It is easy to show that for $w>3$, $E'<E''$. For $w=3$ one can verify that $E'<E''$ for $H(k+3)>7$. Finally, for $3\leq H(k+3)\leq 7$ and $w=3$ the matter can be settled by working directly on the possible hierarchies. ■

From these propositions here are algorithms allowing us to determine the hierarchy giving the minimal entropy cost, without calculating any E_b. It can be proved that for a given value of $P_r>2$, the hierarchy producing minimal entropy is always unique.

Algorithm 1: Given $P_r=H(k+3)$, we generate any Fibonacci-type sequence starting with $H(k-1)$, $H(k)$, $H(k+1)$, $H(k+2)$, $H(k+3)$. We then calculate $p=[-H(k-1)/3]$, and put its value in $\langle H(k-1)+3p$, $H(k)-2p$, $H(k+1)+p$, $H(k+2)-p$, $H(k+3),\ldots\rangle$, which is the sequence giving minimal entropy. That is, $\langle H(k+1)+p$, $H(k+2)-p\rangle$ is the hierarchy looked for.

Example 1: Suppose that $P_r=23$. Produce any Fibonacci-type sequence $\langle H(k-1), H(k), H(k+1), H(k+2),\ldots\rangle$ having 23 as 5th element, such as $\langle -2,9,7,16,23,\ldots\rangle$. Then $-H(k-1)/3=2/3$ so that $p=1$. The sequence we are looking for is given by $\langle H(k-1)+3p$, $H(k)-2p$, $H(k+1)+p$, $H(k+2)-p,\ldots\rangle$, that is $\langle 1,7,8,15,23,\ldots\rangle$. Notice that if we had started with the sequence $\langle 10,1,11,12,23,\ldots\rangle$, then $-H(k-1)/3=-10/3$ so that $p=-3$. This value of p returns the same minimal entropy sequence, and hierarchy $\langle 8,15\rangle$.

Algorithm 2: Given $P_r=H(k+3)$, we generate all the Fibonacci-type sequences having $H(k+3)$ as a member, starting with $H(k-1)$, and such that $H(k+1)<H(k+2)<2H(k+1)$. Then we choose the sequence for which $p=0$.

Example 2: Let $P_r = H(k+3) = 17$. Then the possible Fibonacci-type sequences are

$$\langle 7, 1, 8, 9, 17, \ldots \rangle,$$

$$\langle 4, 3, 7, 10, 17, \ldots \rangle,$$

$$\langle 1, 5, 6, 11, 17, \ldots \rangle.$$

The first element in each row represents the value of $H(k-1)$. The sequence giving $[-H(k-1)/3] = 0$ is $\langle 1, 5, 6, 11, 17, \ldots \rangle$, giving minimal entropy at $P_r = 17$. That is, the hierarchy looked for is $\langle 6, 11 \rangle$.

Example 3: If $P_r = H(k+3) = 11$, the possible hierarchies $\langle H(k+1), H(k+2) \rangle$ are $\langle 5, 6 \rangle$ and $\langle 4, 7 \rangle$. For the former, $H(k-1) = 4$, while for the latter $H(k-1) = 1$. It follows that the hierarchy $\langle 4, 7 \rangle$ has minimal entropy at $P_r = 11$.

6.4. Results and predictions of the model

6.4.1. Patterns that can and cannot exist

According to the principle of minimal entropy production, at plastochrone P_r the spiral pattern that is selected is the one that minimizes the function E_b. Table 6.1 shows the values of P_r at which the different patterns arise. The first three types of patterns are by far the most frequently mentioned in the literature.

The Fibonacci-type sequences that are generated by the model are of the form:

$$\langle p, 2p, 3p, \ldots \rangle, \quad p \geq 1,$$

$$\langle p+1, 2p, 3p+1, \ldots \rangle, \quad p \geq 2,$$

$$\langle p+1, 2p+1, 3p+2, \ldots \rangle, \quad p \geq 1.$$

By eliminating the duplications (like $p = 2$ in the first sequence and $p = 3$ in the second) and rearranging the sequences, according to the interpretative model the only types of spiral patterns that can exist as primitive primordial patterns (not arising from transitions on mature stem) are given by the sequences:

$$J\langle 1, 2, 3, 5, 8, 13, \ldots \rangle, \quad J \geq 1,$$

$$J\langle 1, t, t+1, 2t+1, 3t+2, \ldots \rangle, \quad J = 1 \text{ or } 2, \ t \geq 3, \tag{6.3}$$

Table 6.1. *Sequences that arise as P_r increases*

Sequences	Values of P_r
$\langle 1,2,3,5,8,13,\ldots\rangle$	3, 5, 8, 13
$2\langle 1,2,3,5,8,13,\ldots\rangle$	6, 10, 16
$\langle 1,3,4,7,11,18,\ldots\rangle$	7, 11
$3\langle 1,2,3,5,8,13,\ldots\rangle$	9
$4\langle 1,2,3,5,8,13,\ldots\rangle$	12
$\langle 1,4,5,9,14,23,\ldots\rangle$	14
$5\langle 1,2,3,5,8,13,\ldots\rangle$	15
$\langle 1,5,6,11,17,\ldots\rangle$	17
$6\langle 1,2,3,5,8,13,\ldots\rangle$	18
$\langle 2,5,7,12,19,31,\ldots\rangle$	19
$\langle 1,6,7,13,20,33,\ldots\rangle$	20
$2\langle 1,3,4,7,11,18,\ldots\rangle$	22
$\langle 2,7,9,16,25,\ldots\rangle$	25
$2\langle 1,4,5,9,14,23,\ldots\rangle$	28
$\langle 2,9,11,20,31,\ldots\rangle$	31

[compare with Sequence (2.1)] with the corresponding limit-divergence angles

$$d = 360°[J(t+\tau^{-1})]^{-1}, \tag{6.4}$$

called **the sequences and angles of normal phyllotaxis;**

$$\langle 2, 2t+1, 2t+3, 4t+4, 6t+7, \ldots\rangle, \quad t \geq 2, \tag{6.5}$$

[identical with Sequence (2.2)], with the corresponding limit-divergence angles

$$d = 360[2+(t+\tau^{-1})^{-1}]^{-1}, \tag{6.6}$$

called **the sequences and angles of anomalous phyllotaxis.** For every value of P_r (except the values 2 and 4), one and only one of these systems can arise.

Proposition 6 (the main result). *Patterns expressed by the Fibonacci sequence arise if a cycle is completed early, that is, when plastochrone $P_r <$ 6 (when there are five primordia), or at $P_r = 8$ or 13.*

We only have to look at Table 6.1 to see that this proposition is true. Notice also that for the companion hierarchy of order 3 (Fig. 3.17) there are 5 nodes in the first 3 levels ($T=0, 1, 2$) and $E_b = \log 3240$, while for the hierarchy $\langle 1, 2\rangle$, $E_b = \log 12$, and for the hierarchy $\langle 2, 3\rangle$, $E_b = \log(320/3)$. The cost paid by nature to produce Fibonacci phyllotaxis is minimal. The Fibonacci angle represents the most economical solution of primordial arrangements.

The general fact of observation (Church, 1920b) is the enormous preponderance of the Fibonacci sequence in spiral systems, to the extent that any other sequence may be termed relatively rare and exceptional. As P_r increases, a priority order is established among the patterns, as shown in Table 6.1. A larger value of P_r produces a pattern interpreted to be less frequent. That the Fibonacci sequence is the most frequent at the beginning of the table, coupled with Propositions 1d and 3a, would explain the omnipresence of this sequence in nature.

Whatever the meaning he gives to the word "rhythm," we can read in Church (1920b) that "the special advantages of the Fibonacci rhythm are (1) the maximum compactness of the resultant soma, and (2) its capacity for indefinite growth-extension on the same terms." For the sunflower showing (3, 5) phyllotaxis at the vegetative stage, a cycle is completed at $P_r \leq 3+5$. As the capitulum forms, phyllotaxis can shift to (144, 233) under the determination established at P_r. If the capitulum had to arise at $P_r = 144+233$, then it would be an unlikely structure. That the rhythm is established as late as $P_r = 8$ opens the possibility for the presence of sunflowers expressed by the sequences $2\langle 1, 2, 3, 5, \ldots\rangle$ (divergence of 68.75°) and $\langle 1, 3, 4, 7, 11, \ldots\rangle$ (divergence of 99.5°), as Table 6.1 shows. It is interesting to note that Schoute (1938) found sunflowers showing these sequences: 2.8% of his specimens are cases of the former type, and 14.5% are cases of the latter type.

Proposition 7.

(a) *If* $1/3 < d < 1/2$, *then a sufficient condition for the pattern expressed by the Fibonacci sequence to arise is that* $P_r < 19$.

(b) *If* $1/4 < d < 1/3$, *then the pattern expressed by the first accessory sequence (Lucas sequence) always arises.*

Proof: In case (a), the limit divergence angles corresponding to the sequences of Table 6.1 are all outside the range 120° and 180° when $P_r < 19$ (use Table 2.1). In case (b) the sequence of angles of normal phyllotaxis [Eq. (6.4)] for $J=1$ and $t \geq 2$ tends towards 0° monotonically starting with 137.51°, 99.5°, 77.96°, ... (when $J>1$ the sequence of decreasing angles starts with 68.75°), while the sequence of angles of anomalous phyllotaxis [Eq. (6.6)] starts with 151.1° and increases monotonically towards 180°. It follows that the region [90°, 120°] contains only the angle of 99.5°, corresponding to $t=3$, so that the only pattern that can arise is expressed by the sequence $\langle 3, 4, 7, 11, 18, \ldots \rangle$. ■

The following proposition relates Equations (4.11) and (4.12) to E_b.

Proposition 8. *The function* $\min E_b$ *varies inversely proportional to r (the rise) and to* $\log R$. *It varies proportionally to the plastochrone P, to the phyllotaxis (m, n), and to Richards's area ratio A. In particular, for large P,* $\min E_b \approx P \log 2 + 2 \log \tau$.

Proof: $\min E_b = \log\{2^{m+n}[n(m+n)/m(m-n)]\}$ for a certain Fibonacci hierarchy $\langle n, m \rangle$, $n < m$ (by Proposition 5). For the Fibonacci hierarchies given in Proposition 3c, and for large values of n and m [for which $\min E_b \approx (m+n) \log 2 + 2 \log \tau$], putting the values of $\min E_b$ as a function of $m+n$ shows that $\min E_b$ has a linear relation to $m+n$. Using Equations (4.28), (4.11) and (4.12), or (4.27) gives the results. ■

Notice that the value of the slope of the line in the proof, that is $\log 2 \approx 0.3$, gives an angle of about 17°. We can multiply E_b by a weighting factor $p<1$ so as to reduce the slope and make it equal to the slope of the function L of $m+n$ given by Equation (4.26) (see also Figure 5.1 giving angles of 10° and 13°). This factor p reduces the value of $X(T)$ and makes $X(T)$ grow more slowly while preserving the results of Chapter 6. Using Equations (4.25), (4.26), and (4.27) gives $A \approx \pi L^2/4$ so that again $(\min E_b)^2$ is proportional to A.

6.4.2. Multijugate systems

Spiral systems in plants have two spirals winding in opposite directions, each one going through all the primordia. Only one, the genetic spiral, is retained, going through consecutively produced primordia by the shortest path, i.e., so that the divergence angle is not greater than 180°. Generally speaking, spiral systems may have $J = 1, 2, 3$, or more primordia, at each level of a stem or produced together in the shoot apex. This gives rise to the same number of genetic spirals. Bravais and Bravais (1837) used the term **bijugate** to describe the case of two genetic spirals ($J=2$); hence the term **unijugate** is used for the case of one genetic spiral ($J=1$) and **multijugate** for the case of several spirals ($J>2$). J represents the **jugy** (jugacy or jugacity) of the system. According to the model, the multijugate systems are given by the sequences of normal and anomalous phyllotaxis.

The literature does not give consistent names for the patterns represented by these sequences. I propose the following names, using the vocabulary found in the literature, to unify the concepts: the **main sequence** (Fibonacci sequence), the **accessory sequences** (for normal phyllotaxis with $t>2$), and the **lateral sequences** (for anomalous phyllotaxis). More precisely, for $J=1$ and $t=2$ in Sequence (6.3) we have the

$$\text{main or Fibonacci sequence } \langle 1,2,3,5,8,\ldots\rangle.$$

For $J=1$ and $t=3,4,5,\ldots$ we have the

$$\begin{aligned}&\text{first accessory sequence } \langle 1,3,4,7,11,\ldots\rangle,\\ &\text{second accessory sequence } \langle 1,4,5,9,14,\ldots\rangle,\\ &\text{third accessory sequence } \langle 1,5,6,11,17,\ldots\rangle,\ldots.\end{aligned}$$

The values $J=2$ and $t=2,3,4,\ldots$ give the

$$\begin{aligned}&\text{bijugate main sequence } 2\langle 1,2,3,5,8,\ldots\rangle,\\ &\text{bijugate first accessory sequence } 2\langle 1,3,4,7,11,\ldots\rangle,\\ &\text{bijugate second accessory sequence } 2\langle 1,4,5,9,14,\ldots\rangle,\ldots.\end{aligned}$$

The values $J=3,4,\ldots$ give trijugate, quadrijugate ... systems. The values $t=2,3,4,\ldots$ in Sequence (6.5) give the

$$\begin{aligned}&\text{first lateral sequence } \langle 2,5,7,12,19,\ldots\rangle,\\ &\text{second lateral sequence } \langle 2,7,9,16,25,\ldots\rangle,\\ &\text{third lateral sequence } \langle 2,9,11,20,31,\ldots\rangle,\ldots.\end{aligned}$$

Table 6.2 shows some common sequences with examples. The corresponding divergence angles are given in Table 2.1. The examples in Table 6.2 are based on the fact that at least one contact parastichy pair made with consecutive terms of the corresponding sequence was observed for each case.

6.5. By-products and applications

As a by-product of the model and of the treatment of hierarchies with matrices, L-systems have been bridged with the Perron–Frobenius spectral theory and with Bernardelli–Leslie–Lewis matrix theory used in the biology of populations. The new mathematical results concern the asymptotic behavior of the ratios $f(T+1)/f(T)$, C^T/r^T, and $f(T)/r^T$, where $f(T)$ is the growth function of the L-system generated by the growth matrix

Table 6.2. *Some examples of species exhibiting parastichy pairs in common phyllotactic sequences*

Jugy (J)	Sequences and examples
1	Main ⟨1, 2, 3, 5, 8, ...⟩ *Anthurium crassinervium, Araucaria excelsa, Aspidium felix-Mas, Bellis perennis, Cedrus libani, Cynara scolymus, Euphorbia wulfenii, Helianthus annuus, Opuntia leucotricha, Pinus laricio, Pinus pinea, Saxifraga umbrosa, Sempervivum calcaratum* Normal type for all Phanerogams
	1st accessory ⟨1, 3, 4, 7, 11, ...⟩ *Araucaria excelsa, Echinocactus williamsii, Euphorbia biglandulosa, Monanthes polyphylla, Sedum reflexum*
	2nd accessory ⟨1, 4, 5, 9, 14, ...⟩ *Cereus pasacana, Echinocactus williamsii, Lycopodium selago, Pothos, Strangeria paradoxa*
	3rd accessory ⟨1, 5, 6, 11, 17, ...⟩ *Cereus candicans, Echinopsis tubiflora, Echinopsis zuccarinianus, Lycopodium selago*
	4th accessory ⟨1, 6, 7, 13, 20, ...⟩ *Echinopsis multiplex, Echinopsis tubiflora, Raphia ruffia*
	1st lateral ⟨2, 5, 7, 13, 29, ...⟩ *Cephalotaxus drupacea*
	2nd lateral ⟨2, 7, 9, 16, 25, ...⟩ *Cereus chilensis*
2	Bijugate main 2⟨1, 2, 3, 5, 8, ...⟩ *Cephalaria tartarica, Dipsacus fullonum, Dipsacus laciniatus, Dipsacus silvestris, Pinus pumilio, Podocarpus japonica, Sedum elegans*
	Bijugate 1st accessory 2⟨1, 3, 4, 7, ...⟩ *Echinopsis tubiflora, Raphia ruffia*
3	Trijugate main 3⟨1, 2, 3, 5, 8, ...⟩ *Anthurium crassinervium, Dipsacus laciniatus, Echinopsis multiplex*

Source: Tabulated using Church, 1904b.

C with spectral radius r, $f(T)$ being also the number of nodes in level T of the hierarchy generated by C. This mathematical by-product makes possible a keener analysis of the growth of filamentous organisms, of lineages of cells, and of ramified structures, a subject explored further in Appendix 7.

Interpreting the outcome of the model, Fibonacci phyllotaxis would be the result of an optimization process having to do with maximizing the survival capacities, the fertility, and the flow of energy in the plant. Fibonacci phyllotaxis would maximize the potential energy stored in the cell in the form of ATP and the quanta of energy fixed by photophosphorylation. A land plant with Fibonacci phyllotaxis may be the best suited for economically fixing solar energy, for transforming it for its own growth and organization, and for synthesizing material that can be used by all other nonphotosynthetic organisms (for food production and crop yield). This concerns the functional problem of phyllotaxis, the problem of understanding the purpose served by the orderly arrangements of primordia.

Considering that under certain conditions leaf arrangement could prove to be a limiting factor in productivity (Donald, 1968), it appears that agronomy and agricultural productivity should start to take phyllotactic parameters into account. For Greyson and Walden (1972) "the possibility that the regulation of phyllotaxis could be put to practical advantage in corn breeding is quite real." Since the number, size, and placement of leaves are factors that affect crop yield, "strains possessing variability in these characters are in demand by the plant breeder."

An agronomy of limit systems based on maximizing agricultural production when the conditions of growth are less favorable could be considered. The bases for such agronomy could be laid down by considering more and more unfavorable climatic and growth conditions (e.g., on cold mountain slopes) for a high level of production, or regions altered by a long-term phenomenon of general pollution (acidification, desert formation). The objectives would be to demonstrate that in botany there exists a phyllotactic pattern for inducing the optimal development of land plants, to identify this pattern, and to master its applications. By optimization is meant here the best economical profitability. An applied agriculture with an economical impact could result and could be based on species and varieties adapted to particular climate and soil conditions.

7

Testing the interpretative model

7.1. Searching for quantified observations

In this chapter the predictions made using the model are compared with the data available on spiral systems. A good agreement is found whenever a comparison is possible. In the last decade, theoretical consideration of phyllotaxis has attained a high degree of elegance which contrasts with a lack of quantitative evidence. However, conclusions may be drawn from a comparison between factual and theoretical phyllotaxis, as we will see. These may stimulate future observations, as well as a dialogue between the observer and the modeler.

Theories and models in phyllotaxis are meant to be compared with reality. There are many possible ways to test models. If for example a model is based on the hypothesis that a substance is responsible for the generation of phyllotactic patterns, then this hypothesis can be directly tested by trying to identify the substance and its pathways in the plant. A model for the photorealistic reproduction of plants and patterns is tested directly by a visual comparison between the computer-generated images and the plants themselves. If a model is based on the assumption that the energy costs paid by plants are responsible for the various patterns that can arise, then the model would be tested directly by measuring the wavelengths absorbed by chloroplasts and carotenoids of each species. There might be, however, easier, indirect ways to test hypotheses. Indeed, if predictions come out of a model, then we can use these to test the hypothesis.

We will compare the predictions coming out of the interpretative model, with the data available, and also study the relative frequencies of the various types of spiral patterns in the plant kingdom in general, given that the interpretative model predicts an order of occurrence of the spiral patterns. We will verify whether the order of the respective importance of the spiral patterns in nature corresponds to the order given in Table 6.1.

Table 6.1 gives relative, not quantitative, frequencies. It proposes a priority-order based on increasing entropy and of increasing rareness as we move down the list. A pattern higher in the list is interpreted to be significantly more frequent than any other one lower in the list. We already know that the first type of pattern (main) given by Table 6.1 is more frequent than the next two (bijugate main and Lucas), but we do not know the exact relative importance of these latter two. We will try to determine whether the relative frequencies of occurrence of the sequences $\langle 1,4,5,9,14,\ldots\rangle$ (second accessory sequence) and $\langle 2,5,7,12,19,\ldots\rangle$ (first lateral sequence) are the ones predicted by the model. A pattern is said to be represented by the sequence $\langle n, m, m+n, 2m+n, 3m+2n,\ldots\rangle$, $n<m$, when the secondary numbers of a visible opposed parastichy pair are consecutive terms of the sequence.

We will thus need quantified observations. Despite the large amount of data on phyllotactic patterns, these are not always clearly described for our present purpose. For example, Cronquist (1981) listed and classified the existing genera and families of angiosperms (250,000 species), and leaf arrangement is consistently mentioned in each family described by him. Unfortunately, his "phyllotactic" analysis is typical of most botanical books - very brief and very simple. Cronquist said nothing about Fibonacci patterns, though they are probably what he calls "alternate arrangements." Of the 381 families, 204 are recognized to have alternate-leaved plants exclusively, and an additional 122 to have alternate as well as other leaf arrangements. This is in agreement with what is already known in a more detailed fashion on the relative frequencies of spiral and whorled patterns. The present chapter deals with spiral patterns only.

Other reports are more specific regarding spiral patterns. For example, Camefort (1956) observed that there are such genera as *Cephalotaxus drupacea, Cephalotaxus fortunei,* and *Torreya,* in which the pattern is characterized by the bijugate main sequence $2\langle 1,2,3,5,8,13,\ldots\rangle$. He noticed that in *Taxus baccata, Picea excelsa,* and *Cryptomerica japonica* bijugacy is accidental, the pattern being always described by the main or Fibonacci sequence $\langle 1,2,3,5,8,13,\ldots\rangle$. Camefort reported that the first accessory sequence $\langle 1,3,4,7,11,18,\ldots\rangle$ had been observed by Sterling in *Sequoia sempervirens.* However, these observations are not quantified; we are not informed about the number of observations made on each type of spiral pattern.

Section 7.2 brings together the available data on spiral patterns in plants, scattered through the literature over 160 years. The 12,750 observations on 650 species that are presented, will be considered to be statistically significant, even though they may seem to be limited in number and of unequal values. However, the group of studies we will mention on quantified observations has not been taken at random. These studies are the ones available, and no other study has been reported to the author. Our compilation (Table 7.12 summarizes Tables 7.1 to 7.10) will be compared with the predictions coming out of the interpretative model of Chapter 6.

Camefort observed two specimens of *Cephalotaxus drupacea* displaying the sequences ⟨1, 4, 5, 9, 14, ...⟩ and ⟨2, 5, 7, 12, 19, ...⟩ and a specimen of *Cupressus macrocarpa* displaying the former sequence. He says that he observed neither the third and fourth accessory sequences ⟨1, 5, 6, 11, 17, ...⟩ and ⟨1, 6, 7, 13, 20, ...⟩ nor the second and third lateral sequences ⟨2, 7, 9, 16, 25, ...⟩ and ⟨2, 9, 11, 20, 31, ...⟩ reported in Braun (1831). Using the interpretative model, it can be predicted that certain types of phyllotaxis can exist and that others cannot. The following sequences are reported in the literature: ⟨3, 7, 10, 17, 27, ...⟩, ⟨3, 8, 11, 19, 30, ...⟩, and ⟨3, 14, 17, 31, ...⟩. They will be considered in the source documents for Section 7.3, given that they are not supposed to be generated in shoot apices according to the model. In Section 7.4 I draw conclusions based on the comparison between observations and predictions.

7.2. Data on the frequencies of occurrence of patterns

The data in this section are compiled from the sources mentioned. In particular, Fujita (1938) listed the pairs of secondary numbers he observed in approximately 500 plant species in 100 families of gymnosperms and angiosperms. I have compiled his data in Table 7.1. I have done the same with Zagorska-Marek's (1985) observations on one species, and obtained Table 7.2.

Sterling (1945) reports that to his knowledge bijugacy (2⟨1, 2, 3, 5, ...⟩) and the first accessory sequence (⟨1, 3, 4, 7, ...⟩) have never been mentioned in the literature for *Sequoia sempervirens* (see Table 7.3). Two more apices, not tabulated in the table, were recognized by Sterling as representing the bijugate condition.

Brousseau (1968) analyzed twenty species of California conifers and did not report any exception to the Fibonacci sequence. Crafts's (1943b) observations on ten coniferous species revealed no bijugate patterns and

Table 7.1. *Observations and determination of the frequency of occurrences of various spiral patterns in angiosperms and gymnosperms*

	Angiosperms		Gymnosperms			
Sequence	Veg.	Repr.	Veg.	Repr.	Total	%
$\langle 1, 2, 3, 5, 8, \ldots\rangle$	438	129	45	14	626	93.29
$\langle 1, 3, 4, 7, 11, \ldots\rangle$	1	4			5	0.75
$2\langle 1, 2, 3, 5, 8, \ldots\rangle$		8	5		13	1.94
$\langle 1, 4, 5, 9, 14, \ldots\rangle$		1			1	
$\langle 1, 5, 6, 11, 17, \ldots\rangle$		2			2	
$\langle 1, 7, 8, 15, 23, \ldots\rangle$		2			2	
$\langle 1, 8, 9, 17, 26, \ldots\rangle$		3			3	
$\langle 1, 9, 10, 19, 29, \ldots\rangle$		2			2	
$\langle 1, n, n+1, 2n+1, \ldots\rangle$[a]		15			15	
$\langle 2, 5, 7, 12, 19, \ldots\rangle$			1		1	

[a] $n = 10$ to 18.
Note: Veg. = vegetative; Repr. = reproductive.
Source: Compiled from Fujita (1938).

Table 7.2. *Observations (3,200) on 155 trees of the species* Abies balsamea, *based on contact parastichies made by needle scars*

Sequence	Frequency	%
$\langle 1, 2, 3, 5, 8, \ldots\rangle$	2,994	93.58
$\langle 1, 3, 4, 7, 11, \ldots\rangle$	81	2.53
$2\langle 1, 2, 3, 5, 8, \ldots\rangle$	77	2.41
$\langle 2, 5, 7, 12, 19, \ldots\rangle$	19	0.59
$\langle 1, 4, 5, 9, 14, \ldots\rangle$	10	0.31
$4\langle 1, 2, 3, 5, 8, \ldots\rangle$	8	0.25
$3\langle 1, 2, 3, 5, 8, \ldots\rangle$	3	0.09
$2\langle 1, 3, 4, 7, 11, \ldots\rangle$	3	0.09
$\langle 3, 7, 10, 17, 27, \ldots\rangle$	3	0.09
$\langle 1, 5, 6, 11, 17, \ldots\rangle$	1	0.03
$\langle 2, 7, 9, 16, 25, \ldots\rangle$	1	0.03

Source: Zagorska-Marek (1985).

Table 7.3. *Patterns observed on a random sampling of 22 apices of* Sequoia sempervirens

Sequence	Frequency	%
⟨1, 2, 3, 5, 8, ...⟩	19	86.36
⟨1, 3, 4, 7, 11, ...⟩	2	9.09
2⟨1, 2, 3, 5, 8, ...⟩	1	4.55

Source: Sterling (1945).

Table 7.4. *Patterns for 23 species of conifers with helical phyllotaxis, representing 31 specimens*

Sequence	Species	Frequency	%
⟨1, 2, 3, 5, 8, ...⟩		26	83.87
2⟨1, 2, 3, 5, 8, ...⟩	*Torreya californica*	1	
	Cephalotaxus drupacea	2	
	Larix laricina	1	12.9
⟨1, 3, 4, 7, 11, ...⟩	*Cedrus atlantica*	1	3.23

Source: Namboodiri & Beck (1968).

no first accessory sequences; the frequencies of his observations are unknown. These observations differ from those reported in Table 7.4.

Concerning *Cephalotaxus drupacea* mentioned in Table 7.4, Fujita (1939) reported 141 Fibonacci patterns and 328 bijugate patterns (both included in Table 7.5). For the same species, Fujita (1937) reported 125 Fibonacci patterns and 232 bijugate patterns (tabulated in Table 7.12). The proportions in both cases are about the same. The 44 cases belonging to the first accessory sequence in Table 7.5 represent the species *Cunninghamia laccolata,* for which 572 Fibonacci patterns were found. Tables 7.6, 7.7, and 7.8 show a definite advantage of the first accessory sequence over the bijugate sequence for the species concerned, though in general the advantage is to the latter sequence as we will see. Tables 7.7 and 7.8 show very different percentages for the same species.

In certain species such as *Cephalotaxus drupacea* and *Torreya* and in Rhizophoraceae, patterns different from the Fibonacci pattern prevail,

Table 7.5. *Patterns for 30 species of angiosperms, representing 7,508 observations*

Sequence	Frequency	%
⟨1, 2, 3, 5, 8, ...⟩	7,136	95.04
2⟨1, 2, 3, 5, 8, ...⟩	328	4.37
⟨1, 3, 4, 7, 11, ...⟩	44	0.59

Source: Fujita (1939).

Table 7.6. *Data on 228 seedlings of* Picea abies

Sequence	Frequency	%
⟨1, 2, 3, 5, 8, ...⟩	224	98.24
2⟨1, 2, 3, 5, 8, ...⟩	1	0.44
⟨1, 3, 4, 7, 11, ...⟩	3	1.32

Source: Gregory & Romberger (1972).

Table 7.7. *Patterns for 319* Helianthus *(sunflower) capituli*

Sequence	Frequency	%
⟨1, 2, 3, 5, 8, ...⟩	262	82.13
2⟨1, 2, 3, 5, 8, ...⟩	9	2.82
⟨1, 3, 4, 7, 11, ...⟩	46	14.42
⟨1, 4, 5, 9, 14, ...⟩	2	0.63

Source: Schoute (1938).

Table 7.8. *Data for 140* Helianthus *capituli*

Sequence	Frequency	%
⟨1, 2, 3, 5, 8, ...⟩	133	95.0
2⟨1, 2, 3, 5, 8, ...⟩	1	0.71
⟨1, 3, 4, 7, 11, ...⟩	6	4.29
⟨1, 4, 5, 9, 14, ...⟩	0	0

Source: Weisse (1897).

Table 7.9. *Patterns of the type* $\langle 1, n, n+1, 2n+1, 3n+2, \ldots \rangle$ *for* $n = 4, 5, 6, 7, 8$, *representing 63 observations*

	Patterns				
Species	(4, 5)	(5, 6)	(6, 7)	(7, 8)	(8, 9)
Lysimachia japonica	1	1	13	7	2
Lysimachia clethroides	3	11	3	0	0
Lysimachia fortunei	0	1	8	11	2
Total	4	13	24	18	4

Source: Fujita (1942).

Table 7.10. *Observations made on 211 spadices for 20 species of aroids*

System	Frequency	%
$\langle 1, 2, 3, 5, 8, \ldots \rangle$	96	45.7
$\langle 1, 3, 4, 7, \ldots \rangle$	2	0.95
$\langle 1, 5, 6, 11, \ldots \rangle$	22	10.48
$\langle 1, 6, 7, 13, \ldots \rangle$	8	3.81
$2\langle 1, 3, 4, 7, \ldots \rangle$	2	0.95
$\langle 1, 7, 8, 15, \ldots \rangle$	5	2.38
$\langle 2, 5, 7, 12, \ldots \rangle$	2	0.95
$\langle 1, 8, 9, 17, \ldots \rangle$	1	0.48
$\langle 1, 9, 10, 19, \ldots \rangle$	1	0.48
$\langle 1, 10, 11, 21, \ldots \rangle$	3	1.43
$\langle 1, 11, 12, 23, \ldots \rangle$	1	0.48
$\langle 1, 12, 13, 25, \ldots \rangle$	1	0.48
$\langle 1, 13, 14, 27, \ldots \rangle$	1	0.48
$\langle 1, 14, 15, 29, \ldots \rangle$	2	0.95
$\langle 2, 9, 11, 20, \ldots \rangle$	1	0.48

Source: Davis & Bose (1971).

as we have seen. In other species, such as *Lysimachia* (shown in Table 7.9), the pattern is exclusively different from the usual types.

In Table 7.10 no bijugate pattern and no pattern belonging to the second accessory sequence $\langle 1, 4, 5, 9, 14, \ldots \rangle$ are reported, and 29.52% of the observations represent whorled patterns.

7.3. About aberrant spiral patterns

7.3.1. *List of problematic patterns*

According to the epistemologist Popper (1968), "every good scientific theory is a prohibition, it forbids certain things to happen." Table 7.11 shows four particular sequences reported in the literature a long time ago. According to the interpretative model, they cannot be found in nature as patterns of primordia, that is at the origin. The model does not concern patterns made by mature organs, as these patterns may be the results of all kinds of stresses and disturbances.

Of course, we could say that a model will never be more than an approximation of reality, and forget about these problem cases. And one cannot reject this puzzling matter on the basis that the plants on which these patterns were observed were teratological cases, or that the observations were not scientifically made. We have chosen to examine the reports in which these patterns were described. The discussion will alert the reader to some more difficulties (see Section 5.6) encountered in phyllotactic pattern recognition.

7.3.2. *On the sequence* $2\langle 6,13,19,32,\ldots\rangle$

This sequence is reported to have been observed by Schwendener (1878) on *Dipsacus speciosus*. His figure 91 refers to this case. It is a reconstruction of a specimen of a few millimeters in size. One fails to see the pattern $(26,38)=2(13,19)$ by counting the spirals in the opposed families of spirals shown in the figure. That is probably why Schwendener proposed the alternative sequence $\langle 11,26,37,63,\ldots\rangle$, which he said could be observed on another level of the specimen. From the numbering of the primordia in the figure one sees rather the pattern $2(5,6)$ belonging to the sequence $2\langle 1,5,6,11,17,\ldots\rangle$, followed by a transitional zone where the pattern is difficult to determine. It is clear that a family of 26 spirals is present, but it is impossible to count the spirals in the opposed family. Given that $26=2(13)$ is common to the first two sequences above, we are inclined to expect an opposed family of $42=2(21)$ or $16=2(8)$ spirals, realizing the bijugate condition expressed by $2\langle 1,2,3,5,8,13,\ldots\rangle$, in agreement with Snow (1954), and Church (1904b, see Table 6.2), who observed bijugacy in *Dipsacus silvestris*. On the other hand, Table 6.2 shows a (teratological?) case of trijugacy for *Dipsacus laciniatus*, and Vieth (1964) arrived at the conclusion that the capitulum of *Dipsacus silvestris* is not bijugate, in agreement with Szabo (1930). The pattern in

Table 7.11. *Problematic patterns*

Pattern	Species	Author
$2\langle 6,13,19,32,\ldots\rangle$	*Dipsacus speciosus*	Schwendener (1878)
$\langle 3,14,17,31,\ldots\rangle$	*Monstera deliciosa*	Hofmeister (1868)
$\langle 3,8,11,19,30,\ldots\rangle$	*Grimmia leucophaea* (7/19)	Braun (1831)
	Plantago media (11/30)	Braun (1831)
$\langle 3,7,10,17,\ldots\rangle$	*Pothos*	Schwendener (1878)

Source: Fujita (1937).

this genus is clearly difficult to characterize, and requires a more detailed analysis. This is done in Section 8.6.2.

7.3.3. On the sequence $\langle 3, 14, 17, 31, 48, \ldots\rangle$

This sequence is reported in Fujita (1937) to have been found in *Monstera deliciosa* by Hofmeister (1868). It should thus be found again from an analysis of other specimens of the plant. But Fujita (1937) himself rather recorded for that plant the patterns $\langle 1, n, n+1, 2n+1, 3n+2, \ldots\rangle$ for $n = 14$ to 18. Delpino (1883) reports the values $n = 17$, 18, and 20. Fujita (1942) says he found ten specimens displaying the same type of pattern for $n = 16$ to 18, two specimens showing the pattern $2\langle 1, 8, 9, 17, 26, \ldots\rangle$, one specimen showing the whorled pattern (18, 18), and a specimen with the pattern (15, 18) belonging to the sequence $3\langle 1, 5, 6, 11, \ldots\rangle$. He found no specimen belonging to the sequence $\langle 3, 14, 17, 31, \ldots\rangle$, and to our knowledge this sequence is not mentioned anywhere else than in Hofmeister (1868). It emerged from the phyllotactic fraction 11/31 (about 127.74°) reported in Hofmeister's figure 68. This figure represents a theoretical cylindrical lattice where the pairs (1, 2), (3, 2), (3, 5), (3, 8), and (3, 11) are visible and opposed as well as (3, 14). Using the contraction algorithm of Chapter 2, it is found that these pairs give intervals for the divergence, containing 11/31. A spadice of *Monstera* is reported by Davis and Bose (1971) as requiring further studies.

7.3.4. On the sequence $\langle 3, 8, 11, 19, 30, \ldots\rangle$

Let us now take a closer look at the third puzzling sequence. To my knowledge this sequence is mentioned four times in the literature. Roberts (1984a) made a purely theoretical use of it in an attempt to justify the behavior of the approximately normal curves in Fujita's (1939) paper (see Subsection 5.6.1), which does not even mention the sequence ⟨3, 8, 11, 19, ...⟩. Roberts did not observe this type of pattern. Fujita (1937) reports the sequence reported by Braun (1831), and from what we have seen in Section 5.6.3, with Braun's fractions 7/19 (132.63°) and 11/30 (132°) given in Table 7.11, we cannot infer that we have the sequence ⟨3, 8, 11, 19, ...⟩, even if the angle corresponding to this sequence is 132.18°. Indeed, the computational algorithm of Chapter 2 applied to the two phyllotactic fractions shows that the pairs (1, 2), (3, 2) and (3, 5), corresponding to the Fibonacci sequence, are visible and opposed. The problematic sequence is also reported in Zagorska-Marek's (1985) paper, and is dealt with in Section 5.6.2.

Finally, Delpino (1883) reports that the sequence would have been observed on *Echinocactus erinaceus* by Naumann (1845). In Naumann's report on *Echinocactus* it is noticeable that (1) 35 kinds of *Echinocactus* are correlated with 13 types of sequences among which 12 are acceptable by the model; (2) the problematic sequence ⟨3, 8, 11, 19, 30, ...⟩ is mentioned in relation to the phyllotactic fraction 8/19 corresponding to 151° instead of 132.18° for that sequence; (3) the sequence ⟨2, 5, 7, 12, 19, ...⟩ (corresponding to 151°) is reported for *Echinocactus*; (4) in at least five cases (including the problematic one), there is a clear-cut inconsistency between the phyllotactic fraction obtained and the sequence reported (e.g., 3/7 that is 154° for ⟨1, 3, 4, 7, 11, 18, ...⟩ instead of 99.5°). The case is dismissed. Delpino also mentions that the problematic sequences ⟨3, 16, 19, 35, ...⟩ and ⟨3, 17, 20, 37, ...⟩ would have been observed on a spadice of *Calladium crassipes*, but we have no way to check these assertions. Table 6.2 shows the first and the second accessory sequence for *Echinocactus,* also mentioned by Naumann.

7.3.5. On the sequence ⟨3, 7, 10, 17, 27, ...⟩

Consider now the last pattern in Table 7.11 – ⟨3, 7, 10, 17, 27, ...⟩. We know of six references that mention it. Zagorska-Marek (1985) says she found three plants displaying this pattern, as can be seen in Table 7.2. But for unknown reasons they are not tabulated with all the other cases in her own table. Moreover, we cannot analyze her observations as we did in Section 5.6.2 for the case ⟨3, 8, 11, ...⟩ she reported, given that it is not detailed in her article. However, her lattices of points for ⟨3, 7, 10, 17, ...⟩ and ⟨3, 4, 7, 11, ...⟩ (p. 1846) are identical, as they coincide by superposition. But they have slightly different vertical axes.

The pattern $\langle 3,7,10,17,\ldots\rangle$ is also mentioned by Hejnowicz (1973) who theorizes (p. 225) on the pair (3, 7) in relation to a specimen showing the first accessory sequence $\langle 3,4,7,11,18,\ldots\rangle$. The sequence is also reported by Fujita (1938), who observed it on the reproductive level of a specimen of *Salix viminalis*. But we have no way to verify this specimen. However, Fujita noticed a (2, 3) pattern belonging to the main sequence $\langle 1,2,3,5,8,\ldots\rangle$, on the vegetative level of his specimen.

Schwendener's (1878) figure 41 for *Pothos* (see Fig. 7.1) seems to be the only truly puzzling case. It cannot be discarded like the preceding ones. It represents a natural lattice of 7 spirals in one direction and 10 in another to give a (7, 10) phyllotaxis. The numbering of the primordia is consistent, showing differences of 7 and 10 along the respective spirals. The system (17, 27) is likely to arise; the system (17, 10) is about to arise; and the system (7, 3) just disappeared. From Schwendener's figure we can see the consecutive phyllotactic fractions 2/7, 3/10, 5/17, 8/27, 13/44 corresponding to angles of about 102.9°, 108°, 105.9°, 106.7°, 106.4°, respectively, which get closer and closer to the limit-divergence angle of about 106°, corresponding to the sequence $\langle 3,7,10,17,27,\ldots\rangle$. However, as we move along the genetic spiral there is not a steady rise: at every third or fourth primordium, primordium x is higher than primordium $x+1$. It is not specified whether the pattern characterizes the 75 species of *Pothos*, or whether it corresponds to a single specimen in a particular species. It is interesting to underline that five years later Delpino (1883) reported for *Pothos scanders* Hort. the sequence $\langle 1,15,16,31,\ldots\rangle$ admitted by the model, and pointed out that $\langle 3,7,10,17,\ldots\rangle$ had not been observed yet in nature.

7.4. Conclusions and discussion

7.4.1. Compilation of the data

We can combine the data in Tables 7.1 to 7.10 to produce Table 7.12. Assuming that the data is reliable, we can draw the conclusions in the following subsections.

7.4.2. On the frequency of the pattern $\langle 1,2,3,5,8,\ldots\rangle$

According to Table 7.12 the pattern arises in 91.3% of all the cases. This percentage does not include for example the 6,271 cones collected from 16 species of conifers by Brousseau (1969), given that the 84 exceptions to

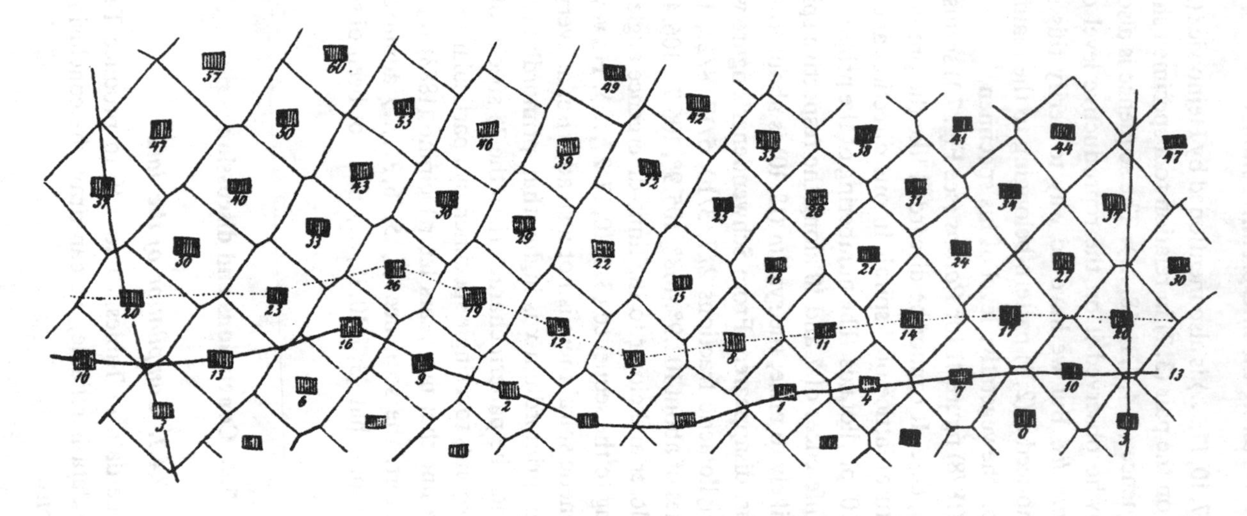

Figure 7.1. A century-old example of a pattern in *Pothos* resuscitated by a recent model. (From Schwendener, 1878.)

Table 7.12. *Data for 8 patterns and 12,750 observations (including the 119 for the very rare spiral patterns not in the table) on more than 650 species*

Pattern	Frequency	%
Main ⟨1, 2, 3, 5, 8, ...⟩	11,641	91.3
Bijugate 2⟨1, 2, 3, 5, 8, ...⟩	666	5.2
First accessory ⟨1, 3, 4, 7, ...⟩	190	1.5
Second accessory ⟨1, 4, 5, 9, ...⟩	17	
Third accessory ⟨1, 5, 6, 11, ...⟩	38	
Fourth accessory ⟨1, 6, 7, 13, ...⟩	32	
Fifth accessory ⟨1, 7, 8, 15, ...⟩	25	
First lateral ⟨2, 5, 7, 12, ...⟩	22	

Source: Compiled from Tables 7.1 to 7.10.

the main sequence are only recorded as "non-standard"; in his study the main sequence arises in 98.7%. This percentage differs from the one in Table 7.4 (83.87%) for 23 species but 31 specimens only. Table 7.12 includes neither the analysis (not quantified) made by Davis (1970) on a few species of palm trees, nor the compilation on pineapple fruits (Section 1.1.3) where all the numbers reported synchronize the Fibonacci numbers. It follows that the usual saying, according to which the Fibonacci sequence arises in 95% of all the cases is realistic, the percentage being probably closer to 92%:

the Fibonacci sequence arises in 92% of the cases.

7.4.3. Relative frequency of occurrence of the sequences ⟨1, 3, 4, 7, 11, ...⟩ *and* 2⟨1, 2, 3, 5, 8, ...⟩

Table 7.12 allows us to say, with a reasonably wide margin of certainty, that the first accessory sequence occurs significantly less than the bijugate sequence. Moreover, in comparative studies of the shoot system in the family of Rhizophoraceae for 16 genera and 120 species, Tomlinson and Wheat (1979) revised current statements about their phyllotaxis. This phyllotaxis is usually described as decussate but is in fact bijugate, the only type of pattern they found. These observations are not tabulated in Table 7.12. Bijugacy would be relatively more important than it appears in the table. It is thus reasonable to conclude that

(1) *The bijugate main sequence* $2\langle 1, 2, 3, 5, 8, \ldots \rangle$ *arises four times more often than the first accessory sequence* $\langle 1, 3, 4, 7, 11, \ldots \rangle$,
(2) *Something like 6% over 1.5% of all cases.*

These are new results.

The relative order of the first three types of patterns in Table 7.12 is the same as in Table 6.1 given by the interpretative model. The model does not propose percentages, but for those three sequences only it shows more than one value of P_r at which these sequences can arise. The predictions of the model for these types of patterns are thus well validated.

7.4.4. Insufficiency of the data available

It is often stated that the sequences $\langle 1, 4, 5, 9, \ldots \rangle$ and $\langle 2, 5, 7, \ldots \rangle$ are very rare in nature. This is in agreement with Table 7.12. But the sequence $\langle 1, 4, 5, 9, \ldots \rangle$ was expected from Table 6.1 to be more frequent than the first lateral sequence and than the third, fourth, and fifth accessory sequences. Table 7.12 indicates just the contrary. Although the pattern data, other than the first usual three, are clearly very limited, we can still draw the following conclusions:

(1) *The accessory sequences* $\langle 1, t, t+1, 2t+1, 3t+2, \ldots \rangle$, $t = 4, 5, 6, 7$, *and the first lateral sequence* $\langle 1, 5, 7, 12, 19, \ldots \rangle$ *have about the same frequency.*
(2) *Each sequence represents about 0.1% of all observations.*
(3) *These sequences with the usual three cover about 99.9% of all observations on spiral patterns.*

7.4.5. Remarks on methodology

Beyond the question of validating or invalidating models, this chapter points to the necessity of careful pattern recognition. Theories and models must fit data, not the other way around. We have seen, however, in Section 5.6, that phyllotactic pattern determination can be a difficult task. In particular, we must be very careful before allowing a single fraction to characterize a pattern. Also, it can be seen for example that the divergence angles $360(t+\tau^{-1})^{-1}/J$, corresponding to the sequences $J\langle 1, t, t+1, 2t+1, 3t+2, \ldots \rangle$ of normal phyllotaxis, tend to become clustered as t or J or both increase. It thus becomes more and more difficult to characterize a pattern from an angle. However, if the divergence angle is close to

137.5°, 99.5°, 78°, or 151°, it is most likely that the sequences involved are those given in Table 2.1 for these angles.

Given a divergence angle or a phyllotactic fraction, the computational algorithm can be used to obtain visible opposed parastichy pairs (m, n), and potential Fibonacci sequences that can be checked on the specimens. When the measured divergence angles significantly gather around the limit-divergence angles in Table 2.1, in an interval such as (90°, 110°), it generally conveys the fact that a particular sequence is involved, and thus that particular visible pairs can be observed, such as (3, 4), (7, 4), (7, 11), ... in the example. The presumed phyllotaxis (m, n) of a system (conspicuous pair) should be checked against divergence angles measured on plants in order to show internal consistency. If for example $m = 14$ and $n = 9$ then, according to Table 2.1, the divergence is expected to be around 78°.

The data must be collected at their origin, in transverse sections of shoot apices rather than on developed mature apices. Also, a wrong estimation of the real axis of a mature shoot will produce a wrong report on the pattern. With a scientific protocol on hand, it is hoped that the cases in Table 7.11 will be reconsidered, given that there is no substantial evidence of their existence, neither as primitive primordial patterns, nor as patterns of organs on mature plants, teratological or regular. It is surprising that the sequences reported so long ago (1831, 1868, 1878, 1883) are not routinely and consistently cited in the literature as the results of observations. An in-depth analysis of the species reported in Table 7.11 should be performed, and the stability or recurrence of the patterns verified. A final conclusion of this study is that ***the sequences prescribed by the interpretative model cover all spiral primordial arrangements.***

8
The interpretative model and whorled patterns

8.1. Multimerous patterns

Section 6.4 described multijugate (spiral) systems, while Section 1.2 introduced the idea of whorled patterns. Richards (1951) classified phyllotactic patterns under three headings: spiral, distichous, and whorled. The whorled or verticillate systems have two (the literature proposes the terms decussation, decussate, dimery, dimerous), three (tricussation, tricussate, trimery, trimerous, tristichy, tristichous), or several (tetramery, pentamery, hexamery, . . .) organs at the same node on the stem, or primordia in the same ring around the center of the shoot apex. This generates vertical rows of organs parallel to the axis of the stem, or rays of primordia in line with the center of the apex, the orthostichies. For uniformity, the suffix "mery" and the prefixes "di," "tri," "tetra," . . . , and "multi" will be used to express whorls, whether they are superposed or alternating.

The terms distichy or distichous (alternating unimery), and decussation or decussate (alternating dimery) are generally used, and will be used here. In a distichous system, general among monocotyledons, the leaves or primordia are distributed on two opposed orthostichies, one leaf at each node. The decussate system displays a pair of leaves at each node, each pair forming a right angle (when the stem is viewed from above) with the pair on the adjacent node of the stem, thus forming four orthostichies. A regular twisting of the orthostichies around the stem brings about secondary spirality, as in spirodistichy and spiromonostichy (discussed in Section 2.4).

Depending on whether the whorls comprise 3, 4, 5, ... superposed or alternating organs or primordia, the systems have 3, 4, 5, ... or 6, 8, 10, ... evenly spaced orthostichies, respectively. Alternating whorls of n organs or primordia, for $n = 1$ to 8 for example, are seen in *Iris*, *Canna*, *Gasteria obtusifolia*, and Gramineae ($n=1$, distichy); *Ligustrum, Aesculus, Clematis, Buxus, Azolla*, and Labiateae ($n=2$, decussation); *Epilobium adenocaulon, Salvinia*, and *Equisetum* ($n=3$); *Lycopodium selago* ($n=4$); *Cucurbita pepo* and the flowers of many dicotyledons ($n=5$); *Raphia ruffia* ($n=6$); *Cereus chilensis* ($n=7$); and *Sedum reflexum* ($n=8$) (examples from Church, 1904b). Superposed whorls are very rare; they occur in such flowers as *Ruta* and *Primula*.

In agreement with the notation used for parastichy pairs in spiral systems, whorls of n organs or primordia are denoted by (n, n), and distichy by (1, 1). Church (1904b) prefers the terms symmetrical ($m=n$) and asymmetrical ($m>n$) to describe the whorled and spiral patterns, respectively. We mentioned in Section 3.5 that in no case is one number in the parastichy pair (m, n) more than twice the other, ranging from (n, n) to $(n, 2n)$. These latter cases are, according to Church (1920b), "relatively so few as to be regarded as undoubtedly secondary, as 'sports' and 'mutants' of the original mechanism."

8.2. Preliminary relationships among the patterns

8.2.1. Multimery versus multijugy

This chapter puts forward the idea that the set of multijugate patterns includes the set of multimerous patterns. All patterns are claimed to be derived, mathematically and evolutionarily, from the subset of multijugate normal and anomalous spiral patterns discussed in Chapter 6. The set of multijugate patterns is subcategorized by the jugacy $J=1, 2, 3, 4, \ldots$, just as the set of distichous and whorled patterns is subcategorized by the number of organs or primordia per node, that is $1, 2, 3, 4, \ldots$. Richards (1951) and Richards and Schwabe (1969) claimed that multimerous systems and multijugate systems are essentially the same.

Richards (1951) presents the famous cases of distichy and decussation as limiting cases of the unijugate and the bijugate main patterns, respectively. Conceptually, distichy can be considered as a spiral pattern with a divergence angle of 180°. More precisely, it is proposed here that distichy is a limiting case of anomalous phyllotaxis given that as t increases in Equation (6.6), d tends towards 180°. The values $t=2, 3, 4, \ldots$ in Equation (6.6), give indeed $d=151.1°$, 158.1°, 162.4°, 165.3°, 167.4°, 168.9°, 170.1°, ..., respectively. Values of d not far from 180°, such as 170.1° corresponding to $t=8$, give spirodistichy, the case of distichy where the two orthostichies are slightly spiraled.

Decussation is defined as a pattern showing an angle of 90° between the orthostichies. It is considered by Richards as the bijugate version of the distichous system. Alternating trimery (whorls of three) is the pattern showing an angle of 60° between the orthostichies; tetramery has an angle of 45°; and whorls of J have an angle of $(360/2J)°$. Working with these orthostichies can be misleading. Indeed, four orthostichies are created in decussation when a primordium is linked to the superposed primordium two whorls apart. But the primordia in a whorl must be linked with the primordia in the consecutive adjacent whorls, thus creating two helices around the stem. The system then appears to be a special limiting case of bijugacy. In the same way, alternating trimery becomes a special case of the trijugate patterns, with three helices in one direction linking all the primordia and three in the opposite direction doing the same. Figure 1.1 shows a case of pentamery, but 5 spirals in one direction and 5 spirals in the opposite direction can be observed.

Williams (1975, p. 38) reports values of Richards's phyllotaxis index (Section 4.4), for orthogonal whorled systems $(n, n) = (1, 1)$ to $(6, 6)$. The respective plastochrone ratios R are equal to 23.083, 2.193, 1.417, 1.217, 1.134, and 1.091. Putting these values of R into the Pattern Determination Table (Table 4.3) and looking for the values of $m+n$ such that γ is closest to 90° (actually γ is seen in each case to be between 87° and 88°) gives $m+n=2, 4, 6, 8, 10$, and 12, which are the values of $n+n$ for the six whorled systems above. Whorled patterns behave like multijugate patterns, and Table 4.3 can be used for multimerous patterns as well. Following Richards, Williams also considers whorled systems as multijugate systems.

8.2.2. Schoute's false whorls

Schoute (1936) reminds us that at the very beginning of investigations into phyllotaxis, Schimper and others asserted that whorls did not exist, all leaf whorls having a definite succession of their members. Investigations with the scanning electron microscope confirm that the organs in a whorl arise in a more or less rapid sequence, followed by a longer plastochrone. Rutishauser and Sattler (1987) suggest that a whorl results from a single primordium breaking into a number of smaller units after it has been initiated. For Schoute there are true and false whorls and his survey shows that in vascular plants *the great majority of all whorls are false whorls* (e.g., in Equisetineae, Casuarinaceae, Labiatae), that true whorls are scarce, and that floral whorls "probably always are false whorls."

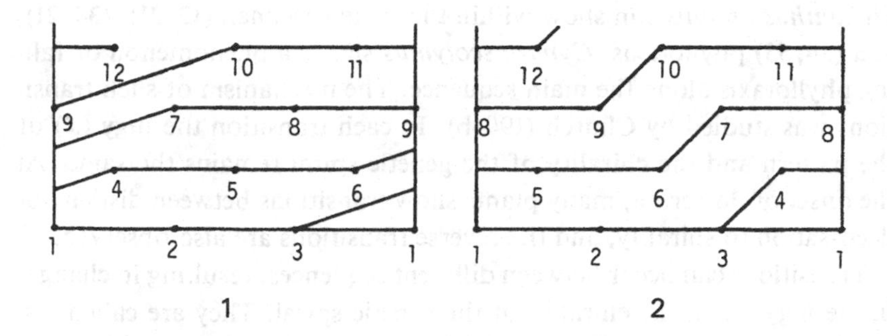

Figure 8.1. Diagrams from Schoute (1938) showing how trimerous alternating whorls can be transformed into spiral systems. Though the points are positioned identically in both diagrams, by linking and numbering them in the ways shown we obtain (1) a (2, 3) system belonging to the main sequence, and (2) a (3, 4) system belonging to the first accessory sequence.

Following Naegeli, Schoute (1938) traced whorled phyllotaxis back to spiral patterns.

Figure 8.1 shows how a cylindrical lattice of alternating trimery is interpreted by Schoute to be a spiral pattern, based on the examination of specimens of *Helianthus annuus.* In Figure 8.1(1) we have the systems (2, 3) and (5, 3), while the same points in Figure 8.1(2) show the systems (1, 3) and (4, 3). In Figure 8.1(1), the mean divergence angle is seen to be equal to $(120+120+180)°/3=140°$, while in Figure 8.1(2) it is $(120+120+60)°/3=100°$. Schoute says that trimerous whorls might also arise from the trijugate main sequence $3\langle 1,2,3,5,8,\ldots\rangle$, but that he knows of no instance of this case.

8.3. Transitions between patterns

8.3.1. Continuous and discontinuous transitions – natural and induced

Within a particular spiral system transitions can occur where the phyllotaxis, expressed by consecutive terms in a sequence, changes to other consecutive terms of the same sequence. Rising and falling phyllotaxis are examples. These transitions arise by a continuous change in the rise *r.* It is thus natural to use the term "continuous transition" for packings of primordia, already used in the literature to describe cylindrical packings of circles and spheres (Harris and Erickson, 1980). For example,

Helianthus annuus can show within the same specimen (13, 21), (34, 21), and (34, 55) phyllotaxis. *Cynara scolymus* shows a phenomenon of falling phyllotaxis along the main sequence. The mechanism of such transitions was studied by Church (1904b). In each transition the jugy (*J*) of the pattern and the chirality of the genetic spiral remains the same. At the onset of flowering, many plants show transitions between distichy or decussation to spirality, and the reverse transitions are also observed.

Transitions can occur between different sequences, resulting in changes in the jugy, or in the chirality of the genetic spiral. They are called discontinuous transitions. For example (from Church, 1904b), *Podocarpus japonica* shows (5, 8) (main sequence) apices varying to 2(3, 5) (bijugate main sequence), and weaker axes varying from (3, 5) (main sequence) to (3, 4) (first accessory sequence). *Cereus chilensis* can rise from (7, 8) (fifth accessory) to (7, 9) (second lateral).

There are also discontinuous transitions between whorled and spiral patterns. For example, *Echinopsis eyriesii* shows the patterns (7, 8), (7, 7), (8, 8), (6, 7) and 3(2, 3), and *Echinopsis tubiflora* the patterns (6, 7), (5, 7), (6, 6) and 2(3, 4). The scales of the fruits of *Raphia ruffia* vary between (6, 6), (6, 7), (7, 7) and 2(3, 4). *Lycopodium selago* shows the patterns (2, 2), (3, 3), (4, 4), (4, 5), (5, 6), and (5, 5) (called decadent by Church).

More recent empirical studies (e.g. Meicenheimer, 1982, 1987a) on various plant species confirm that discontinuous transformations occur naturally. A famous example is the mature *Bryophyllum tubiflorum*, which shows no less than five different patterns on the same specimen (Gomez-Campo, 1974, Hellendoorn and Lindenmayer, 1974). The mutant *Zea mays* (Greyson and Walden, 1972) shows the normal distichous pattern in the lower 1/3 or 2/3 of the leaves, and a decussate or spiral arrangement in the upper thirds.

Also experiments (e.g., by Snow and Snow, 1935; Vieth & Arnal, 1961; Schwabe, 1971, 1984; Maksymowych, Cordero & Erickson, 1976; Erickson & Meicenheimer, 1977; Meicenheimer, 1981; Mingo-Castel et al., 1984; Marc & Hackett, 1991), with various artificial interferences on the meristem (e.g., surgery, radiations, injections of hormones such as benzyladenine and gibberellin), lead to the conclusion that the same genome can produce a variety of patterns, and shifts from whorled to spiral patterns and vice versa. That one plant can show two types of patterns is indicative of the kinship between them. The phenomenon of phyllotactic transition indicates that the various patterns do not much differ from one another.

8.3.2. The mechanism of transition

Transitions to symmetry (whorl) are obtained (and abundantly illustrated by Church, 1904b) by equalization of the number of curves in either direction. Church observed that these transitions take place with the addition or loss of parastichies, one at a time, with consequent alteration of the pattern in other respects. For example, a (6, 7) construction (e.g., Cactaceae) would attain true symmetry with whorls of (7, 7) and (6, 6) by adding either a long parastichy, or by losing a short one, respectively. In all such cases the original genetic spiral effect would vanish. By adding a short parastichy it would appear as $(6, 8) = 2(3, 4)$, a case of bijugy.

Conversely, the addition of curves to a symmetrical system, as in *Hippuris*, *Casuarina*, and *Equisetum* (wholly secondary whorls), would give a secondary regression to a spirality, "though in these highly specialized types the occurrence is so rare as to pass as a 'freak.'" Finally, Church argues that excessive variations of these types soon lead to "hopeless irregularity," to chaotic or unordered phyllotaxis.

Now, it is well known that the cycle of primordium formation by an apical meristem comprises the following events. The meristem enlarges, and a part of it forms a new primordium. The surface area of the meristem reduces abruptly, and then enlarges again. With the production of each new primordium, the center of the apical dome moves instantaneously toward a new position. Green and Baxter (1987) used this movement to describe a simple geometrical activity of the meristem for four categories of phyllotaxis: whorled, distichous, trijugate, and spiral. It is easy to imagine that the figure made by, or on which is, the center of the meristem as the primordia arise is respectively a point, a segment, a polygon, and a circle. I suggest that there are geometrical transitions between these figures, as there are biological transitions between the phyllotactic patterns.

Let us look at the case of the polygons and circles. Consider for example a shoot whose pattern is defined by the main sequence and the Fibonacci angle. Looking at the mature stem we might be able to recognize easily 3 orthostichies. This corresponds to a phyllotactic fraction of 1/3, i.e., to an approximation, equal to 120°, of the divergence angle. Looking along the stem towards the apex a better approximation can be obtained with a fraction of 2/5, corresponding to 144°, and to 5 orthostichies. Continuing in this way the next fractions will be 3/8, then 5/13, ..., corresponding to 135°, 138.46°, ..., respectively, which get closer to 137.5°. For the same plant we may thus imagine that the center of the dome moves on the vertices of polygons, with 3, 5, 8, 13, ..., sides, respectively. For example, *Euphorbia wulfenii* showing 8 orthostichies and 135°, and 3 with 120°, is along those lines. In the same manner, in the *Euphorbia biglandulosa* studied by Green, the numbers of orthostichies correspond to the numbers in the first accessory sequence, but the various polygons determined by the movement of the center of the apical dome have their vertices on a circle. In a plant showing spirality, there are as many orthostichies as approximations to the divergence angle. The larger the number of orthostichies, the better the approximation we get of the divergence, and the closer the polygon approximates a circle.

For natural transitions, the underlying geometrical activity of the apex may have to do with the facts that the circle for spiral patterns (1) is a polygon with many sides; (2) reduces to a point (that is, to whorled patterns) when the radius decreases; and (3), is a special case of an ellipse which by flattening becomes the segment of distichy. What happens in the apical meristem when phyllotaxis changes? The transitions between patterns are in particular the results of an interplay between the variations of R or r (the rise) and the variations of d (the divergence angle), due to the plasticity of the growing point. When d is relatively stable, changes in R or r bring continuous transitions. They occur via symmetrical expansion or contraction of the stem circumference. Theoretically, discontinuous transitions can occur via an asymmetry introduced in some way in the system. Following Church (1904b), for Meicenheimer and Zagorska-Marek (1989) the discontinuous transitions could be brought about by a local increase or decrease of the circumference of the apex, making possible the adjunction or rejection of a parastichy in a family. We will come back to the subject of transitions in Chapter 10 where crystallographic phyllotaxis is introduced, and a likely mechanism of transition will be proposed, based on the results of Chapter 4.

8.4. The primitiveness of spirality

8.4.1. Evolutionary levels in pattern generation

The most highly perfected condition of phyllotaxis is especially obvious in the structure of flowers, and also in specialized assimilating shoots (*Hippuris*, Labiateae, Oleaceae, Equisetineae, Dasycladeae, Characeae) presenting the case of alternating whorls (Church, 1904b, p. 142). In *floral phyllotaxis* (called anthotaxis), progressing from the most primitive angiosperm group (Magnoliidae and Hamamelididae) to the highly advanced one (Asterideae), whorled patterns become more strongly favored, and the prevalence of Fibonacci numbers of floral organs, even in whorled flowers, seems to be due to the spiral onset of many flowers (see Endress, 1987).

It may be difficult to state that at the evolutionary level of flowers, spiral floral phyllotaxis is primitive and whorled phyllotaxis is advanced, if transitions can be observed to go in all directions, and if in the earliest known fossil flowers both spiral and whorled arrangements occur. In coal-measure flora of the Equisetalean species, whorls were very common; even in fossil algae there are examples of whorls. However, all Phanerogams, Cryptogams, and the bulk of the algae, with whorled series of lateral members, either commence with an asymmetrical (spiral) condition or show traces of it in subsequent development (Church, 1904b; 1920b). Kubitzki (1987) stresses the view that the dimerous, trimerous, and pentamerous conditions have their origins in spiral anthotaxis, and underlines the impossibility of a return to spiral phyllotaxis from trimery.

Applying the theory of recapitulation or the Haeckel law to the numerous patterns observed in *Acacia* apices, Rutishauser (1986) reached the conclusion that three evolutionary levels may be distinguished. Spiral phyllotaxis is considered as primitive; fasciculate (half-whorls), peculiar, and scattered phyllotaxes are considered as intermediate; and verticillate (whorls) phyllotaxis is considered as derived. Thus following Church, the peculiar patterns would be expressions of the transitions from spiral to whorled pattern.

Many botanists, including Church (1904b, 1920b) and Corner (1981), have claimed that spirality is primitive in the biological sense, and that the other systems are derived from earlier conditions of spiral organization: "the varied phases and complications of whorled and anomalous systems, among higher plants, are wholly secondary and derivative" (Church, 1904b). According to Church (1920b), comparative morphology would show that whorls (symmetry) can be regarded as secondary specializations from an ancestral asymmetrical (spiral) condition, the alternation of whorls being in fact the strongest evidence in this connection. Spirality is clearly not universal, but spiral systems expressed by the main or Fibonacci sequence are so overwhelmingly preponderant that "the deduction is well warranted that Fibonacci phyllotaxis may be regarded as phylogenetically primitive . . . . Fibonacci symmetry is, in fact, one of the most archaic of somatic factors. It is difficult to trace anything phyletically more remote" (Church, 1920a) than the asymmetry expressed by the main sequence, which is the most common and the most widely distributed.

"This undoubtedly affords the clue for its extreme persistence, and indicates that it is the more elementary and most probably the original case; whorled systems appeared as an afterthought" (Church, 1904b). For Church, asymmetry (spiral) rather than symmetry (whorl) is the primitive condition, in the absence of any reason to the contrary; and symmetry is a secondary phenomenon physiologically independent from the first but phylogenetically related to it. Spiral patterns would be mechanistically, morphologically, and developmentally the simplest, and evolution generally progresses from simplicity to complexity.

Church thought that all types of phyllotactic patterns, other than those expressed by the sequence $\langle 1,1,2,3,5,8,\ldots\rangle$, could throw light on the causes that tend to induce symmetry. Church (1904b) sees the accessory patterns $(m, m+1)$ [such as (4, 5), (5, 6), (6, 7), ...], observed in *Echinocactus williamsii* ($m=3$), *Stangeria paradoxa* ($m=4$), *Echinopsis tubiflora* ($m=5$), *Echinopsis multiplex* ($m=6$), *Echinopsis eyriesii* ($m=7$), and *Richardia africana* ($m=8$), as signs of a nearer approximation to symmetry, as an approach to equality in the number of parastichies. The slightest deviation from absolute symmetry produces an apparent spiral effect. The phenomenon can be observed in *Lycopodium*, and *Raphia*, and is well illustrated in Church's diagrams (1904b). Whorls are only the limiting cases, in a sense which is clarified in Section 8.5.

For Church the primary "intention" of nature thus took the form of spiral patterns to which all other types of phyllotaxis are phyletically related. Expressing himself somewhat disparagingly, he speaks about "academic morphologists contemplating the plant for over a century, in complete ignorance of its phyletic origin in submarine environment" (1920a). Supported later by Corner, Church (1920a) emphasized that the whole process of pattern generation in higher plants must have started in the sea, with principles of growth such as overtopping. The latter principle is known to be the ancestor of the principle of spiral growth. It is one of the evolutionary principles of Zimmermann's telome theory which would be at the basis of the existence of all vascular plants (Section 3.2).

8.4.2. Methodological consequences on modeling

The primitiveness of spirality in the evolution of plants is empirically supported and conceptually defendable, but nevertheless it is an assumption, however well grounded. The methodological consequences of this assumption will now be examined.

We said above that the variety of patterns represents quantitative differences of the same mechanism. This evidence brought the theorists to address, at the same time, the problems of generation of spiral patterns, of whorled patterns, and of transitions. One of the reasons for the difficulties of making a complete model comes from the failure to recognize distinct steps in the process of modeling phyllotactic patterns, corresponding to distinct steps in their evolutionary generation. If spirality is primitive, then a mathematical model must first be able to reproduce all spiral patterns. The other types of patterns are by-products of evolution from spiral patterns; they are secondary patterns that may require a different mechanism or an additional control. To explain the transitions from spiral patterns to other types of patterns is, from this point of view, a secondary problem.

The patterns must be generated in a sequence corresponding to the order of their appearance in the real world, that is, to their relative frequencies of occurrence. This suggests the following **evolutionary classification of phyllotactic patterns**, corresponding to a presumed sequence of evolutionary events. It is proposed that the main categories of patterns in their general phylogenetic order are:

Normal phyllotaxis with $J=1$
Normal phyllotaxis with $J=2,3,4,\ldots$
Anomalous phyllotaxis ($J=1$)
Alternating whorls
Superposed whorls

First, the challenge is thus to contain all the spiral patterns within a model. Second, it is to extend the model so that transitions to whorls are adequately mastered, in a way that integrates the two steps. It is clear that by separating the components of a problem, it becomes easier to solve.

The model in Chapter 6 corresponds to the first methodological step proposed here. The spiral patterns controlled by the sequences (6.3) and (6.5) are obtained. This model suggests that the various types of spiral (multijugate) patterns arose in their relative frequencies because of their ability to maximize energy under the condition that increasing numbers of primordia are required to produce the same rhythm.

8.5. Fundamental relationships among the patterns

8.5.1. Generating alternating multimery from multijugate normal systems with $t=2$ – the first hypothesis

We must show how the angles of 90° (decussation), 60° (trimery), 45° (tetramery), etc., proceed from spiral patterns. Two proposals are sum-

Table 8.1. *Generating alternating whorls from normal patterns [sequence (6.3) with $t=2$ and $J=2, 3, 4, \ldots$ representing the jugy, and divergence (6.4)], or from anomalous systems [sequence (8.1) with $J=2, 3, 4, \ldots$, and divergence (8.2)]*

Whorl of	Divergence	Divergence (6.4)		Divergence (8.2)		
$J=$	in whorl	Jugy J	$t=2$	J	$t=4$	$t \gg 4$
2	90°	bi	68.75°	2	81.2°	90°
3	60°	tri	45.84°	3	54.1°	60°
4	45°	tetra	34.38°	4	40.6°	45°
5	36°	penta	27.50°	5	32.5°	36°

Note: The symbol ≫ means "much larger than."

marized in Table 8.1. The first explains how alternating whorls evolved from normal patterns expressed by the bijugate, trijugate, tetrajugate, ... main sequences, and having the divergence angles shown in the fourth column of the table [angles in (6.4) with $J=2, 3, 4, 5$, and $t=2$; the angles for t larger than 2 are irrelevant because they are smaller than the latter angles]. The angles in the fourth column are not so far from the divergence angles in the corresponding whorls (column 2) which are larger. That is the angle in (6, 4), namely $360/(2J+J\tau^{-1})$, is close to $360/2J$ the angle for the whorl of J.

This interpretation would no doubt please Richards. Church (1904b) as well might agree with that point of view. He states that the distinction between a truly decussate (2, 2) system and the bijugate variant (2, 4) is often indistinguishable to the eye. "From the point of view that a decussate system represents a doubled construction, (2, 2) = 2(1, 1), the possibility of the secondary reversion to the doubled spiral construction implied in bijugate systems is very apparent." The case in which J is larger than 1 was recognized earlier by Bravais and Bravais (1837) as a sort of intermediate condition giving both whorls and spirals at the same time. Using *Epilobium angustifolium* as an example, Church (1904b) claimed that multijugate systems may now be viewed from the standpoint of transitions to symmetrical (whorled) constructions, and that multijugy ($J=2, 3, 4, \ldots$) is "a distinct sign of symmetrical construction."

If alternating whorls come from normal phyllotaxis with $t=2$, then the angles in column 4 of Table 8.1 must increase in order to become those of column 2. It is likely that *during evolution rules of pattern formation have been refined, strengthened, or weakened.* One of these is Hofmeister's rule which applies very often. According to this rule, new primordia arise on the shoot apex in positions farthest from the margin of the bases of the neighboring primordia already present. Biologically, the angles in alternating whorls might then result from those in multijugate normal systems with $t = 2$, by the evolutionary expansion towards more space for each primordium. Then *alternating whorls corresponds to a strengthening of Hofmeister's rule,* while *superposed whorls result from a weakening of this rule.*

In a spiral system such as the one expressed by the main sequence, the control of the divergence angle is very loose for the conspicuous pairs (1, 2) and (2, 3) given that the angle is in the ranges [0°, 180°], and [120°, 180°], respectively (Chapter 2). As phyllotaxis rises along the sequence, the margin of fluctuations for the divergence shrinks. Indeed, higher-order conspicuous pairs such as (8, 13) or (13, 21) give the intervals [135°, 138.46°] and [137.14°, 138.46°] close to the Fibonacci angle. In these cases, the divergence is perfectly controlled. Assuming that the strengthening of Hofmeister's rule has been genetically encoded in some cases during evolution, then the divergence angle became maximized. Spiral systems expressed by low secondary numbers can be cradles for all the possibilities, including erratic patterns.

8.5.2. Alternating multimery derivation from anomalous systems – the second hypothesis

The second proposal to explain alternating whorls now follows. This second hypothesis may look more appealing, given that we do not even need to imagine a supplementary evolutionary constraint (e.g., a strengthening of Hofmeister's rule). Assuming that multijugy derives from unijugy, alternating whorls may be considered to derive from anomalous patterns expressed by the Sequence (6.5), in the way already suggested for distichy, that is, from the angle in (6.6) with t becoming large. Let us consider a specimen showing a divergence $d = 170°$. The spiral pattern expressed by the sequence $\langle 2, 17, 19, 36, \ldots \rangle$ [$t = 8$ in Sequence (6.5)] is involved. But practically speaking this pattern looks like distichy. It is in fact spirodistichy with two slightly spiralled orthostichies.

Now let us imagine patterns expressed by the sequence

$$J\langle 2, 2t+1, 2t+3, 4t+4, 6t+7, \ldots \rangle \tag{8.1}$$

[where for $J=1$ we have Sequence (6.5)]. For $J=2,3,4,\ldots$ we have hypothetical anomalous bijugy, trijugy, tetrajugy, ... The divergence angle in Equation (6.6) is then divided by J to give the divergence corresponding to Sequence (8.1):

$$d = 360[2+(t+\tau^{-1})^{-1})]^{-1}/J. \tag{8.2}$$

It is easily shown that for any fixed value of J, as t becomes large the divergence becomes the one in a whorl of J primordia, as expressed in the last column of Table 8.1. In this way, alternating multimerous systems are the expressions of multijugate anomalous systems ($J>1$).

As an example, let us consider a 2(9, 11) spiral pattern [$t=4$, $J=2$ in Sequence (8.1)]. Then the corresponding divergence angle would be around $162.4°/2 = 81.2°$. This angle is close to 90°, so that practically speaking the system conforms to a decussate system; we have a spirodecussate system with slightly spiraled orthostichies.

8.5.3. Superposed whorls are normal multijugate systems

Alternating whorls are found in the vast majority of symmetrical plant constructions. For Church (1904b) alternation is the normal and primitive condition for whorls. He states that superposed whorls are exceedingly rare, their presence in the vegetative shoot is doubtful (Church, 1920b). They occur in flowers and are in such cases commonly regarded as of secondary origin, as a special case of the more general asymmetrical form of a spiral vortex (Church, 1904b). It is proposed here that superposed whorls derive from normal spiral systems in the following way.

Take for example a normal spiral pattern with $t=9$ and $J=3$. This is a trijugate system with a divergence angle of $37.4°/3 = 12.5°$, an angle not far from 0°, the divergence for superposed whorls of 3. Table 8.2 shows divergence angles in multijugate normal patterns [Sequence (6.3)] with $t=9$, and with t much greater than 9. The normal spiral system looks like a system of superposed whorls where the J orthostichies are slightly spiraled. But as t increases in normal multijugate systems, the divergence angle [in Equation (6.4)] approaches 0°, so that the orthosthichies become perfectly straight.

8.5.4. Summary of the model and existence of predicted patterns

The model of Chapter 6 is based on the hierarchical representation of phyllotactic patterns. In a hierarchy three important characteristics can

Table 8.2. *Generating superposed whorls from normal patterns*

Whorl of $J=$	Divergence in whorl	Seq. (6.3) Jugy J	Divergence (6.4) for $t=9$	Divergence (6.4) for $t \gg 9$
2	0°	bi	18.7°	0°
3	0°	tri	12.5°	0°
4	0°	tetra	9.4°	0°
5	0°	penta	7.5°	0°

Note: The symbol ≫ means "much larger than."

be defined – the stability $S(T)$ of the pattern, its complexity $X(T)$, and its rhythm w. The entropy E_b is a function of these parameters, expressed by an equation. The model makes it possible to compute the entropy cost of every spiral pattern and to compare the costs. The pattern chosen at plastochrone P_r at which a cycle is completed is the one that minimizes the entropy function. Multijugate patterns result from a rhythm of period 2 set at different values of P_r.

Preliminary relationships and resemblances between multijugate and multimerous patterns have been proposed. Chapter 8 is based on the assumption that an understanding of why the phyllotactic patterns arise involves a phylogenetic perspective. Reliable and acceptable grounds for understanding multimerous patterns have been established, which brought forward a refinement (Section 8.4) in Church's evolutionary classification of the patterns. Figure 8.2(1) shows how in evolution spiral systems were transformed smoothly into whorls.

Fundamentally, whether we retain the first proposal for alternating whorls (Section 8.5.1) or the second (Section 8.5.2), there is just one type of phyllotactic patterns (the multijugate patterns) and two types of spiral systems (the normal and the anomalous systems). Multimerous systems are spiral systems bearing special names as t takes values greater than say 5 [see Fig. 8.2(1)]. Multimerous systems are in turn subdivided into two classes – the superposed whorls and the alternating whorls represented by the right-hand side of the two curved dotted lines in Figure 8.2(1). The figure shows that as t increases in Sequences (6.3) and (6.5), spiral systems gradually become spiro-whorled and then whorled systems. In the diagram, the value of J determines the scale from 0° to 180°/J for the divergence angle, for cases of J-mery and J-jugy, superposed or alternating. The figure considers the second hypothesis (Section 8.5.2) for

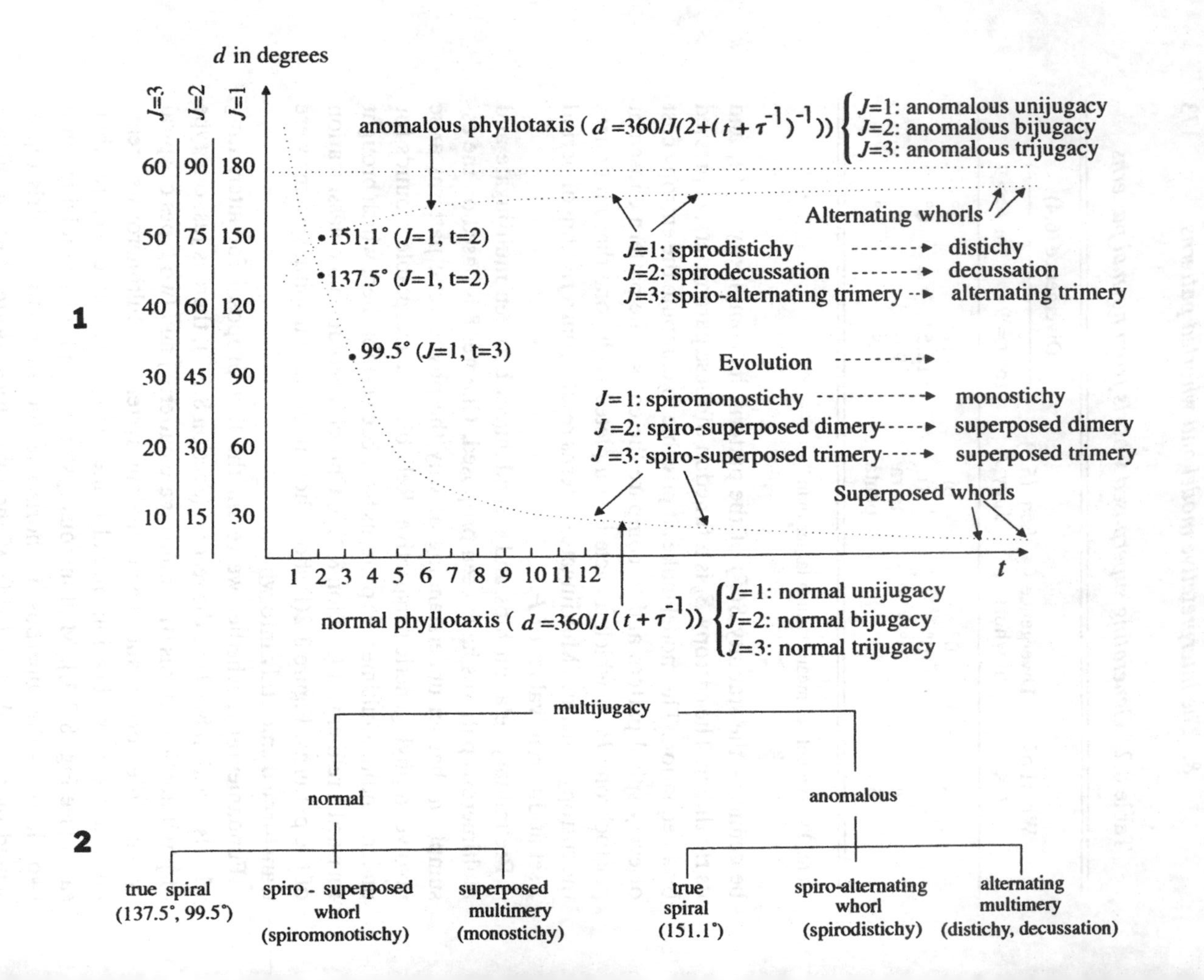
1
d in degrees
J=3
J=2
J=1
60 90 180
50 75 150
40 60 120
30 45 90
20 30 60
10 15 30
1 2 3 4 5 6 7 8 9 10 11 12
t
anomalous phyllotaxis (d =360/J(2+(t + τ^-1)^-1))
J=1: anomalous unijugacy
J=2: anomalous bijugacy
J=3: anomalous trijugacy
151.1° (J=1, t=2)
137.5° (J=1, t=2)
99.5° (J=1, t=3)
Alternating whorls
J=1: spirodistichy
J=2: spirodecussation
J=3: spiro-alternating trimery
distichy
decussation
alternating trimery
Evolution
J=1: spiromonostichy
J=2: spiro-superposed dimery
J=3: spiro-superposed trimery
monostichy
superposed dimery
superposed trimery
Superposed whorls
normal phyllotaxis (d =360/J (t + τ^-1))
J=1: normal unijugacy
J=2: normal bijugacy
J=3: normal trijugacy
2
multijugacy
normal
anomalous
true spiral (137.5°, 99.5°)
spiro - superposed whorl (spiromonotischy)
superposed multimery (monostichy)
true spiral (151.1°)
spiro-alternating whorl (spirodistichy)
alternating multimery (distichy, decussation)

explaining alternating whorls. Figure 8.2(2) is a simplified version of Figure 8.2(1).

The limiting cases (multimery) are more improbable structures relative to the prominent spiral systems. Each spiral system has an entropy cost. So, decussation, for example, has an entropy cost which depends on the approximation of the spiral system to true decussation, with forms of spiro-decussation. With increasing t in anomalous systems ($J=2$, for decussation), the divergence angle increases and approaches 90°, and the entropy cost increases as spiro-decussation approaches more closely a true decussate system, which is a relatively very expensive system.

Figure 8.2(1) predicts the existence of patterns such as monostichy, where all primordia or leaves would be on a single orthostichy parallel to the axis of the plant. This seems to be an impossible pattern. I learned afterwards that it really exists. It is found, for example, in *Utricularia* (R. Rutishauser, personal communication). The diagram also predicts the existence of superposed dimery, which is apparently common in pinnate leaves.

Notice that for every t the divergence angle is much closer to 180° for anomalous systems, than to 0° for normal systems [e.g., for $t=8$, $d=170°$, and $d=42°$ in Eqs. (6.4) and (6.6), respectively]. In other words, the upper dotted curve in Figure 8.2.1 is closer to the upper asymptote, than the lower dotted curve is to the lower asymptote. This would explain the relative abundance of distichy, decussation, and alternating whorls, in general, with respect to superposed whorls.

The model can be used to propose a dynamic and unifying view of phyllotaxis. The usual classifications [e.g., whorled and spiral patterns, or symmetrical and asymmetrical patterns (Church, 1904b), or spiral and nonspiral patterns (Green, 1991)] sanction the dichotomy between phyllotactic patterns, consecrate the static view of them, and do not show the

Caption for Figure 8.2
(1) Entropy model of Chapter 6 reconsidered. The divergence angles are isolated points on the two dotted curves. Multimery is a special case of multijugy. Normal phyllotaxis [Sequence (6.3) and Divergence (6.4)] and anomalous phyllotaxis [Sequence (6.5) and Divergence (6.6)] gradually changing to whorled phyllotaxis via spiro-whorled phyllotaxis, by the increase of t, for three cases of whorls (choose the value $J=1$, 2, or 3 and read the graph consequently). Using the second hypothesis (Section 8.5.2), distichy, decussation, and alternating whorl are seen to be special cases of anomalous phyllotaxis. Superposed whorls are special cases of normal phyllotaxis. The figure highlights three famous divergence angles. Between the whorls and the usual cases of spiral phyllotaxis are cases such as spiro-distichy and spiromonostichy. (From Jean, 1988c.) (2) Simplified version of Figure 8.2(1) with examples for the various cases.

unity behind the diversity. In the model it is proposed that the variety of phyllotactic patterns are simply variations on the theme of the rhythmic production of organs. Fundamentally, there are multijugate patterns only. Among them special categories of patterns can be discriminated. And if spirality was not primitive, then Figure 8.2(1) can be read from right to left – that is, from multimery to spirality – instead of from left to right.

8.6. Predictive and explanatory values of the interpretative model

8.6.1. *Analysis of the phyllotaxis of* Dipsacus

In Section 7.3.2 I emphasized the difficulty of pattern recognition in the genus *Dipsacus*, generating theoretical problems considered by many morphologists. Barabé and Vieth (1990) deduced from the interpretative model, a coherent interpretation of the growth and phyllotaxis of *Dipsacus silvestris* in its normal and biastrepsis states. Biastrepsis appears for example in *Rubia*, *Saponaria*, *Galium*, and *Mentha*. It consists of an alignment of leaves on an abnormally twisted stem [Fig. 8.3(1)]. It is in *Dipsacus silvestris* that this phenomenon has been the most studied. The leaves of this plant normally show decussation [Fig. 8.3(2)].

In the case of the twisted stem the authors observed a mean divergence angle of 151.3°, while for the regular stem they observed a mean divergence angle of 75.4°. They used Equations (6.4) and (8.2) to deduce the possible values of t:

$$t = (360\tau - dJ)/dJ\tau, \tag{8.3}$$

$$t = (2dJ + dJ\tau - 360)/\tau(360 - 2dJ), \tag{8.4}$$

where J and t are integers. For the regular stem we must have $J = 2$, given that the system is decussate (this excludes the possibility $J = 1$ and $t = 4$ given by (8.3)). The results are as follows:

For 75.4°, the regular stem: $t = 1.76$ ($J = 2$, normal phyllotaxis),
$t = 1.96$ ($J = 2$, alternating whorl).

For 151.3°, the twisted stem: $t = 1.76$ ($J = 1$, normal phyllotaxis),
$t = 2.03$ ($J = 1$, anomalous phyllotaxis).

Given that 1.96, and 2.03 are the values nearest to an integer ($t = 2$ giving the first lateral sequence) Barabé and Vieth conclude that the plant in its regular and twisted forms is controlled by the sequence $J\langle 2, 5, 7, 12, 19, \ldots\rangle$

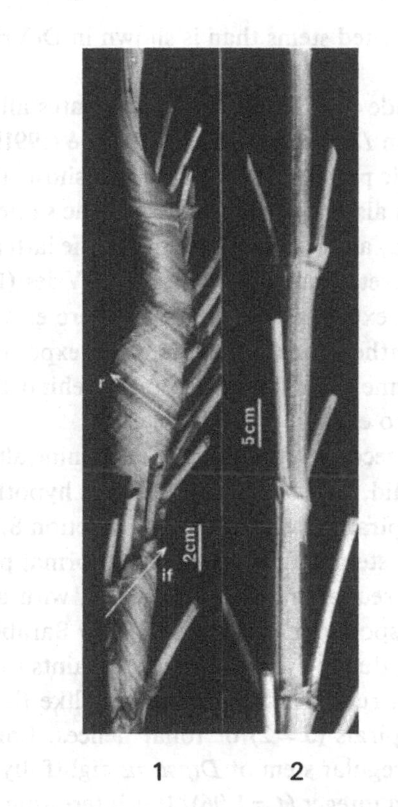

Figure 8.3. (1) Portion of twisted stem of *Dipsacus silvestris*. (2) Portion of normal stem from same plant. (From Barabé & Vieth, 1990.)

with $J = 2$ and $J = 1$, corresponding to forms of spiro-decussation (regular stem) and spirodistichy (twisted stem), respectively.

The authors then compare their interpretation with those of Plantefol (1963) and DeVries (1920) in order to find out whether they agree with their own. DeVries explained that the leaves are already arranged in a spiral pattern before the beginning of the torsion, and that they show a phyllotactic fraction of 5/13. For DeVries biastrepsis was explained by the transition from bijugacy to unijugacy. For Plantefol this phenomenon was explained by the presence of one foliar helix, while the regular stem shows two such helices. Barabé and Vieth's interpretation is not only in agreement with that of Plantefol and DeVries, but with other observations and interpretations. It allows them to analyze known facts in a more coherent framework than permitted by other studies. Barabé and Vieth's results show a greater formal unity between the phyllotaxes of the regular and the twisted stems than is shown in DeVries's and Plantefol's interpretations.

The authors conclude that Jean's model integrates all the phyllotactic variability observed on *Dipsacus silvestris*. Barabé (1991b) gives an overview of the phyllotactic patterns in the genus and shows that all the variability in the genus can also be accounted for by the same model. This includes normal, twisted, and trimerous stems. In the latter case the author proposes a new interpretation of the work of DeVries (1920) in the light of the model. Barabé expresses the view that there exists a complementarity between the mathematical model and the experimental model of foliar helices, the former explaining the unity behind the diversity and the latter attempting to explain the differences.

Instead of using the second hypothesis for explaining alternating whorls, as Barabé and Vieth did, we can resort to the first hypothesis that whorls derive from normal spiral patterns with $t=2$ (Section 8.5.1). This would mean that the regular stem is fundamentally a normal pattern expressed by the bijugate main sequence $2\langle 1, 2, 3, 5, 8, \ldots \rangle$, with a subsequent adjustment of the corresponding angle of 68.75° to Barabé and Vieth's reported value of 75.4°, due to evolutionary constraints such as a strengthening of Hofmeister's rule. This interpretation, like the preceding one, implies two genetic spirals ($J=2$) or foliar helices. Barabé and Vieth's interpretation of the regular stem of *Dipsacus* rightfully comes from the value of t closest to an integer ($t=1.96$). It is interesting to note however that $t=1.76$, the alternative value mentioned by the authors for both the regular and twisted stem, (1) gives a positive answer to Vieth's (1964) question, namely "Is the capitulum of *Dipsacus* bijugate?" (where the term bijugate here means the bijugate main sequence), (2) supports Church's statement of bijugacy as can be seen from Table 6.2 for *Dipsacus silvestris*, (3) indicates that the twisted stem is controlled by the Fibonacci sequence, in agreement with Plantefol (1963) and DeVries (1920), and (4) substantiates the conclusion that the abnormality is explained by a transition from $2\langle 1, 2, 3, 5, 8, \ldots \rangle$ to $\langle 1, 2, 3, 5, 8, \ldots \rangle$.

Barabé, Brouillet and Bertrand (1991) used the same method for the analysis of the phyllotaxis of four species of *Begonia*. They registered high divergence angles, in the range 150° (for *Begonia holtonis*) to 178° (for *Begonia coccinea*). On the basis of Figure 8.2(1), they concluded that the phyllotaxis is anomalous, expressed by the sequence $\langle 2, 2t+1, 2t+3, \ldots \rangle$ where $t = 2, 3, 4, 5, \ldots$, and more precisely spirodistichous. They noticed that the plastochrone ratio is generally greater than 1.75 (except for *Begonia foliosa* where $R \approx 1.47$), a value recognized by Rutishauser (1982)

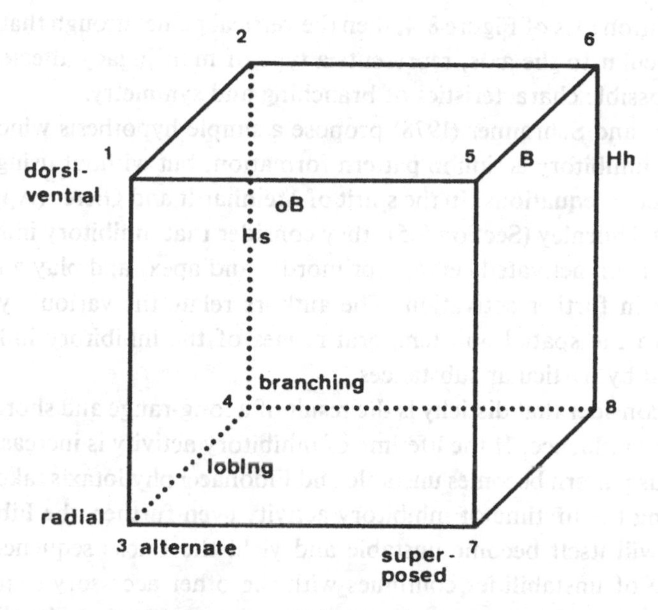

Figure 8.4. Sattler's cubical space (1988a) in floral morphology. Plane 1-2-4-3 represents alternate positioning, plane 5-6-8-7 superposed positioning, and plane 1-5-7-3 represents lack of branching, etc. The point Hh (androecial members of *Hypericum hookerianum*), for example, combines superposition with branching and a symmetry somewhat intermediate between radial and dorsiventral.

as a characteristic value in spirodistichous patterns. Looking at Table 4.3 it can be seen then that γ must be very small. This is the case, given that the observations gave values of γ between 1° and 28°.

8.6.2. Correlation with other models

Figure 8.2(1) can be correlated with Sattler's (1988a) cubical space (Fig. 8.4), which summarizes in a simplified three-dimensional diagram his dynamic multidimensional approach to floral morphology. Each of the three dimensions of the cube represents a submodality of growth: symmetry (from radial to dorsiventral), degree of branching (from lack of branching via lobing, to branching), and positioning (from alternate to superposed). Each point within the cube represents one particular combination of the "states" of the three submodalities. The alternation–superposition axis is made explicit by Figure 8.2(1), which shows the whole range of intermediate patterns between alternation and superposition via the spiro-alternating whorls, the regular normal and anomalous spiral systems, and the spiro-superposed whorls. If one takes a point on the alternation–

superposition axis of Figure 8.4, then the vertical plane through that point, perpendicular to the axis, represents a type of multijugacy affected with all the possible characteristics of branching and symmetry.

Richter and Schranner (1978) propose a simple hypothesis which supposes an inhibitory action in pattern formation, but without using diffusion–reaction equations. In the spirit of Meinhardt and Gierer (Appendix 8) and of Thornley (Section 9.5), they consider that inhibitory influences emanate from activated centers (primordia and apex) and play a regulatory role in further activation. The authors relate the various types of pattern to the spatial and temporal ranges of the inhibitory influences generated by particular substances.

They consider that distichy is the result of a long-range and short-living inhibitory influence. If the lifetime of inhibitory activity is increased, the distichous pattern becomes unstable and Fibonacci phyllotaxis takes over. Increasing the lifetime of inhibitory activity even further, the Fibonacci pattern will itself become unstable and yield the Lucas sequence. This sequence of unstabilities continues with the other accessory sequences. There is thus a systematic sequence of situations beginning with distichy. The same reasoning can be made starting with decussation instead of distichy, leading to the various types of bijugy, via alternating dimery, and leading to superposed dimery. Richter and Schranner suggest that a distichous pattern would turn into a decussate one if the lifetime of inhibitory action is increased and the spatial range reduced at the same time. Figure 8.5 is a summary of their suggestions. They conclude that a detailed quantitative investigation along the lines of the Gierer–Meinhardt theory would be required in order to more carefully examine their phase diagram.

Richter and Schranner do not mention the lateral sequences, superposed whorls, and "spiro-whorls" shown in Figure 8.2(1). But they can be included in their reasoning by correlating Figure 8.5 with Figure 8.2(1). Figure 8.6 is an extension of Figure 8.5. Starting, for example, with distichy ($J = 1$) and increasing the lifetime of inhibition while keeping the spatial range constant, corresponds to following the upper curve in Figure 8.2(1) to its point of intersection with the lower curve, via spiro-distichy and the lateral sequences, leading to Fibonacci and accessory sequences, and ultimately to monostichy via spiromonostichy. The horizontal axis in Figure 8.5 is a vertical axis in Figure 8.2; increasing the lifetime is decreasing the divergence. Also, decreasing the spatial range of inhibition is increasing the jugy J. Every transition between patterns can be explained along these lines.

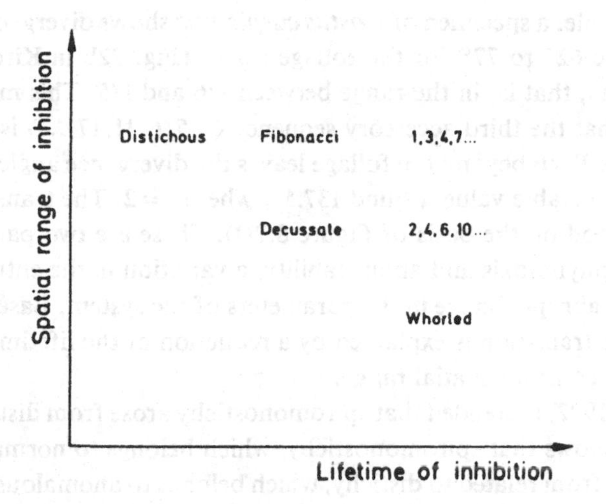

Figure 8.5. Phase diagram relating the various types of phyllotaxis to two main characteristics of the inhibiting substance – its lifetime and spatial range. (From Richter & Schranner, 1978.)

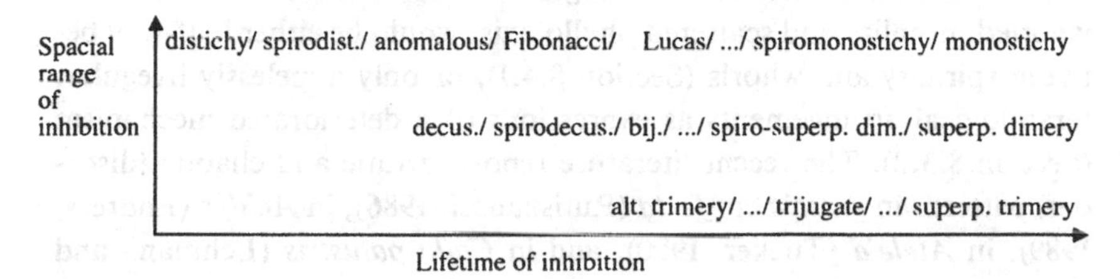

Figure 8.6. Revision of Richter and Schranner's proposal (Figure 8.5) in the light of the interpretative model, going from alternating whorls to superposed whorls by following the two curves in Figure 8.2(1) via their junction, for three values of J. The patterns underlined are those mentioned in Figure 8.5.

8.6.3. Perturbed patterns

Kirchoff and Rutishauser (1990) reviewed the literature on costoid phyllotaxis, and brought forward new data on three species of *Costus*. For these authors spiromonostichy in *Costus* and other Costaceae is characterized by low divergence angles, often as low as (30°–) 50°. In Section 2.4 we have seen that spiromonostichy in *Costus* with $d \approx 47°$ is a special case of normal phyllotaxis with $J = 1$ and $t = 7$. If the divergence angle differs from that angle in other areas of a specimen, then other sequences of integers can be involved.

For example, a specimen of *Costus cuspidatus* shows divergence angles in the range 62° to 77° for the foliage leaves (Fig. 22b in Kirchoff and Rutishauser), that is, in the range between 1/6 and 1/5. This means that $t = 5$ and that the third accessory sequence $\langle 1, 5, 6, 11, 17, \ldots \rangle$ is involved in this area. Then beyond the foliage leaves the divergence angle abruptly changes to a stable value around 137.5°, where $t = 2$. The transition can be understood on the basis of Figure 8.2(1). These are two parent cases of normal phyllotaxis and an unstability, a variation in the entropy level induced an abrupt change in the parameters of the system. Based on Figure 8.6, the transition is explained by a reduction of the lifetime of inhibition, in a constant spatial range.

Hirmer (1922) pretended that spiromonostichy arose from distichy. Figure 8.2(1) shows that spiromonostichy, which belongs to normal phyllotaxis, is far from related to distichy, which belongs to anomalous systems. For other reasons, Kirchoff and Rutishauser (1990) also reject Hirmer's hypothesis. The model of Chapter 6 thus offers a framework for the analysis of costoid phyllotaxis.

Patterns that are neither multijugate nor multimerous, such as half-whorled, peculiar and scattered phyllotaxis, would be either half-way between spirality and whorls (Section 8.4.1), or only hopelessly irregular, teratological arrangements as expressions of a deteriorated mechanism (Section 8.3.2). The recent literature reports erratic and chaotic (disorder) patterns in *Acacia conferta* (Rutishauser, 1986), in *Achlys* (Endress, 1989), in *Ateleia* (Tucker, 1990), and in *Calla palustris* (Lehmann and Sattler, in press). However, a closer look at these plants points to regular patterns.

In *Calla* multimery seems to be the rule, but as suggested by Lehmann and Sattler, homeosis (replacement of tepal primordia with stamen primordia of a different shape) and crowding of floral buds transformed it into an irregular pattern.

Ateleia has flowers lacking four of the five petals. The initiation of the solitary petal precedes initiation of stamens. Tucker (1990) states "that disturbance in petal initiation sets the stage for further drastic perturbation of stamen initiation."

Achlys is the only one of the 16 genera of Berberidaceae that is perianthless. Endress (1989) says that there are no whorls and no regular divergence angles of around 137.5° or 99° "as would be normally expected in spiral flowers." The floral apex of *Achlys* produces stamens in unordered and variable arrangements, while in the regular related taxa there is whorled arrangement of all floral organs. Endress (1989) speculates that petals control orderly initiation of stamens.

On adult shoots of *Acacia conferta*, Rutishauser (1986, his figure 4) says that the first leaves show helical phyllotaxis but that afterward the generative helix vanishes and phyllotaxis "looks chaotic" with a slight tendency toward whorls. According to him, this irregularly scattered pattern would not come from stem elongation or from a helical or verticillate phyllotaxis that becomes hidden or disturbed later.

The diagram drawn by Hellendoorn and Lindenmayer (1974) of *Bryophyllum tubiflorum* (mentioned in Section 8.3.1) also shows what appears to be a case of chaotic phyllotaxis. But the authors apparently see it differently. They state that the specimen exhibits several kinds of phyllotactic patterns (5–7), that their computer simulation shows that these changes are due either to changing apical size or to changing rates of production of an inhibitor on the apical surface, and that observations support the latter assumption. The problem with *Acacia*, in particular, might be one of a similar nature. There might be true and false chaotic patterns.

The ABPHYL (*ab*normal *phyl*lotaxis) syndrome in *Zea mays* (Greyson & Walden, 1972; Greyson et al., 1978) refers to altered patterns of leaf initiation arising through "subtle modifications of normal cellular processes," and "genetically based causative factors" (as in *Arabidopsis* flowers, Yanofsky et al., 1990).

It is suggested here to use the expression **perturbed pattern**, instead of abnormal, irregular, teratological, aberrant, scattered, or chaotic pattern. When the perturbation is small, the mechanism is self-correcting and confines the divergence angle around a certain value. When it is a bit stronger, as in *Zea* or in artificially induced changes (Section 8.3.1), the perturbation produces a shift to another standard pattern with an eventual reversion back to the initial one. The term "perturbed" relates to the dynamic view that an initial multijugate pattern has been modified during evolution by more or less known biological factors, and that order may be present in what is statically called irregular, aberrant, deformed, and abnormal patterns. The case of *Dipsacus* (Section 8.6.2), showing "abnormalities" known as biastrepsis and trimerous nodes, is a good example. For Sachs (1991) irregular initiations are rare exceptions that could represent early stages in the evolution of new morphogenetic controls.

The interpretative model explains regularity and evolutionary transitions between spiral and whorled patterns. The stability of a divergence angle indicates no change in the cyclic process of pattern formation. Regular patterns can be altered by means of one thousand circumstances. Random fluctuations in entropy produce true chaotic patterns. It is believed that perturbed patterns stress rather than contradict the principles that apply to the generation of multijugate patterns.

9

Convergences among models

9.1. Basic morphology of phyllotactic patterns

9.1.1. Packing efficiency; the noble numbers

A central morphological feature of phyllotaxis is spirality (parastichies, genetic spirals). Thornley (1975b) used an exponential genetic spiral to describe the centric representation of primordia. Ridley (1982a) used the parabolic spiral, introduced in phyllotaxis by Vogel (1979) under the name of cyclotron spiral (the path followed by an elementary particle accelerated in a cyclotron is a parabolic spiral). Hernandez and Palmer (1988) generated the sunflower capitulum using exponential and Archimedian genetic spirals. Their program, which respects the natural development from rim to center (Section 1.4.2), allows the user to vary the sizes of the head and seed number, and the spacing between the seeds. Using computer graphics and the Fibonacci angle, Dixon (1992) studied particular genetic spirals called "loxodromes," those spirals cutting the generators of surfaces of revolution (cone, cylinder, sphere) by a constant angle. Depending on the rate of advance of the spiral across the surface of revolution, he distinguished between grey and green spirals (grey spirals being special cases of green spirals), and obtained concho-spirals and equiangular spirals.

The generative spiral, generally given in polar coordinates by $r = a\theta^b$, is parabolic when $b = 1/2$, golden when $b = 1/\tau$, Archimedian when $b = 1$, reciprocal for $b = -1$, logarithmic (equiangular) when $r = ae^{b\theta}$ ($b = \cot\phi$, where ϕ is the constant angle made by the tangent at any point of the spiral with the ray from the pole), or exponential with $r = a\log b\theta$. Only in the case of the logarithmic spiral is the plastochrone ratio constant.

The central fact about spirality in phyllotactic patterns is the overwhelming presence of the golden ratio τ, which appears as a constant of nature [the ratio of the expressions in Eqs. (4.7) to (4.10) is τ]. Phyllotaxis is a consequence of the occurrence of τ, regardless of the growth of the plant in either size or length. Growth determines the conspicuous parastichy pair, and contributes to stress the presence of τ by the phenomenon of rising phyllotaxis, obtained by a steady decrease in the internode distance (the rise) following an increase in size of the apical dome with respect to the size of the primordia.

Out of the Bravais–Bravais biologically plausible mathematical assumptions in the cylindrical lattice, namely:

The divergence angle d is an irrational number.

By moving upward along the vertical axis of the cylindrical lattice [see Fig. 2.3(2)] the neighbors of this axis, located on two infinite asymptotic polygonal lines, are alternately on each side of the axis and get closer to it.

Coxeter (1972) deduced the set C of limit-divergence angles expressing normal phyllotaxis [Eq. (6.4) involving τ]. The set $\mathbf{J}$ of limit-divergence angles of normal and anomalous phyllotaxis [see Eq. (6.6) also incorporating τ], obtained from Jean's model of Chapter 6, have continued fractions terminating with $1/\tau$ (Appendix 1 at the term "continued fraction" introduces the subject). In these cases the alternation of the neighbors of the vertical axis generally starts from a certain point of the lattice. Some approaches (e.g., Ridley, 1982a; Rivier, 1988) stress that phyllotaxis is the solution of a structural variational principle giving maximal uniformity to the patterns and requiring divergence angles whose continued fractions terminate as $1/\tau$, that is, with 1's only. The numbers whose continued fractions terminate in this way have been called **noble numbers**. The set of noble numbers is given by $N = \{(a+b\tau)/(c+d\tau)$: a, b, c, d are integers, $ad-bc = \pm 1$, $a<b$, $c<d\}$. We have, with a strict inclusion, $C \subset \mathbf{J} \subset N$.

Ridley (1982a, 1986, 1987) introduced the teleonomic consideration of packing efficiency to deal with the packing of seeds in the sunflower head, centered at points along a genetic spiral. They are considered to be of equal area and distributed at a constant divergence angle. He defined packing efficiency by

$$\eta = \frac{\text{mean square distance between nearest neighbors}}{\text{the average area per point}},$$

$$\eta = A/M,$$

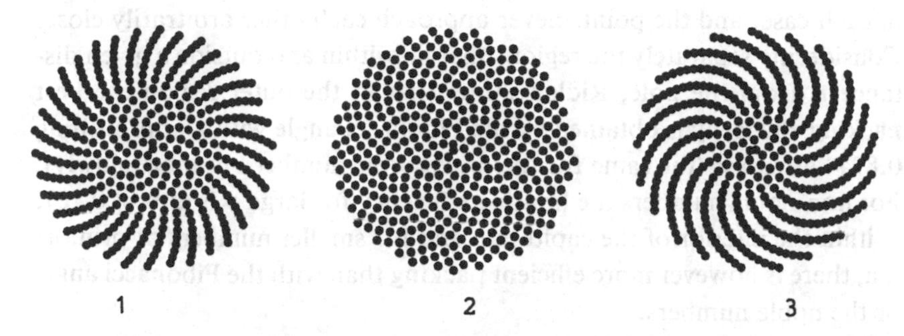

Figure 9.1. Depiction of the critical role of the divergence angle in packing efficiency; the three patterns differ only by a slight value of the divergence angle d. In (1) it is equal to 137.3°, in (2) to 137.507764...° (the Fibonacci angle corresponding to the noble number τ^{-2}), and in (3) to 137.6°. Notice that in the case of the Fibonacci angle a parastichy pair is visible (not just one family of spirals), and the points are evenly distributed in the disk. In the other two cases, η (see the text) falls to zero while moving outward, although the area per point is still the same. (From Prusinkiewicz & Lindenmayer, 1990.)

where A is the area of the capitulum containing M seeds. The larger η is, the more uniformly the points are distributed over the capitulum. Ridley showed that for a constant divergence angle the cyclotron spiral distributes more evenly the points of the spiral lattice. Moreover, when the divergence angle chosen is the Fibonacci angle, packing efficiency is maximized, and $\eta \approx 0.8169$. With the other limit-divergence angles of phyllotaxis, the efficiency is the same but it is approached more slowly than with the Fibonacci angle.

Dixon (1983) illustrated the failure to close-pack points on a disk with $1/\pi$, $1/e$, $1/\sqrt{5}$, and $\tau^{-2}/2$. More generally, outside the set of noble numbers the pattern becomes very strongly spiraling because it is dominated by a set of long parastichies (the other set is not apparent), or the parastichies become rays in the outer portion of the disk (see Fig. 9.1). This figure shows that a difference as little as 0.1° in the divergence angle results in improbable structures. Even for the rational approximation 34/89 (angle of about 137.5281°) to the Fibonacci angle (about 137.5077°; difference of less than 0.03°) along the cyclotron spiral, we obtain in the outer region of the lattice, 89 radii where the points become arbitrarily close as we move away from the pole.

For irrational non-noble divergence angles such as $\sqrt{2}-1$ (corresponding to an angle of about 149.11°) and a cyclotron spiral, we can observe 29 spirals in the center and 70 near the rim, but no set of opposed spirals in each case, and the points never approach each other arbitrarily close. Considering separately the regions that lie within and outside a given distance p from the pole, Ridley shows that in the outer region the best packing efficiency is obtained by the Fibonacci angle with η approaching 0.8944 as $p \rightarrow \infty$ (the same for the other noble number but more slowly). For non-noble numbers the limiting efficiency for large p is at best 0.707. Within the interior of the capitulum and for smaller numbers of primordia, there is however more efficient packing than with the Fibonacci angle or the noble numbers.

The set N of noble numbers is precisely what Marzec and Kappraff (1983) obtained from a slightly different point of view. It is their set ϕ_G of divergence angles proved to distribute the leaves around the stem of a plant more uniformly than the neighboring angles can do. More precisely, the authors demonstrated that the Fibonacci angle distributes the points on a circle more uniformly than is possible with any other angle. Otherwise stated, if a is an irrational number and $[a]$ its fractional part, the n points $[a], [2a], [3a], \ldots, [na]$ split the segment $[0, 1]$ into $n+1$ pieces. The next point $[(n+1)a]$ always falls into one of the largest segments, splitting it into two. It happens that the two segments are so unequal that one is twice or more than twice as large as the other. We then say that the point causes a "bad break." Then an appropriate economy principle is: no "bad break" appears. Knuth (1973) proved the theorem that the only value that satisfies the economy principle is the Fibonacci angle. For all numbers a, except for $1/\tau$ and $1/\tau^2$ (which are equivalent: their sum is 1), nonuniform splits arise; that is, "bad breaks" appear. The Fibonacci angle divides the largest segment into the golden ratio, and it gives the most uniform distribution of the n points in $[0, 1]$. Marzec and Kappraff (1983) showed that noble numbers, whose continued fractions end as $1/\tau$ (no intermediate convergent from a certain term; see "continued fraction" in the Glossary), eventually lead to the most uniform distribution. This uniform distribution is a mathematical version of the botanical economy principle that all leaves should have equal access to sunlight and should not be overcrowded by new ones.

9.1.2. Self-similarity

Outside the set of noble numbers the structures generated are not self-similar. Following Occelli (1985) and Rivier (1986), Rothen and Koch (1989a) clearly illustrated the meaning and role of self-similarity in phyllotactic patterns. They showed that when the value of the rise r changes

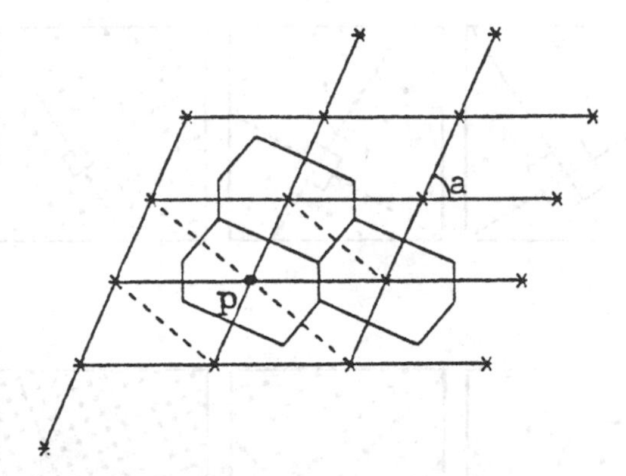

Figure 9.2. A visible opposed parastichy pair in a regular cylindrical lattice determines a set of congruent parallelograms filling the space completely, without overlapping. Each point of the lattice, such as point p, belongs to six triangles, such that six sides meet at point p. Drawing the perpendiculars in the middle of these sides determines the Voronoi polygon, such as the scales of pineapples. The diagram shows the three families of parastichies determined by the scales in contact. When the parallelogram is a rectangle, the Voronoi polygon is a rectangle instead of a hexagon. Three hexagons meet at triple points. The set of hexagons fill the plane without overlapping.

in the cylindrical lattice (the divergence angle being constant), the Voronoi polygon (the Dirichlet domain or Wigner–Seitz cell) does not change in form, exclusively when the divergence d is a noble number.

Now, what is a Voronoi polygon? Given a point p of the cylindrical lattice, such a polygon encloses by definition all points of the plane nearest to p than to any other point of the lattice (see Fig. 9.2). A Voronoi polygon is a natural unit cell of a structure. The noble numbers give shape invariance or self-similarity of the Voronoi polygons around the points of the lattice, as is illustrated in Figure 9.3(1). When the divergence angle is not a noble number, the shape of the polygon changes when r changes, as Figure 9.3(2) shows. Non-noble divergences generate structures that are not self-similar (Rivier et al., 1984). This is especially remarkable when the divergence is a rational number, as Figure 9.3(3) shows.

It is particularly important that the Voronoi polygon stays similar to itself when one considers the natural continuous transitions observed in daisies or sunflowers showing a system (m, n) in the center of the head or capitulum, $(m, m+n)$ in the middle portion of the capitulum, and $(2m+n, m+n)$ in the outer portion, corresponding to changes in r (rising phyllotaxis). In such capituli there are circular defect lines (see Fig. 9.4). These defect lines can be observed in Figures 1.2 and 1.4. Self-similarity

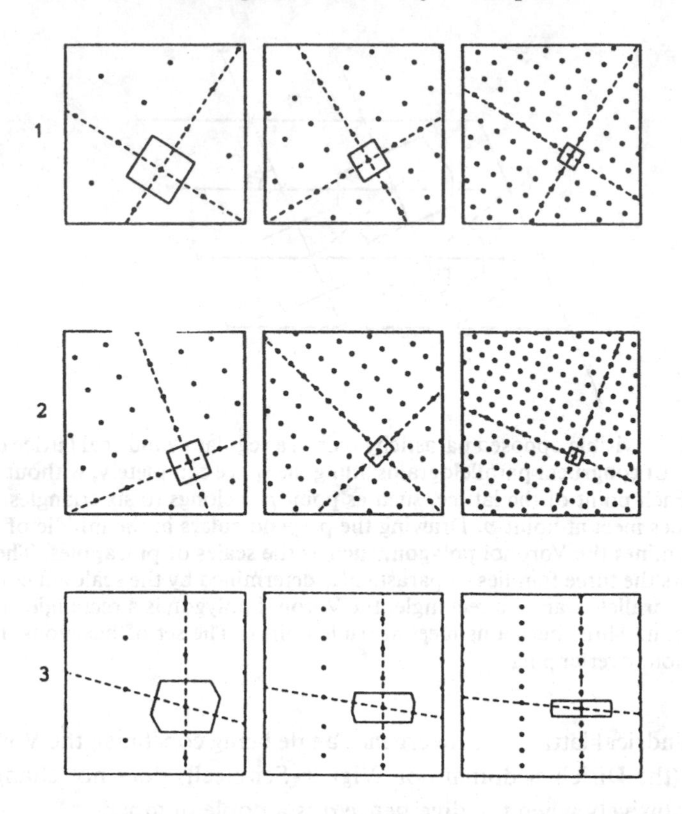

Figure 9.3. Three cylindrical lattices differing only by the value of their divergence angles. (1) Built with $d = \tau^{-1}$, a noble number, the decrease of the rise r does not change the shape of the Voronoi polygon, but only its size and orientation, while the system rises from (2, 3) to (5, 3) to (5, 8) phyllotaxes. (2) The lattice is built with $d = \sqrt{2} - 1$ whose continued fraction ends with 2's only. At the second step the polygon is not a square as in the other two cases. The system rises from (2, 5) to (7, 5) to (12, 5) phyllotaxes. (3) Voronoi polygons in a lattice of points with $d = 1/3$. (From Rothen & Koch, 1989a.)

means that we need only one algorithm to produce both small and large capituli, that is, the same code to raise their phyllotaxes.

The golden ratio τ expresses the symmetrical proportions involved in the generation of phyllotactic patterns. The problem of pattern generation thus consists in devising principles that will put the systems on tracks defined by the golden ratio, as the minimality principle found for example in Adler's model does. Now why does nature prefer self-similar structures and efficient packing? It is a question which brings us deeper into

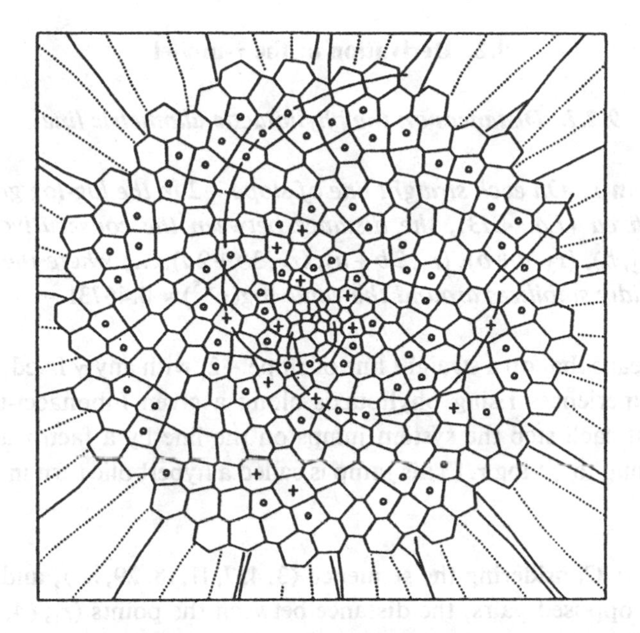

Figure 9.4. A picture of a daisy flower obtained by drawing Voronoi polygons (florets) about the points on a genetic spiral. It shows two circular defect lines. They are made with pentagonal and heptagonal florets marked respectively with ○ and +. They allow the transition from (13, 21) phyllotaxis in the middle part of the capitulum, to (34, 21) phyllotaxis in the outer part. The figure shows the direction of various families of parastichies. (From Rothen & Koch, 1989a, originally obtained by Occelli, 1985.)

the causes. It brings us back to primitive land plants; to Lyell's principle of evolutionary continuity according to which a large part of cellular organization is very ancient and is common to all organisms; and to principles of optimal design (Chapter 6). But for the moment let us not inquire further along these lines of causality. Instead of trying to understand why τ has a key role in the phenomenology, let us take for granted the importance of τ in the morphogenesis of spiral patterns. The minimality condition that will be introduced, the τ-model, is an algorithm which may not seem to have much to do with morphogenesis, but it will allow us to order the set N of noble numbers. The condition incorporates a general notion of rhythm. If this rhythm is established at plastochrone m (when m primordia exist), then an element of N is selected by the minimality condition, which maximizes packing efficiency of the primordia for this value of m.

9.2. Derivation of the τ-model

9.2.1. Distances on the phyllotactic allometric line

Proposition 1. *On each straight line of slope -2 in the log-log grid (Eq. 4.11), γ fixed (Eq. 4.13), the distance between the consecutive points (r_1, a), (r_2, b), $(r_3, a+b)$, $(r_4, 2b+a)$, $(r_5, 3b+2a), \ldots$, where the r_i's are rises, rapidly stabilizes around the value* $\log(\tau^{\sqrt{5}}) \approx 0.46731$.

This means that on a straight line of slope -2, with any γ fixed, when a system experiences rising phyllotaxis along a given Fibonacci-type sequence, at each step the system jumps on the line by a factor approximately equal to $\sqrt{5} \log \tau$. (This jump is called a hyperbolic turn in Appendix 9.)

Example 1: Considering the sequence $\langle 3, 4, 7, 11, 18, 29, \ldots \rangle$, and a fixed γ for the opposed pairs, the distance between the points $(r_1, (4,7))$ and $(r_2, (11,7))$ in the log-log grid is given by $\sqrt{5}\log(18/11) \approx 0.47825$.

Example 2: One goes from the orthogonal conspicuous pair $(5, 3)$ to the orthogonal conspicuous pair $(5, 8)$ by a jump of $\sqrt{5}\log(13/8) \approx 0.47148$ in the log-log grid, on the line $r = 1.8944(m+n)^{-2}$.

Proposition 2. *Given the integer m, the value of $n < m$ for which the distance between (r_m, m) and (r_n, n) in the log-log grid, for any given γ, is the closest to $\sqrt{5}\log\tau$, is such that $n < m \leq 2n$ and is given by $n = (m\tau^{-1})$, the parentheses denoting the closest integer to $m\tau^{-1}$.*

Proposition 3. *Given the integer m and the angle γ, if point $n = k_1$ is such that $\sqrt{5}\log(m/k_1\tau)$ is closest to 0, then for the integer $m+n$ and the same γ, the number $\sqrt{5}\log[(m+n)/k_2\tau]$ is closest to 0 for $k_2 = m$.*

9.2.2. Minimality principles

Minimality criterion: Given m, for any γ, a point $n < m$ is said to satisfy the minimality criterion if the value of the function of k, $\sqrt{5}\log(m/k\tau)$, $k < m$, is closest to 0 for $k = n$. The value $M(m) = \sqrt{5}\log(m/n\tau)$ is then called the minimum of the function. The point n is unique.

By Proposition 3, if n satisfies the minimality criterion for m, then m satisfies the minimality criterion for $m+n$, and $m+n$ satisfies the minimality criterion for $2m+n$ and so forth.

Proposition 4. *Given a term m of the Fibonacci-type sequence $\langle a, b, a+b, 2b+a, \ldots\rangle$, if the minimality criterion is satisfied for the adjacent member $n<m$ in the sequence, then the criterion is satisfied for any point p greater than m in the sequence by the number $q<p$ and adjacent to p in the sequence. The consecutive minima $M(F_{k+1}b+F_k a)$ are alternately positive and negative and decrease in absolute value toward* 0.

Minimality condition: At plastochrone m, when m primordia exist, the minimality condition states that nature requires the primordia to show the visible opposed parastichy pair $(m-n, n)$ such that n satisfies the minimality criterion.

It is known (Appendix 4) that each one of the two extensions $(a, a+b)$ and $(a+b, b)$ of a visible opposed pair (a, b) is a visible pair, and that one of them is opposed. Given that the consecutive minima in Proposition 4 are decreasing toward zero while alternating in signs, the consecutive alternate extensions of the conspicuous pair $(m-n, n)$, that is (m, n), $(m, m+n)$, $(2m+n, m+n), \ldots$ are visible and opposed. This means that if the system starts to experience the minimality condition at plastochrone m, the system enters into a loop for the first time, generating conspicuous pairs along the Fibonacci-type sequence $\langle m-n, n, m, m+n, \ldots\rangle$.

9.3. Ordering the noble numbers

For each m the minimality condition gives a unique n. Table 9.1 gives the sequences obtained from Proposition 4 for various values of m. Chapter 2 gives the divergence angles corresponding to the sequences. The order of the noble numbers corresponding to the angles (divide these angles by 360 to obtain approximations of these numbers) has to do with the figures different from ones at the beginning of the continued fraction of those numbers. For example, the continued fraction corresponding to the sequence $\langle 1, 4, 5, 9, 14, 23, \ldots\rangle$ is $[0; 4, 1, 1, 1, \ldots]$. And before the appearance of $1/\tau$ in the continued fraction of this angle, that is before the rhythm starts, the minimality condition does not yield that sequence.

Indeed, consider the value $m=9$ in the sequence $\langle 1, 4, 5, 9, 14, 23, \ldots\rangle$. For consecutive numbers greater than 9 in the sequence, the minimality

Table 9.1. *Values of n satisfying the minimality criterion for given values of m > n, and for γ fixed*

m	n	Visible opposed pair $(n, m-n)$	Divergence	Sequence
3*+	2	(2, 1)	137.51°	⟨1, 2, 3, 5, 8, ...⟩
4	3	(3, 1)	99.5°	⟨(1), 3, 4, 7, 11, ...⟩
5*+	3	(3, 2)	137.51°	⟨1, 2, 3, 5, 8, ...⟩
6*	4	2(2, 1)	68.75°	2⟨(1), 2, 3, 5, 8, ...⟩
7*+	4	(4, 3)	99.5°	⟨(1), 3, 4, 7, 11, ...⟩
8*	5	(5, 3)	137.51°	⟨1, 2, 3, 5, 8, ...⟩
9*	6	3(2, 1)	45.84°	3⟨(1), 2, 3, 5, 8, ...⟩
10*	6	2(3, 2)	68.75°	2⟨(1), 2, 3, 5, 8, ...⟩
11*+	7	(7, 4)	99.5°	⟨(1), 3, 4, 7, 11, ...⟩
12	7	(7, 5)	151.1°	⟨(2, 5), 7, 12, 19, ...⟩
13*	8	(8, 5)	137.51°	⟨1, 2, 3, 5, 8, ...⟩
14*+	9	(9, 5)	77.96°	⟨(1, 4, 5), 9, 14, ..., ⟩
15	9	3(3, 2)	45.84°	3⟨(1), 2, 3, 5, 8, ...⟩
16*	10	2(5, 3)	68.75°	2⟨(1), 2, 3, 5, 8, ...⟩
17*+	11	(11, 6)	64.08°	⟨(1, 5, 6), 11, 17, ...⟩
18	11	(11, 7)	99.5°	⟨(1), 3, 4, 7, 11, ...⟩
19*+	12	(12, 7)	151.1°	⟨(2, 5), 7, 12, 19, ...⟩
20	12	4(3, 2)	34.38°	4⟨(1, 2), 3, 5, 8, ...⟩
21	13	(13, 8)	137.51°	⟨1, 2, 3, 5, 8, ...⟩
22*	14	2(7, 4)	49.75°	2⟨(1, 3, 4), 7, 11, ...⟩
23	14	(14, 9)	77.96°	⟨(1, 4, 5), 9, 14, ...⟩
24	15	3(5, 3)	45.84°	3⟨(1), 2, 3, 5, 8, ...⟩
25	15	5(3, 2)	27.5°	5⟨(1, 2), 3, 5, 8, ...⟩
26	16	2(8, 5)	68.75°	2⟨(1), 2, 3, 5, 8, ...⟩
27	17	(17, 10)	106°	⟨(3, 7, 10), 17, 27, ...⟩

Note: The asterisk (*) in Table 9.1 corresponds to a value of m for which the τ-model and the interpretative model (Chapter 6) give the same value of n (Section 9.4). The consecutive plus signs (+) correspond to the ordered values of m for which the divergence angles obtained from these two models are the consecutive divergence angles obtained from Marzec's entropy model (Section 9.5).

condition is satisfied by this sequence, but for the first four it is not. Indeed, for $m=9$, $n=6$ gives the minimality condition, for $m=5$ the condition is satisfied for $n=3$, and for $m=4$, it is satisfied for $n=3$ (the sequences then obtained can be read in Table 9.1). This explains why there are parentheses around 1, 4, and 5, in the sequence on the line where $m=14$ in Table 9.1. Of course, these parentheses do not mean that the visible oposed parastichy pairs (1, 4) and (5, 4) are excluded by the algorithm. When the sequence ⟨1, 4, 5, 9, 14, ...⟩ arises in a plant, at whatever plastochrone m it may be, say $m=14$ for example, then all consecutive alternate extensions of (5, 9) are visible opposed pairs, together with the contractions of (5, 9), which are (1, 4) and (5, 4).

9.4. The τ-model and the interpretative model

Given $m = P_r$, the τ-model and the interpretative model give the same value of n, the same sequence, and the same divergence angle for $m = 3, 5, 6, 7, 8, 9, 10, 11, 13, 14, 16, 17, 19, 22, \ldots$ (see the starred items in Table 9.1). Among these divergence angles are the usual values 137.5°, 99.5°, 78°, 68.75°, and 151.1°, corresponding to the vast majority of types of spiral phyllotaxis (more than 98% according to Table 7.12). Moreover, up to $m = 26$ the two models give the same angles, but in a slightly different order. When $P_r = 4$, the minimal entropy condition gives no value of n given that in the first two levels of a phyllotactic hierarchy it is impossible to have 4 nodes if level one must contain one double node.

As a result of his observations, Wright wrote (1873, p. 399) that "each leaf of the cycle is so placed over the space between older leaves nearest in direction to it as always to fall near the middle, and never beyond the middle third of the space, or by more than one sixth of the space from the middle, until the cycle is completed." This means that the divergence angle d is generally such that $1/3 < d < 1/2$, that is between 120° and 180°. This case is dealt with in Proposition 5 below. Proposition 6 deals with the case where $1/4 < d < 1/3$, that is, the interval [90°, 120°] that includes the well-known divergence angle of 99.5°.

Proposition 5. *If* $1/3 < d < 1/2$, *then a sufficient condition for the Fibonacci pattern to arise under the minimality condition of the τ-model is that* $m < 12$ (*in the interpretative model the pattern arises when* $P_r < 19$; *see Proposition 7a of Chapter 6*).

Proposition 6. *If* $1/4 < d < 1/3$, *then a sufficient condition for the pattern governed by the sequence* $\langle 1, 3, 4, 7, 11, \ldots \rangle$ *to arise under the minimality condition of the τ-model is that* $m < 27$ (*with the interpretative model this pattern always arises whatever be the value of* P_r; *see Proposition 7b of Chapter 6*).

Looking at Tables 9.1 and 9.2, it is easily seen that Propositions 5 and 6 hold. The sequences in Table 9.2 are practically not observed in nature (Table 7.12), so that the differences between the minimal entropy model and the τ-model are very scarce as far as the priority order of the patterns

Table 9.2. *Values of $n < m = P_r$ giving the minimal entropy condition for given values of m, and γ fixed; the divergence angles and sequences of integers corresponding to the pair (n, m) is shown*

m	n	Visible opposed pair $(m-n, n)$	Divergence	Sequence
4	—	—	—	—
12	8	4(2, 1)	34.38°	4⟨1, 2, 3, 5, 8, ...⟩
15	10	5(2, 1)	27.5°	5⟨1, 2, 3, 5, 8, ...⟩
18	12	6(2, 1)	22.9°	6⟨1, 2, 3, 5, 8, ...⟩
20	13	(13, 7)	54.4°	⟨1, 6, 7, 13, 20, ...⟩
21	14	7(2, 1)	19.64°	7⟨1, 2, 3, 5, 8, ...⟩
23	15	(15, 8)	47.25°	⟨1, 7, 8, 15, 23, ...⟩
24	16	8(2, 1)	17.19°	8⟨1, 2, 3, 5, 8, ...⟩
25	16	(16, 9)	158.1°	⟨2, 7, 9, 16, 25, ...⟩
26	17	(17, 9)	41.77°	⟨1, 8, 9, 17, 26, ...⟩
27	18	9(2, 1)	15.28°	9⟨1, 2, 3, 5, 8, ...⟩

Note: Table considers only the values of m for which the values of n differ from the one given by the τ-model in Table 9.1.

is concerned. In the τ-model this order is deduced by assuming the importance of τ in the phenomenology. In the minimal entropy model the order is deduced by assuming that phyllotactic systems are hierarchies with simple and double nodes having a rhythm, that is, repetitive cycles completed at various values of the plastochrone P_r.

Both models considered in the present section point to the usual types of patterns, but starting from $m = 27$, the types of patterns generated are different. For example, the sequence ⟨3, 7, 10, 17, 27, ...⟩ does not occur in the minimal entropy model but it does occur in the τ-model. The patterns predicted by various models are discussed in Sections 9.5.3 and 9.7. The idea of comparing models is not to show that one is better than the other but to show similarities of results under dissimilarities of assumptions.

9.5. Marzec entropy and diffusion model

9.5.1. Phyllotaxis as a dissipative structure

Marzec's study (1987) uses a one-dimensional model first developed by Thornley (1975a), where the rise r is eliminated from the problem. In this model the leaves on the stem are abstractly treated as points on a circle, each point acting as a morphogen source. Phyllotaxis becomes a dissipative structure, that is, a nonequilibrium system sustained by a flow of matter and energy through its boundaries by the creation (negentropy) and diffusion and decay (entropy) of the morphogen according to a diffusion-reaction partial differential equation. The nature of the morphogen is not defined (its identification represents the main problem of diffusion approaches to phyllotaxis).

Thornley's model is dynamic, setting a new leaf primordium at the position of the absolute minimum of the concentration field established by the existing primordia, one by one, until a constant divergence $\Delta\mu$ emerges. The strength of each primordium as a morphogen source is decreased by a factor $\beta < 1$ after each plastochrone. Then, setting a constant divergence angle for about 100 primordia and using an idea of perturbation, Thornley attempts to reproduce that angle. He finds bands of angles (78–85°, 97–101°, 105–110°, 132–134°, 136–143°, and 146– 180°) containing some of the observed divergence angles. He determined that when $\beta > \frac{1}{2}$, only the angle τ^{-2} is possible, this angle being approached for values of β near 1.

Marzec's treatment is first of all static. It establishes the set of noble numbers N (those numbers whose continued fractions do not have intermediate convergents after a finite number of terms). They produce uniform spacing of leaves. Any angle in the set possesses a purely geometrical minimization property (Marzec & Kappraff, 1983, Theorem 9, p. 216) by virtue of which a more even distribution of points (leaf primordia) on the circle is produced, each leaf being set so that all the leaves have the maximum space (no "bad breaks"; see end of Subsection 9.1.1). The results of this minimization principle are then applied directly to the minimization of a static concentration function. Marzec uses an informational entropy function defined by the formula

$$S = -2\pi\langle C\rangle \int_0^{2\pi} p(\mu)\ln p(\mu)\, d\mu,$$

where $\langle C\rangle$ is the mean concentration and $p(\mu)$ is a morphogen concentration field established by the leaf primordia, depending on the constant divergence $\Delta\mu$. The entropy of the morphogen distribution is considered to measure the spacing of the leaves. Considered as a function of $\Delta\mu$, it is found that S is a maximum, and the rates of entropy production are minimal at a $\Delta\mu$ which asymptotically approaches, as β approaches unity, one of the members of N.

9.5.2. Relationships between the two entropy models of phyllotaxis

"In rough order of their appearance as β increases, the most prominent members of N are 137.5°, 99.5°, 78.0°, 64.1°, 151.1°, 54.4°, 158.1°, and 106.4°" (Marzec, 1987). This is akin to what happens in the τ-model and in the interpretative model for $J=1$ (unijugy). The angles obtained consecutively in Marzec's model correspond to the increasing values $P_r=3$ and 5 (main sequence, 137.5°), 7 and 11 (first accessory sequence, 99.5°), 14 (second accessory sequence, 78°), 17 (third accessory sequence, 64.1°), 19 (first lateral sequence, 151.1°), 20 (fourth accessory sequence, 54.4°), 25 (second lateral sequence, 158.1°), respectively, under the minimal entropy condition (see Tables 9.1 and 9.2) and to the value $m=27$ (106.4°) under the minimality condition of the τ-model. The angle of 106.4° does not correspond to any of those given by the interpretative model. The three models show the same order of generation for the values 3, 5, 7, 11, 14, 17, and 19 (the values marked with a plus sign in Table 9.1), corresponding to the main divergence angles observed.

Both the dynamic and the static approaches have been considered in Jean's model of Chapter 6. We have seen that the search for a suitable Equation (6.2) was preceded by the consideration of Equation (6.1) not incorporating a notion of rhythm, and corresponding to a dynamic approach. Unfortunately, with Equation (6.1) (and rhythm $w=2$) only the pattern defined by the Fibonacci sequence arises, as in Thornley's model for $\beta > \frac{1}{2}$. Similarly, for low values of β the only solution of the dynamic problem in Marzec's model is τ^{-2}. Why is that? Marzec conjectures that this robustness of τ^{-2} underlies its predominance in plants.

9.5.3. Priority order in phyllotactic systems

The consecutive values for the divergence obtained from the τ-model, Marzec's model, and the interpretative model are given in Table 9.3. Given the actual state of the results, the comparison can be done for unijugate patterns only ($J=1$). The table includes the only other model that proposes a priority order of phyllotactic patterns. This is the XuDong, Bursill and JuLin (1989) model of equal disk on a parabolic spiral lattice for normal phyllotaxis with $J=1$ and a constant divergence angle. Let us say a few words about it.

The total areas covered by aggregates containing identical numbers (100) and sizes of disks were investigated as a function of the divergence angle d. For each d in the range $(0, 180)$, the maximal radius of each

Table 9.3. *Priority order of divergence angles (in degrees) according to various models*

Model	Divergence angles											
Marzec	137.5,	99.5,		78,	64.1,		151,	54.4,		158,	106	
Interpretative	137.5,	99.5,		78,	64.1,		151,	54.4,	47.3,	158		
XuDong et al.	137.5,	99.5,		78,	64.1,	84,		54.4,				131.5
τ-model	137.5,	99.5,	151,	78,	64.1,						106	

aggregate was determined. Then peaks were seen to arise representing the least-dense packings (for example $d = 120$ gives an absolute maximum). The priority order for the closer-packed structures is seen to be given by a series of decreasing minima in the plot, obtained at 137.51° (absolute minimum), 99.5°, 77.96°, 64.08°, 84.0°, 54.40°, 131.5°, 140.3° and 110.0°, respectively. The values in small characters correspond to fractional values of t in the sequence of normal phyllotaxis [Sequence (6.3)]; these values did not give rise to countable numbers of visible spirals when the structures were plotted. The angle of 151.1° (anomalous phyllotaxis) seems to be farther in the list. If the number of disks is increased from 100 to 500, or if a growth law for the disks is introduced, the priority order remains the same.

It is meaningful that each model successfully predicts the priority order of occurrence of the predominant spiral patterns observed on botanical specimens. In other words these four models, developed from a variety of assumptions, predict approximately the same priority. This points to a very simple mechanism common to the models. Underneath the diversity of assumptions lie a common simple structure (e.g., a Fibonacci hierarchy of primordia) and a common minimization principle (a principle of optimal design). It is interesting to note that the τ-model favors the first lateral sequence (151°) over the second accessory sequence (78°). The data in Chapter 7 have not allowed us to draw a definite conclusion regarding the priority order in nature for these two sequences.

9.6. Adler contact-pressure model

9.6.1. The maximin principle and its consequences

Adler (1974, 1977a) elaborated a model in the cylindrical lattice, based on the idea of mutual pressures of primordia, that is, on the assumption that the system grows until the primordia experience contact pressure. Then the centers of the primordia move by growth so as to maximize the minimal distance between these primordia. More precisely, Adler's maximin principle assumes that the rise r decreases as the plastochrone P increases, and it states that: If contact pressure starts at $P = P_c$, then for $P \geq P_c$ and for the rise $r(P)$, d has this uniquely determined value for which the distance $d(0, n)$ (a function of d) from leaf 0 to its nearest neighbor n is maximized. In the absence of contact pressure, $d(P)$ and $r(P)$ are independent from one another, but at $P = P_c$, for r given, $d = d(r(P))$.

The contact-pressure model gives particular conditions for the systems (1, 2), (2, 3), (3, 5), (2, 5), and (7, 5) to arise, and general formal conditions under which a system follows a contact-pressure path. Then rising phyllotaxis occurs, (m, n) being replaced by $(m, m+n)$, $m > n$. When $r \geq \sqrt{3}/38$ (arbitrary P) or when $P < 5$ (arbitrary r), $\max d(0, n)$ has only one maximum for which we can have phyllotaxis (2, 1) or (3, 2). The main result in Adler's model states that if d is between 1/3 and 1/2, a sufficient condition to have Fibonacci phyllotaxis is that either $P_c < 5$ or $r(P_c) \geq \sqrt{3}/38 \approx 0.0456$. This means that if d is between 1/3 and 1/2 Fibonacci phyllotaxis is unavoidable if contact pressure begins before leaf 5 appears, when there are 5 leaves (0, 1, 2, 3, 4). Compare with Proposition 6 of Chapter 6, wherein it is not necessary to suppose that d is between 1/3 and 1/2.

In Adler's model we have phyllotaxis (2, 1) if r is in the interval $[\sqrt{3}/14, \sqrt{3}/6] \approx [0.1237, 0.2888]$, and phyllotaxis (3, 2) if r belongs to the interval $[0.0456, 0.1237]$. For $P = 5$ and $r < \sqrt{3}/38$, $\max d(0, n)$ has two maxima at which the phyllotaxis can be either (5, 3) or (5, 2), so that the phyllotaxis expressed by the sequence $\langle 2, 5, 7, 12, \ldots \rangle$ becomes possible. For $P > 5$ and $r < \sqrt{3}/38$, the function has two or more maxima, and many types of phyllotaxis are possible.

This model cannot be used to explain the initial placement of the primordia, and it assumes a spiral arrangement. It explains rising phyllotaxis in general terms, by proving that the system will experience rising phyllotaxis along the sequence $\langle a, b, a+b, 2b+a, 3b+2a, \ldots \rangle$ if at time P_c the system is expressed by any two consecutive terms of this sequence, and if contact pressure is maintained. Thus we may have the normal phyllotaxis path, expressed by the sequence of normal phyllotaxis, and another phyllotaxis path, expressed by the sequence $\langle t, 2t+1, 3t+1, 5t+2, \ldots \rangle$ which coincide with the sequence of anomalous phyllotaxis [Sequence (6.5)] only when $t = 2$.

9.6.2. Comparison with the minimality condition of the τ-model

Assuming orthogonal intersection of the conspicuous pair and a divergence angle in the interval $[1/3, 1/2]$, Fibonacci phyllotaxis is unavoidable in the τ-model if $m<6$ (when five primordia exist; Proposition 5), for which $r \geq 1.8944/(2+3)^2 \approx 0.0758$. This value [obtained from the allometry Eq. (4.15)] is close to the mean of the end points of the interval given above for the (2, 3) pattern in Adler's model, that is $(0.0456+0.1237)/2 \approx 0.0844$. Under the same conditions, phyllotaxis (1, 2) is obtained if $r \geq 1.8944/(1+2)^2 \approx 0.2105$, a value approximately equal to the mean of the end points of the interval given for the (1, 2) pattern in Adler's model, that is $(0.1237+0.2888)/2 \approx 0.206$.

Adler (1977a) gives a set of conditions by which the system (2, 5) can become (7, 5) under contact pressure. Among these conditions is that r be in the interval $[\sqrt{3}/78, \sqrt{7}/84] \approx [0.0222, 0.0315]$. In that interval the opposed parastichy pair (2, 5) has an angle of intersection γ greater than 90° (for $\gamma = 90°$, the system (2, 5) has, by Equation (4.15), $r \approx 0.0386 > 0.0315$), and the system (2, 5) is replaced by the system (7, 5) for $r = \sqrt{3}/78$. This latter system is replaced by the system (7, 12) for the value of $r = \sqrt{3}/218 \approx 0.0079$. It follows that in Adler's model the system (7, 5) is orthogonal for $r \approx (0.0079+0.0222)/2 \approx 0.0151$. In the τ-model, at $m=12$ the system is (7, 5), and it is orthogonal when $r \approx 0.0132$, a value close to 0.0151. The above value $r = 0.0222$ corresponds to an angle of intersection γ of the conspicuous pair (7, 5), of about 63° [from Eq. (4.11)]. I also obtained the same angle of 63° in a discussion of Mitchison's computer simulation (Section 5.2.1). This angle corresponds to the point where a system (m, n) is replaced by a system $(m+n, m)$.

Another interesting parallel that we can draw between the two models is the following. In the τ-model, given m, a small interval of values for $r \approx r_m/m$ is obtained from Equation (4.11) giving $r_m = p(\gamma)/m^2$, for a γ fixed. An n being also given by the minimality condition, such that the pair $(m-n, n)$ is conspicuous, an interval of values for the divergence angle d is obtained (for example, by the contraction algorithm of Chapter 2). In the (d, r) phase space, the *τ-model confines the point (d, r) in a small rectangle.* As the minimality condition is maintained, the system experiences rising phyllotaxis, and the rectangle shrinks and moves towards the point $(d, 0)$ where d is the limit-divergence angle corresponding to the sequence $\langle m-n, n, m, m+n, 2m+n, \ldots \rangle$. Contact pressure gives a conspicuous pair $(m-n, n)$ for which $d(0, m-n) = d(0, n)$, so that *Adler's model confines the point (d, r) on an arc of circle inside the rectangle* mentioned above. If contact pressure is maintained, the point (d, r) moves on a sequence of arcs of circles, phyllotaxis rises, and r decreases towards 0 on the contact-pressure path.

9.7. Fujita's a priori spiral patterns

According to Fujita (1937) the types of spiral patterns to be found in nature are, a priori, expressed by the sequences

(1) $$J\langle 1, t, t+1, 2t+1, 3t+2, \ldots\rangle, \quad t \geq 1, \quad J \geq 1,$$
with a divergence angle of $(J(t+\tau^{-1}))^{-1}$,

(2) $$J\langle p, pa+1, (1+a)p+1, (1+2a)p+2, \ldots\rangle, \quad a \geq 2, \quad p \geq 2, \quad J \geq 1,$$
with a divergence angle of $(J(p+(a+\tau^{-1})^{-1}))^{-1}$,

(3) $$J\langle p, ap-1, (1+a)p-1, (1+2a)p-2, \ldots\rangle, \quad p \geq 3, \quad a \geq 3, \quad J \geq 2,$$
with a divergence angle of $(J[(p-1)+[(a-1)+\tau^{-1}](a+\tau^{-1})^{-1}])^{-1}$.

The first of Fujita's sequences is the already-known sequence (6.3) of normal phyllotaxis produced by the interpretative model. In Fujita's second sequence, the cases $p=2$ and $a=2, 3, 4$ have been observed. They correspond to the lateral sequences (6.5) of the interpretative model with $t=2, 3, 4$. Now the cases having a probability of existing according to the interpretative model are those where $p=2$ and $a \geq 2$, so that it is possible that the cases where $a=5, 6, 7, \ldots$ will be found in nature. The third a priori sequence of Fujita cannot exist according to the interpretative model.

Fujita's a priori sequences permit the existence of a wide variety of patterns. Fujita's set F of divergence angles form a proper subset of the set N of noble numbers. But we obviously do not need that many angles, and thus that many sequences in phyllotaxis. This possibility of so many angles might be an inherent limit for a model. As we stated earlier from Popper, a good scientific theory forbids certain things to happen. On the other hand, a model that cannot be used to reproduce patterns that are known to exist shows another type of limitation. The set **J** of angles permitted by the interpretative model form a proper subset of the set F, so that $C \subset \mathbf{J} \subset F \subset N$. We know from Chapter 7 that the set **J** covers all observable spiral cases. It is conjectured that the τ-model produces (and orders) all the elements of N (certainly it produces the elements of **J**).

For unknown reasons, Adler considers the sequence $J\langle t, 2t+1, 3t+1, 5t+2, \ldots\rangle$, $t \geq 2$, as corresponding to Fujita's second sequence with $p=t$ and $a=2$. In this sequence, putting $n = F_{k+1}t + F_{k-1}$ and $m = F_{k+2}t + F_k$, the divergence is in the bounds F_{k+1}/n and F_{k+2}/m so that at the limit the divergence is $d = (J(t+\tau^{-2}))^{-1}$. This divergence decreases monotonically toward zero as t increases, just like the divergence corresponding to the sequence of normal phyllotaxis. The divergence corresponding to anomalous phyllotaxis in the interpretative model increases toward 180°, 90°, 60°, ... as t increases, depending on the values $J = 1, 2, 3, \ldots$, respectively. This makes possible the existence of alternating whorls (distichy: 180°; decussation: 90°, ...) and of spiro-alternating whorls. The sequence of normal phyllotaxis makes possible the existence of superposed whorls, such as monostichy, via spiromonostichy (Chapter 8). Adler's a priori sequence does not make possible the existence of alternating whorls [at least in the manner seen in Figure 8.2(1)], and does not cover the lateral sequences (6.5) known to exist (e.g., $\langle 2, 7, 9, 16, \ldots\rangle$; see Table 6.2).

Epilogue

An important part of the author's theory of phyllotaxis (see the introduction to Part II) is the model of pattern generation presented in Chapter 6 and developed in Chapter 8. The model proposes an entropy-like function and is used to interpret the phenomenon of phyllotaxis functionally, in terms of its ultimate effect assumed to be the minimization of the entropy of plants. A principle of optimal design is used to discriminate among the phyllotactic spiral lattices represented by the control hierarchies. Every control hierarchy can be attributed a complexity $X(T)$, a stability $S(T)$, a rhythm w, and a cost E_b (a real-valued function defined on the set of hierarchies and called the bulk entropy). Algorithms derived from the model can be used to identify the spiral pattern giving minimal cost when a cycle of primordial initiation is completed (at P_r plastochrones). For example, the costs of the patterns $\langle 1, 2\rangle$, $2\langle 1, 2\rangle$, and $\langle 3, 4\rangle$ are approximately equal to 1.08, 1.98, and 2.83, respectively. According to the model, the Fibonacci pattern is the most stable and the least complex, thus its overwhelming presence.

The model is developed for spiral patterns, but takes into account whorled patterns as well. It is used to generate the regular patterns and shows that multimery is a special case of multijugy. Whether spirality is primitive or not, the model supports the idea that the patterns must be modeled in an order comparable to their evolutionary order of appearance in nature. An evolutionary order has been sketched, going from Fibonacci patterns to normal and anomalous spiral patterns, and to alternating and superposed whorls. This model can be used to deal with the perturbed patterns (Section 8.6.3).

The set **J** of divergence angles (a subset of the set N of noble numbers) is obtained from the model, together with predictions that can be directly compared to the data available (Chapter 7). With the model we can predict the existence of apparently improbable patterns, and say which types of patterns can and cannot exist. We can also settle difficult problems of pattern recognition in phyllotaxis, thus complementing the allometry-type model used in Part I for practical pattern recognition. Comparisons have been made in Chapter 9 with other models built on entirely different frameworks. The priority orders of spiral patterns, given by the Marzec entropy model, the τ-model, the XuDong et al. model, and the interpretative model, have been shown to be very similar to one another. It is believed that a model can be tested by the predictions, the by-products and the new interpretations we can obtain from it.

Part III will show that primordial patterns observed in plants can be observed in other areas of nature as well. Pattern formation in plants must therefore transcend, in some way that remains to be clarified, the strictly botanical substratum. Consequently, models of pattern generation in phyllotaxis must not be based too much on strictly botanical devices or material, such as diffusion of phytohormones. They must introduce more formal, systemic concepts that are able to explain similar patterns wherever they arise, and that can be given particular meanings depending on the area where they are applied. Part II is the expression of the belief that entropy is one such concept, and Section 6.1.2 underlines the fact that ideas of entropy–energy are gradually imposing themselves on the field of phyllotaxis.

Part III
Origins of phyllotactic patterns

Introduction

The problem of the origins of phyllotactic patterns can be dealt with at many different levels. In the study of higher plants Bolle's theory represents a preliminary level, and Zimmermann's theory a second level which proposes to go back to the ancestral land plants. Church's perspective on phyllotaxis proposes to go back to the sea to understand the primary meaning and function of the patterns found in higher plants. The first three levels have been considered in Chapters 3 and 8.

We have seen in Part I that phyllotactic patterns are branched structures and hierarchies. They were present at the beginning of the evolution of land plants, and are found in brown algae as well. Branching is one of the most fundamental growth processes, present in the structure of the sunflower, one of the most striking example of phyllotaxis (see Section 3.3) and of floral evolution. But these branched structures are also seen in minerals and animals, so that it is not surprising to see phyllotaxis-like patterns in other areas of nature, as Part III will show.

Chapter 10 presents my application of phyllotactic methods and of the theorem of Chapter 2 to protein crystallography, and makes predictions of what should be an exact representation of patterns of amino acid residues in polypeptide chains. The chapter examines the work of contemporary botanists and crystallographers who understood the relevance of the crystallographic paradigm in phyllotaxis. It is mainly devoted to showing the success of crystallographic phyllotaxis or phyllotactic crystallography.

Systems research, whose importance I have always stressed, necessarily brings interdisciplinary comparisons of forms and functions. Chapter 11 shows that comparative morphology in general is relevant for the understanding of phyllotaxis. This chapter deals with Meyen's morphogenetical parallelism on the isomorphisms between forms and patterns in nature in general, representing the fourth level of investigation on the origins of phyllotactic patterns.

Lima-de-Faria's autoevolutionism offers a fifth level of investigation, which can help to better explain the origins of forms and functions in biology in general. This is a new approach to evolution, a recent alternative to neo-Darwinism, although still a minority opinion at the fringes of official circles of thoughts and of the mainstream. Autoevolutionism states that the origins of forms and functions in biological structures must be sought at the levels of evolution that preceded biological evolution, and that left their "imprints" on the latter. That phyllotaxis is now considered by many as generalized crystallography is a way to state and to recognize that there is a physical organization of matter that operates functionally in the genesis of phyllotactic patterns. In order to find the best explanation for phyllotactic patterns it is argued in Chapter 11 that we must take into consideration the fundamental properties of space–time, created by the evolution of elementary particles, chemical elements, and minerals.

By remaining close to neo-Darwinism, the importance generally given to the gene prevents vision beyond it, and thus it is not possible to go beyond Zimmermann's and Church's theories (though of course these two authors are not necessarily genocentric). Beyond level 3, morphogenetical parallelism in nature, and the tenets of autoevolutionism that regard the gene as of secondary importance, allow us to make significant steps toward solving the mystery of phyllotaxis. The reason is that in doing so we become able to link the existence of phyllotactic patterns to the general laws of nature, and the structure of the universe. Among these laws are those of branching (Chapter 3), of allometry (Chapter 4), of homology (Chapters 10, 11), of gnomonic growth (Section 11.3), and of emergence (Section 11.4).

Chapter 12 comprises a reflection on the actual status of the subject of phyllotaxis, which is presented as a multidisciplinary field of research. Considering the general dissatisfaction generated by old lines of thought (Section 12.1), this chapter stresses the need to develop new avenues of thinking. It once again emphasizes that phyllotactic patterns come from more universal laws than it is usually postulated, laws that had initially nothing to do with living matter.

10
Exotic phyllotaxis

10.1 Historical meeting again

Past chapters in this book very often mention Bravais and Bravais who successfully worked on the problem of phyllotaxis in the 1830s (Bravais–Bravais theorem, lattice, formula, etc.). One of the reasons for this success might be that these brothers were a botanist and a crystallographer. In those times living organisms were generally perceived as living crystals. That the crystallographic paradigm was shortly abandoned after the Bravais brothers may explain the relatively poor development of the subject of phyllotaxis in the second half of the nineteenth century. Van Iterson (1907) introduced geometrical methods of analysis of phyllotactic patterns based on symmetry theories of crystal structures. His work received very little attention and was revived to some extent in the second part of the century by botanists and mathematicians dealing with phyllotaxis in terms of contact circles on cylinders. The fact that crystallographers have rejoined the effort very recently might be an unconscious collective historical recognition that the discipline was from its very beginning under a good omen, and that the intuition of the initiators of the cylindrical treatment of phyllotaxis put the subject on the right track.

For crystallographers all aspects of the challenge of phyllotaxis are excluded except the geometrical aspect, and phyllotaxis is identified with the study of spiral lattices. For example, using digitized coordinates they made on 39 sunflower-seed packings and curve-fitting tools, Ryan et al. (1991) showed that *Helianthus tuberosus* (12–15 mm in diameter) can be best fitted with the golden spiral, while *Helianthus annuus* (25 cm in diameter) seems to require a combination of the golden and logarithmic spirals.

The study made by Rothen and Koch (1989b) uses a logarithmic spiral along which primordia are considered as tangent nonoverlapping circles. The logarithmic spiral is given more precisely by $r = r(0)R^{\theta/2\pi d}$, and the lattice determined by the spiral is the discrete set of points $r(n) = r(0)R^n$ and $\theta(n) = 2\pi nd$, where $n = \ldots, -1, 0, 1, 2, 3, \ldots$, $d = (\theta(n+1) - \theta(n))/2\pi$ is the constant divergence angle, and $R = r(n+1)/r(n)$ is the constant plastochrone ratio.

Crystallographers are rediscovering in their own terms already-known results. For example, Rothen and Koch (1989a,b) redefined the treatment of packing of spheres and continuous fractions in phyllotaxis. Among their results, starting with d and R arbitrarily fixed they found conditions for which the circle centered at k is tangent to the circles centered at $k+n$ and $k+m$ giving the parastichy pair (n, m). Thus, given d all pairs (n, m) can be obtained, and given (n, m) an interval of values for d is obtained. For close packing of circles (when each circle has six neighbors instead of four), d and R are uniquely determined by n and m, and the third tangent circle is $m+n$ (reminiscent of Adler's contact-pressure model of circles in the plane, 1977a). If (n, m) is replaced by $(m+n, m)$, we have rising phyllotaxis and continuous transitions leading to noble numbers. Rothen and Koch obtain a figure showing R as a function of d [combine for example Eqs. (4.2) and (4.4) for various values of m and n to obtain such a function], giving the familiar cascade of possible transitions between patterns found in many studies [e.g., Maksymowych & Erickson, 1977; Veen & Lindenmayer, 1977; see Fig. 12.1(2)]. Rothen and Koch's papers also overlap with those of van Iterson (1907), Ridley (1982a), Marzec and Kappraff (1983), and Jean (1988a).

In Section 9.1.2 we described the original developments by Rothen and Koch on the idea of self-similarity. The treatment proposed by XuDong et al. (1988), using digitized data from botanical specimens, Fourier transforms, and power spectra, is striking (see Fig. 10.1). Beyond these efforts, a generalization of classical crystallography is taking place along the lines developed by Mackay (1986), and phyllotaxis is now considered to be an important subject of study, somewhat similar to crystal structure analysis (see Appendix 1 under the term "crystallography"). Thus, the spiral-lattice structure in the seed arrangements of the sunflower becomes a natural example of a **quasicrystal** which lacks translational and rotational periodicities. Crystallographers necessarily have a pattern theory to share, which is why they are bringing a fresh view on phyllotaxis, the study of patterns on plants.

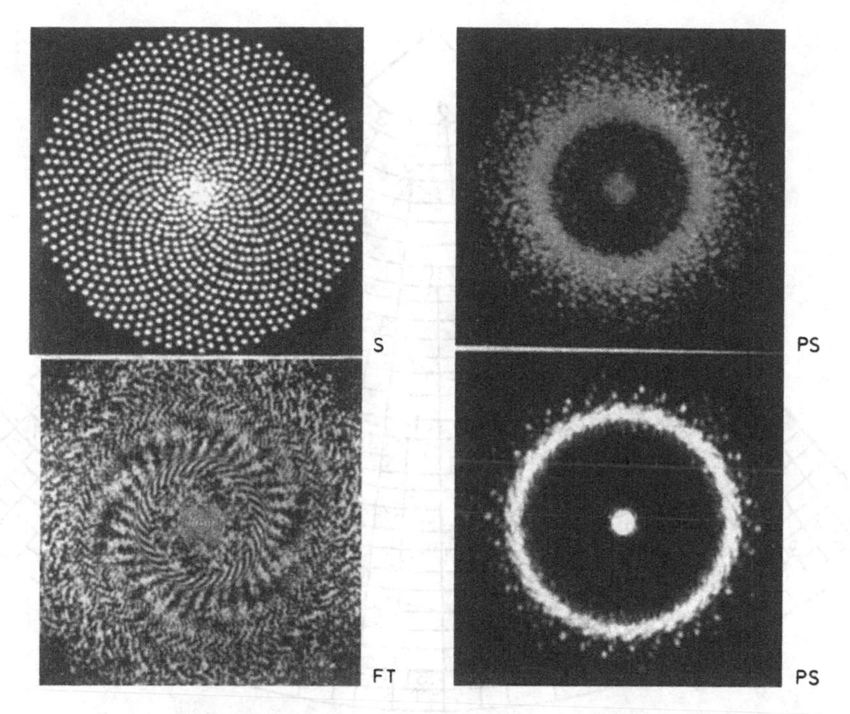

Figure 10.1. Spiral lattice model (with golden generative spiral and Fibonacci angle), its Fourier transform and power spectra (two different contrasts) generated by computer simulation. (From XuDong et al., 1988.)

10.2. Phyllotaxislike patterns

10.2.1. In biology

Spiral patterns can be observed in such living systems as *Hydra* and *Botryllus* where the overall process of column growth and bud formation is strictly homologous to that of leaf primordia (Berrill, 1961, pp. 228, 423). The following diagrams represent various structurally similar systems, beginning with a quadrant of a jellyfish showing a pattern of tentacles (Fig. 10.2). In phyllotaxis this pattern is described as representing a 4(2, 3) visible opposed parastichy pair. Indeed in each of the 4 quadrants, there are two families of spirals containing 2 and 3 spirals, respectively. Each family goes through all the tentacles, and winds in opposite directions with respect to the center of symmetry; at every intersection of two spirals there is a tentacle, and considering any spiral, the tentacles are

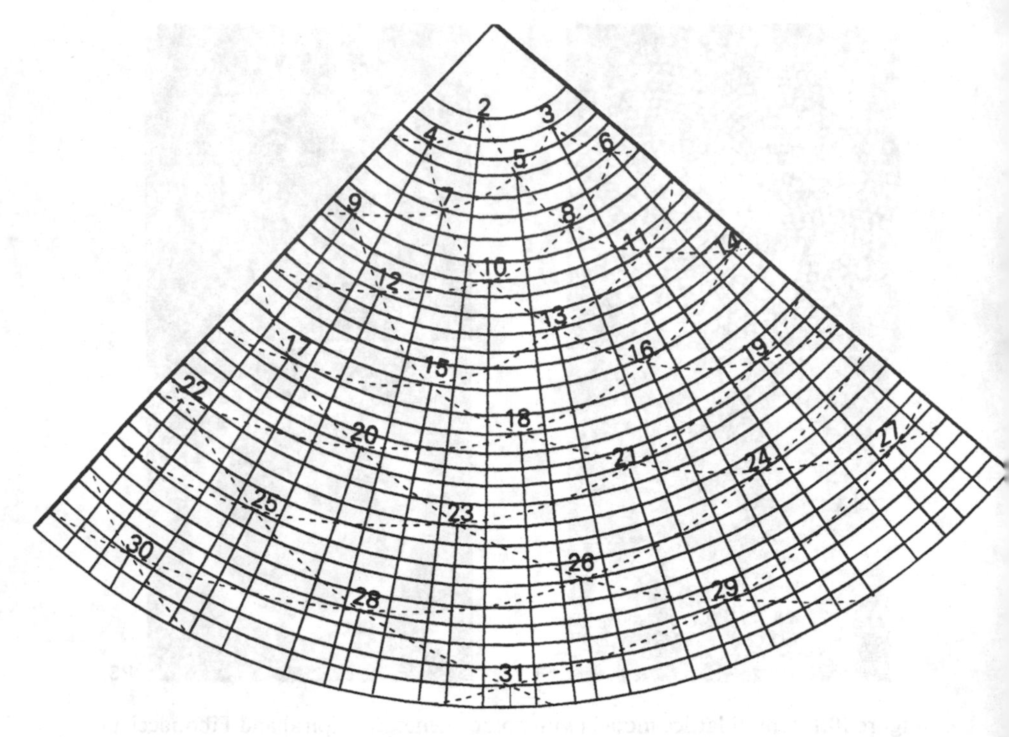

Figure 10.2. Schematic (centric) representation of the arrangement of ex-umbrellar tentacles in a quadrant of *Olindias formosa*. The order of development of tentacles in *Gonionemus* also conforms with this scheme. Diagram shows a (2, 3) visible opposed parastichy pair. (From Komai & Yamazi, 1945.)

linked from 2 to 2 or from 3 to 3. The pair (3, 5), not depicted in the diagram, is also a visible opposed pair.

One very good illustration of phyllotaxis is the pattern in capituli of Compositae (e.g., daisy, sunflower, chrysanthemum) such as the one shown in Figure 10.3(1). The capitulum pattern has a striking resemblance to that of the diatom in Figure 10.3(2). In the Division of Phycophyta (algae) we find the very small, unicellular, brown diatoms, and the very large brown *Fucus*. We have seen in Chapter 3 (e.g., Fig. 3.7 & 3.13) that the ramified structure of *Fucus* is linked to the structure of the capitulum.

10.2.2. In cylindrical crystals

Microtubuli, bacterial flagella and pili, capsids of viruses, fibers of chromatins, and tubular assemblies of enzymes are examples of tubular arrays

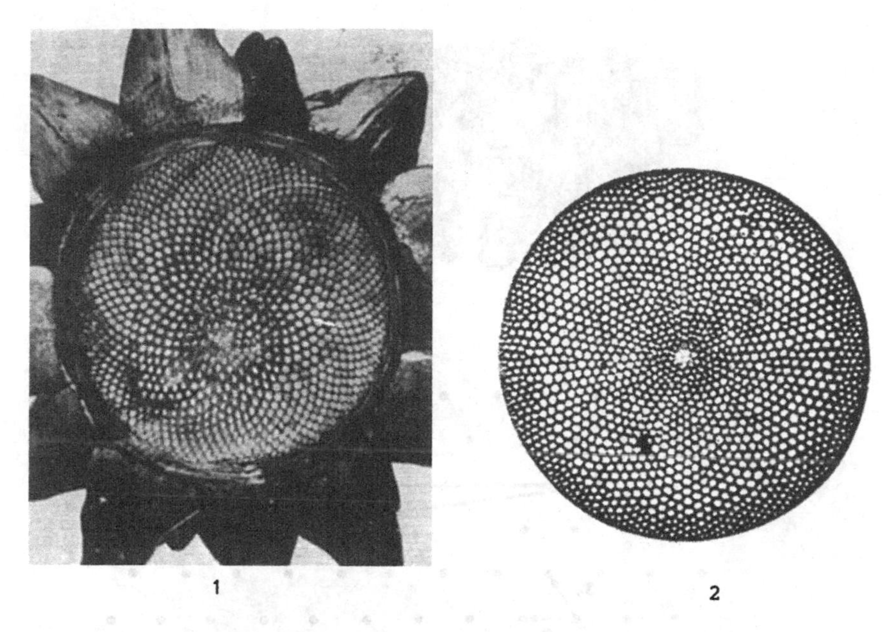

Figure 10.3. (1) Capitulum of *Cynara scolymus* shaved down to the ovaries of the disk-florets to show the phyllotactic system (55, 89). (From Church, 1904b.) (2) Valve of *Coscinodiscus*, a bacillariophyta universally distributed in freshwater, seawater, and soil. (From von Denffer et al., 1971.)

of spheres. A detailed analysis has been made of the geometry of these arrays (e.g., van Iterson, 1907; Erickson, 1973a). They are examples of what Harris and Scriven (1970) term "cylindrical crystals," such as the one shown in Figure 10.4. These structures apparently have in common the fact that they consist of protein subunits of a small number of species and are capable of self-assembly. They can show transitions in patterns, analogous to edge dislocations in crystals. Harris and Erickson (1980) made an extensive analysis of the properties of such structures largely from a crystallographic point of view. They use the botanical term "parastichy," and they introduced the terms "continuous" and "discontinuous contractions" to describe what in these idealized tubular representations are obviously analogous to phyllotactic transitions. That is why we used these terms in Section 8.3.1.

Figure 10.5(1) is a photograph of the submicroscopic cylindrical structure of a bacteriophage. It presents, as Erickson (1973a) demonstrated using the methods and terms of phyllotaxis, the arrangements 4(2, 3) and 4(3, 5) of its hexamers [Fig. 10.5(2)]. He showed also that the flagellum

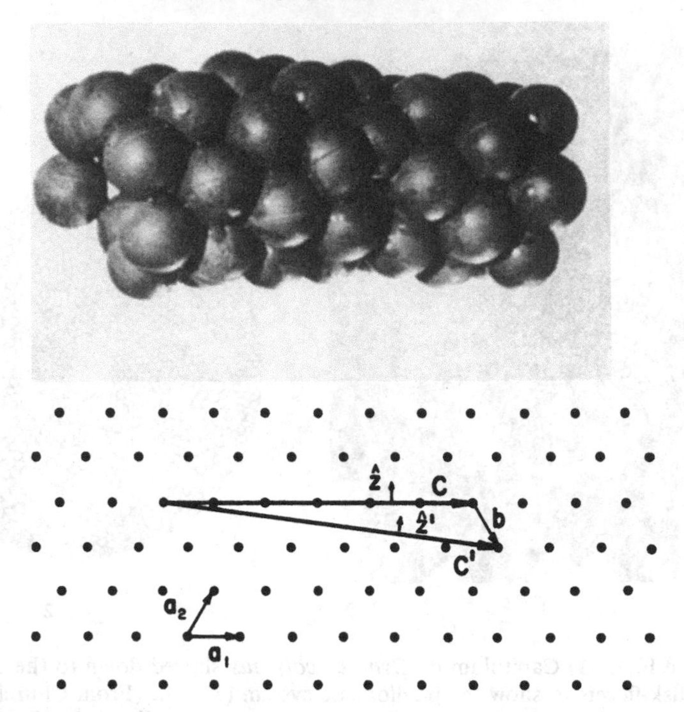

Figure 10.4. Dislocated cylindrical crystal made from rubber balls and its planar representation. (From Harris & Scriven, 1971.)

of the bacteria *Salmonella* presents the patterns 2(2, 3) and 2(3, 5), and that tobacco mosaic virus displays the visible opposed pairs (1, 16) and (17, 16) [Fig. 10.5(3)]. Many microtubules display the cylindrical packing patterns (6, 7) and (7, 13) of protein monomers.

10.3 Structure of polypeptide chains

10.3.1. Protein crystallography and systems research

The elucidation of the three-dimensional structure of protein molecules is a landmark in molecular biology. Knowledge of their structure comes from the interpretation of X-ray diffraction patterns. To determine a molecular structure, an X-ray beam is sent through a crystal of the protein molecule. It is diffracted (scattered) and a detector (e.g., a film) registers the pattern of diffraction, such as the one shown in Figure 10.6(1). This photograph of myoglobin is a bidimensional section of a three-dimensional pattern of points. The mathematical analysis of the diffraction is very

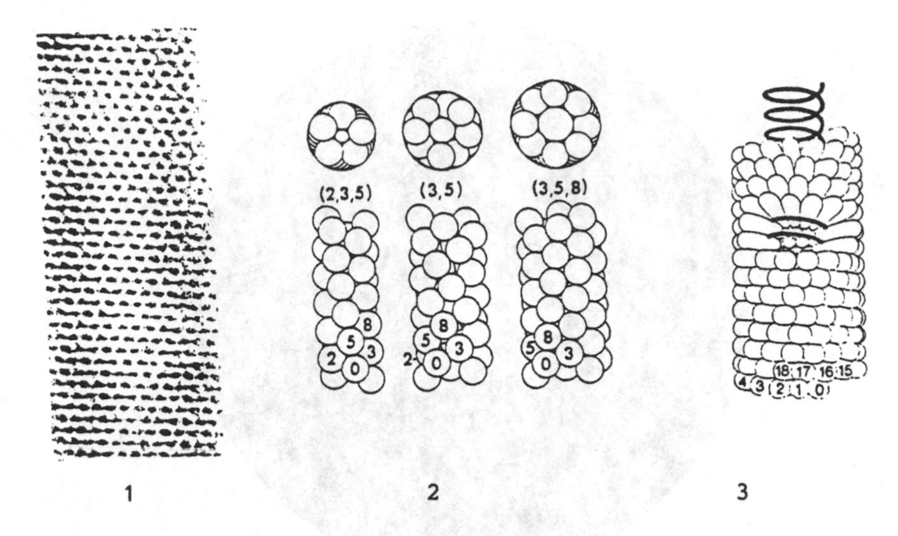

Figure 10.5. **(1)** Optically filtered image of the polyhead of a bacteriophage showing a 4(2, 3, 5) arrangement of its hexamers. (From DeRosier & Klug, 1972.) **(2)** Packings of spheres on cylinders, with which Erickson (1973a) determined the "phyllotaxis" of such patterns using published high-resolution electron micrographs and optical diffraction studies. **(3)** Model of protein coat of tobacco mosaic virus showing the triple points pattern (1, 16, 17). (From Lauffer & Stevens, 1968.)

complex because of the large number of atoms in the molecule, which provide thousands of diffraction spots. The intensity of each spot is measured, and the problem is then to reconstruct an image of myoglobin from these intensities. This is done using Fourier series, by considering the phases of the diffracted X-rays. An electron-density map [Fig. 10.6(2)], comparable to the level curves in geography, is traced using a high-speed computer. The final Fourier synthesis contains about one billion terms. It gives the density of the electrons in a great number (25,000) of regularly spaced points in the crystal. The next step is to interpret this chart. The three-dimensional distribution is represented by a series of parallel sections placed one above the other.

Proteins consist of one or many polypeptide chains, each possessing 100 or more amino acid residues. Frey-Wyssling (1954) reported angles of 5/18, 8/29, and 13/47 between consecutive amino acid residues in α-polypeptide chains (poly-*l*-alanine, poly-γ-methyl-*l*-glutamate). He noticed that these fractions belong to the sequence

$$\langle 1/3, 1/4, 2/7, 3/11, 5/18, 8/29, 13/47, \ldots \rangle \qquad (10.1)$$

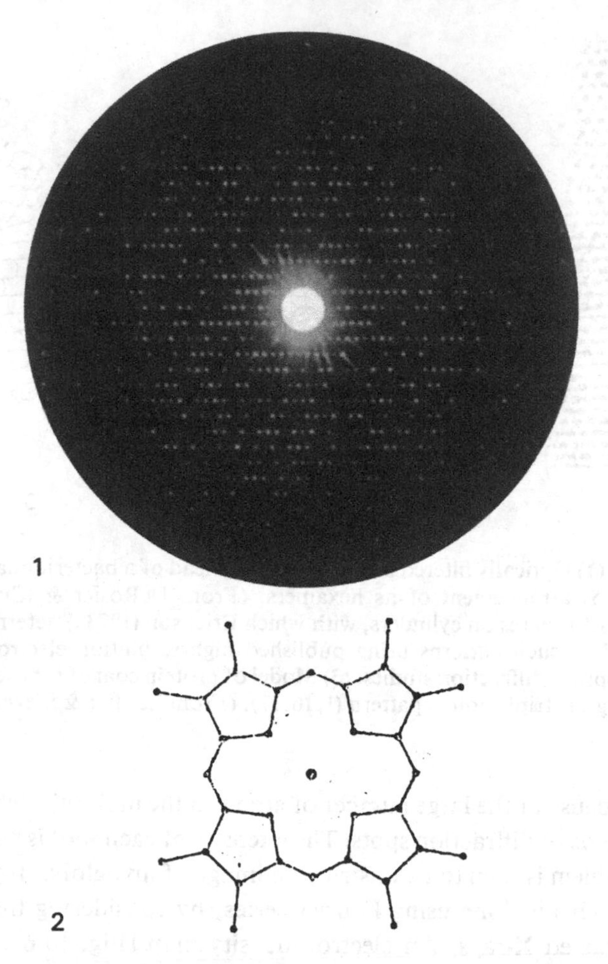

Figure 10.6. (1) X-ray photograph of a normal crystal of myoglobin, showing a pattern of spots. (2) Section from an electron-density map of myoglobin, with its skeleton group superimposed and the iron atom in the middle. (From Kendrew, 1961.)

where each term is the Farey sum of the two terms immediately preceding it. Pauling and Corey (1951) reported a divergence of 3/11 in poly-γ-benzyl-*l*-glutamate.

Abdulnur and Laki (1983) proposed a graphic representation of the α-chain of tropomyosin found in rabbit skeletal muscle (Fig. 10.7). The authors also illustrated with a similar map the α-helical region of influenza virus hemagglutinin HA2 chain. In Figure 10.7 the constant angle

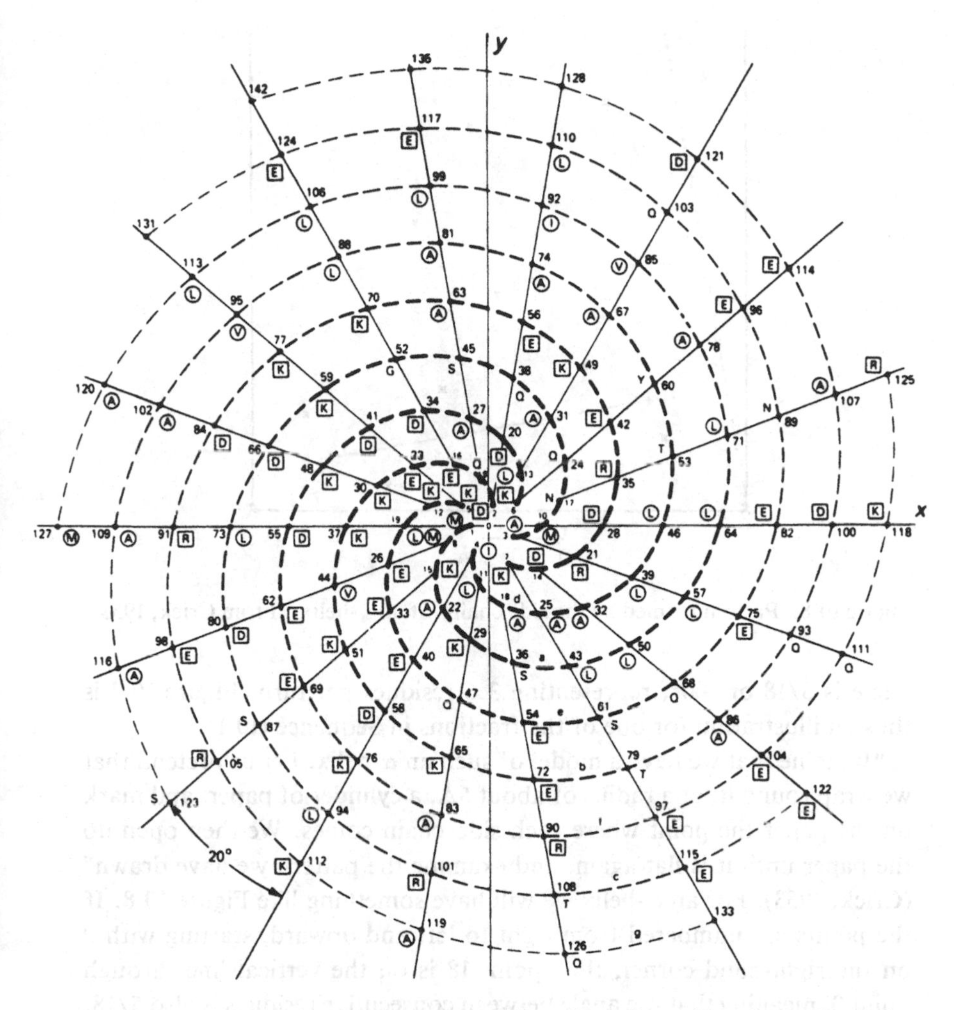

Figure 10.7. Two-dimensional representation of relative orientations of α-helix residues. The numbers represent the amino-acid residues, and the letters the peptides. (From Abdulnur & Laki, 1983.)

between consecutive residues can be calculated in a variety of ways. For example, between residues n and $n+1$ there are five rays and 20° between each ray. Another method essentially consists in looking for a phyllotactic fraction. Two superposed residues, on any of the 18 rays coming from the center of the figure, are located. Then we go from one to the other via all the intermediate residues and by the shortest path. By doing so we make 5 turns around point 0 and meet 18 residues. It follows that the

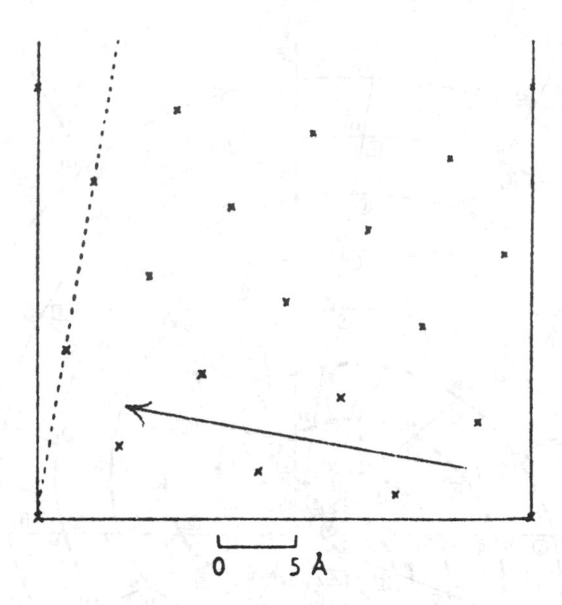

Figure 10.8. Pattern formed by the side chains of an α-helix. (From Crick, 1953.)

angle is 5/18 or 100°, representing 3.6 residues per turn. Figure 10.7 is thus an illustration for one of the fractions in Sequence (10.1).

"Imagine that we have a model of such an α-helix. Let us pretend that we wrap round it, at a radius of about 5Å, a cylinder of paper, and mark on the paper the point where each side-chain comes. We then open up the paper until it is flat again, and examine the pattern we have drawn" (Crick, 1953). For an α-helix we will have something like Figure 10.8. If the points are numbered from right to left and upward, starting with 0 on the right-hand corner, then point 18 is on the vertical line through point 0, meaning that the angle between consecutive residues is also 5/18.

The two representations (Fig. 10.7 & 10.8) are essentially the same, though visually very different. In one case we have what is called in phyllotaxis the centric representation of the chains, and in the other we have the cylindrical representation, as it is obvious from Crick's description. This means that the geometrical representation of the chains has not been improved in the intervening 30 years, even if the literature in the 1950s refers to "angles of about 20°," and of "about 3.6 residues by turn." These authors do not mention phyllotaxis where well-established results can contribute to the analysis of the patterns determined by the residues in the chains, as an example of what systems research is about.

10.3.2. Mathematical analysis of protein lattices and predictions

The authors in the references of Figures 10.7 and 10.8 retained the fraction 5/18, while there are many more fractions in Sequence (10.1). In Figure 10.7 there are families of spirals linking the residues by 11 and by 29, but for biological reasons the authors chose to draw the family of 7 spirals linking the residues by 7. In Figure 10.8 the pair (3, 4) is conspicuous, points 3 and 4 being the nearest to 0. We come up with the Sequence (10.2) of integers, that is Schoute's first accessory or the Lucas sequence of phyllotaxis

$$\langle 1, 3, 4, 7, 11, 18, 29, \ldots \rangle, \tag{10.2}$$

where the consecutive terms are the denominators of the fractions in Sequence (10.1).

The accuracy of the interpretation of the X-ray diffraction patterns can be improved. The properties of the spiral or cylindrical lattice are now well known from phyllotaxis. The fundamental theorem of Chapter 2, more precisely a particular form of it [Sequence (2.1) with $J = 1$ and $t = 3$], shows the link between Sequences (10.1), (10.2), and the data. The fractions in Sequence (10.1) are the end points of the intervals given by the theorem (see Problem 2.2). The limit-divergence angle is $d = (3 + 1/\tau)^{-1}$, that is, approximately 0.28, which when multiplied by 360° gives 99.51°. This angle differs by half a degree from the value proposed by protein analysts. There is thus a good agreement between the theoretical value and the experimental approximation. But at the molecular level a difference of 0.5° could be much more meaningful than in Figure 9.1 where it was observed that a difference of less than 0.1° leads to improbable structures.

The theorem of Chapter 2 allows us to predict what should be the exact representation of the position of amino acid residues in α-polypeptide chains, hopefully opening the way for new observations. In Figure 10.7 we can see that the pairs (3, 1), (3, 4), (7, 4), (7, 11), (18, 11), and (18, 29) are visible and opposed. The points 1, 3, 4, 7, 11, 18, 29 alternate on each side of the negative y axis. But the rhythmic alternation of the terms of Sequence (10.2) breaks off after 29. The terms of Sequence (10.2) should not only alternate on either side of the axis but they should get closer to it. On the contrary, in Figure 10.7, points 29, 47, 76, and 123 are on the same side of the axis and go away from it. A correct representation would show the residues distributed along the counterclockwise (genetic) spiral by a (divergence) angle of approximately 99.51°.

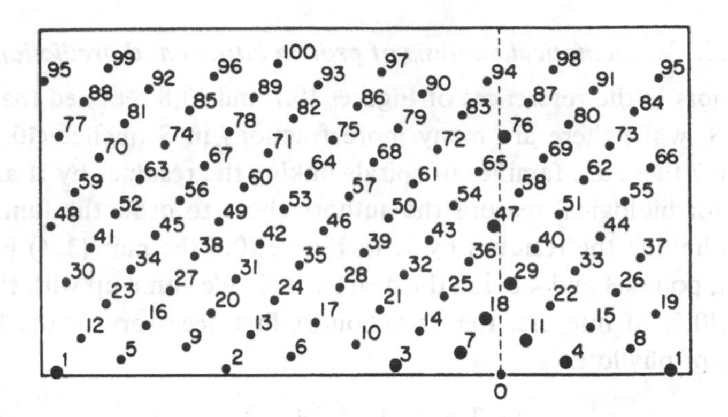

Figure 10.9. Proposed geometrical model for α-polypeptide chains, corresponding to Crick's interpretation of the relative arrangement of amino acid residues in the chains. (From Jean, 1985a.)

Figure 10.9 is the cylindrical representation of the model proposed here, corresponding to a (7, 11) conspicuous parastichy pair. The figure shows the required alternation of the members of (10.2) along the vertical axis. The coordinates of point 1, determining the regular lattice, are

$$(d,r)=((3+1/\tau)^{-1},1/200).$$

The coordinates of point n are $(n(3+1/\tau)^{-1}-[n(3+1/\tau)^{-1})],n/200)$, where $[\cdot]$ means "the integer nearest to." The abscissa is what is called in phyllotaxis the secondary divergence of the point. The unit used for the divergence is the horizontal distance between two representations of the same point (e.g., 1 or 95). Putting $r=1/50$ would bring points 3 and 4 nearest to 0, for a (3, 4) conspicuous pair as in Figure 10.8. Note that point 18 is not on the vertical axis anymore as in Crick's drawings. Equation (4.5) transforms the cylindrical representation into the spiral lattice corresponding to Figure 10.7. The 18 rays in Figure 10.7 should be 18 spirals, as in Figure 1.5. The vertical axis in Figure 10.9 corresponds to line *OC* in Figure 1.5.

Whether it be the study of proteins or phyllotaxis, the two fields of research involve a similar packing problem. The divergence angle proposed for Figure 10.9 induces self-similarity and packing efficiency (Section 9.1). This bridge between the structure of proteins and patterns in higher plants could prove to be more fundamental than described here.

Marvin (1989, 1990) used phyllotactic methods to analyze the symmetry of the filamentous bacteriophage called *Inovirus*. He believes that the

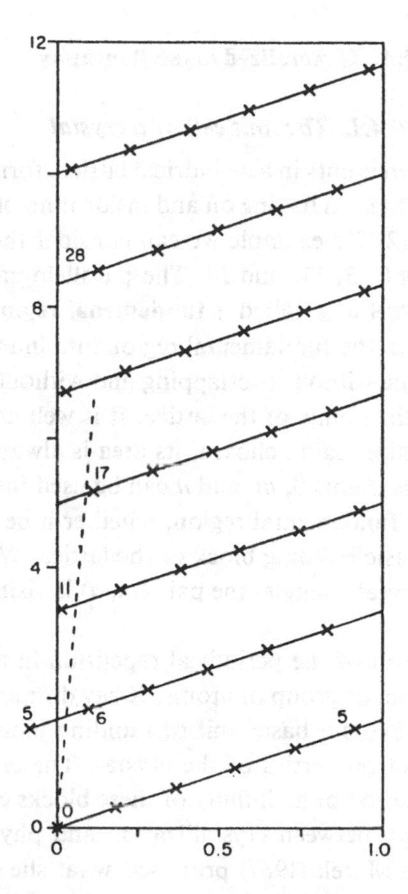

Figure 10.10. Cylindrical lattice representing the symmetry of *Inovirus*. Dotted line called the phy-line (for phyllotaxis line) represents the axis of α-helix associated with protein unit 0. Solid line is called the basic helix (corresponding to the genetic spiral). There are 5.618 units per turn (called the phy-turn). This corresponds to the limit divergence angle of normal phyllotaxis $(5+\tau^{-1})^{-1}$. (From Marvin, 1989; reprinted by permission of the publishers, Butterworth Heinemann Ltd, ©.)

symmetry of α-helix subunit arrangement in the virion is related to the symmetry of leaf arrangement in plants. The projection of the α-helix axis onto the cylindrical lattice is shown to run between the units indexed 1, 5, 6, 11, 17, 28, ... of the third accessory sequence (see Fig. 10.10). As I predicted in Figure 10.9 for the first accessory sequence, the units alternate from one side to the other of the projected α-helix axis, and they approach increasingly close to the axis. Marvin concluded that *Inovirus* symmetry is a generalization of phyllotaxis symmetry.

10.4. Generalized crystallography

10.4.1. The unit cell of a crystal

Let m and n be any two points in a cylindrical lattice, forming with points 0 and $m+n$, a parallelogram having on and inside it no other point of the lattice. In Figure 2.3(2) for example we can consider the points 0, 5, 8, and 13, or the points 0, 5, 13, and 18. The parallelogram made by the points 0, m, n, and $m+n$ is called a **fundamenal region of the lattice**. Translations transform the fundamental region into infinitely many such regions filling the plane without overlapping and without interstices, and their vertices are all the points of the lattice. It is well known that whatever be the fundamental region chosen, its area is always the same. The triangle made with the points 0, m, and n can be used just as well to generate the lattice. The fundamental region, whether it be a parallelogram or a triangle, is the basic building block of the lattice. When 0, m, and n determine a fundamental triangle, the pair (m, n) is visible (as defined in Appendix 4).

A crystal is the result of the periodical repetition in the three-dimensional space of an atom or group of atoms. X-ray diffraction of any crystal shows that there exists a basic unit or building block having all the physical and chemical properties of the crystal. The crystal is thus the result of the superposition of an infinity of these blocks creating a lattice. By virtue of similarity between crystallization and phyllotactic pattern formation, Zagorska-Marek (1987) proposed what she called the PTU, for "phyllotaxic triangular unit," an homology to the primitive unit cell of crystallography. The PTU is simply the fundamental region made by the points 0, m, and n, where (m, n) is the orthogonal conspicuous, or the contact, parastichy pair of the cylindrical lattice. This fundamental region represents a group of leaf primordia which upon repetition, as in crystals, yields the observed phyllotactic pattern. In Figure 10.11, the PTU is the triangle made by the centers 0, m, and n of the primordia represented as tangent circles.

10.4.2. Multimery, multijugy, and transitions revisited

The PTU is considered to be the element common to all phyllotactic patterns $J(m, n)$, spiral $(n<m)$ or whorled $(m=n)$, depending on the angle μ defined in Figure 10.11. When $\mu=0$ we have J-mery (Section 8.1). When $\mu>0$ we have J-jugy (Section 6.4.2). Figure 10.12, which shows the close

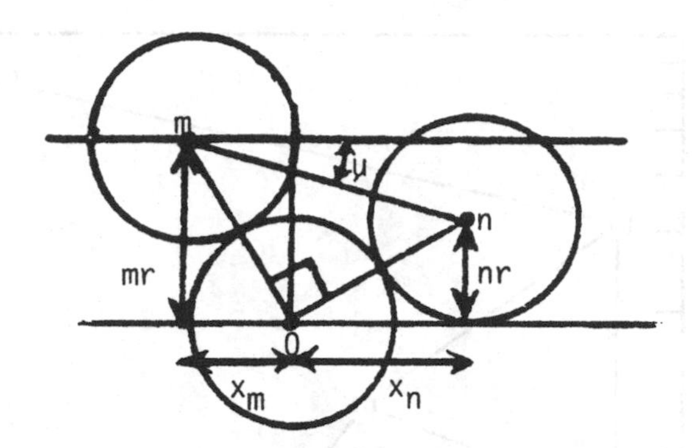

Figure 10.11. The PTU for an orthogonal (m, n) pattern of tangent circles, $n<m$, in a cylindrical lattice; r represents the rise, and the secondary divergences of points n and m are represented by x_n, and x_m, respectively. The base of the fundamental region is defined by the $(n-m)$-parastichy and is oriented at angle μ relative to the horizontal level in the lattice.

relationships between the various patterns, is based on Richards's formula [Eq. (4.30)] or correspondingly to Equation (4.6) with $\gamma=90°$, and on the trivial formula $\tan\mu=(m-n)r/(|x_m|+|x_n|)$. In the figure the phyllotactic patterns $J(m, n)$ are represented as points in the μ-r phase space. The lines interconnecting the points represent elementary edge dislocation pathways which transform the various patterns into one another via changes in the factors. According to the diagram, all types of transitions are theoretically possible, though not all of them have been found in nature. The diagram supports the conclusion of Section 8.5.4 that multimery is a special case of multijugy.

To explain discontinuous transitions, the possibility of a local change in the circumference of the apex has been invoked in Section 8.3.2. The question whether such a change is possible is answered in Zagorska-Marek (1987), by making reference to wedge disclinations present in bubble-raft crystals (Ishida & Iyama, 1976) floating on the surface of water (see Figure 10.13). These rafts are considered by her to be a good model for the group of cells making up the top of the apical dome. Like any close-packing of spheres, the array of bubbles has sixfold rotational symmetry, each bubble being surrounded by six other bubbles. This can be altered by disclination, that is, by removal or insertion of a wedge 60° wide (equilateral triangle of bubbles). Then the rotational symmetry of the structure changes to fivefold or sevenfold, and the central bubble has five or seven neighbors. Similar wedge disclinations would be present at the shoot apex when the number of initials changes.

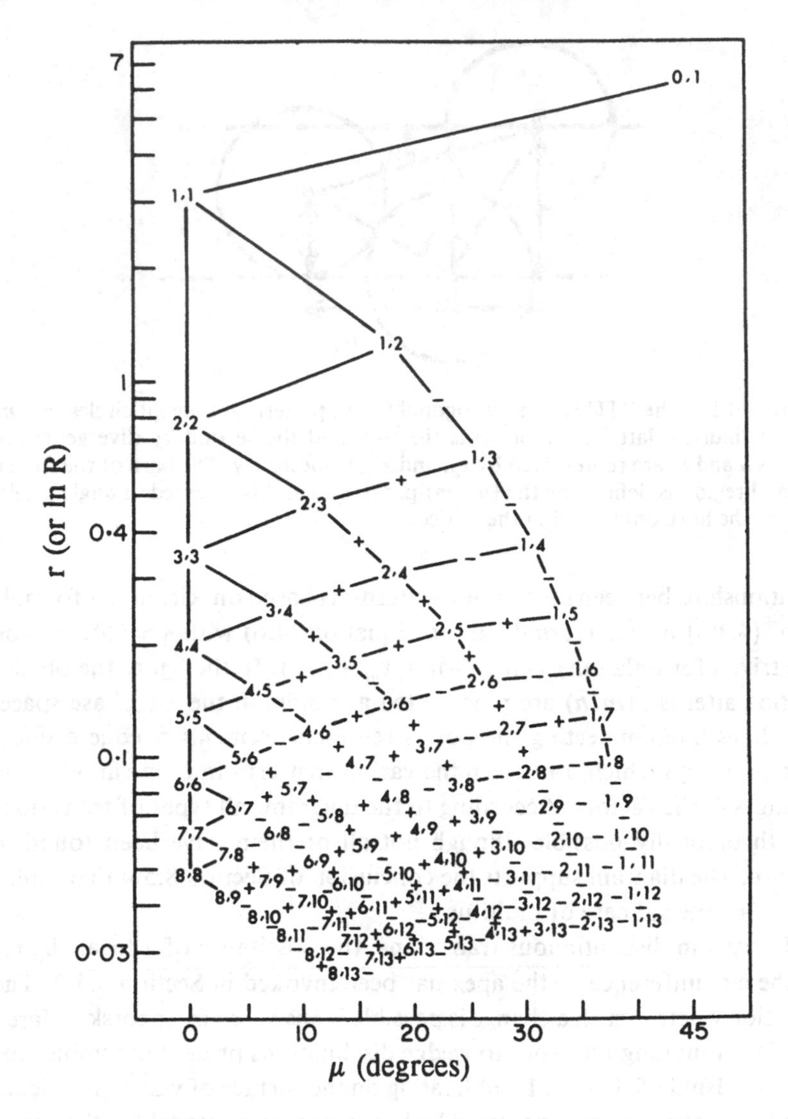

Figure 10.12. Possible changes in contact phyllotactic patterns that can arise through elementary dislocation events. Relationship between r or $\ln R$, and the angle μ of the unit cell for patterns of phyllotaxis modeled as orthogonal contiguous circles in the plane or on the cylinder. The plus and minus signs indicate whether the chirality of the generative spiral reverses (+) or remains the same (−) during the transformation from one pattern to another. (From Meicenheimer & Zagorska-Marek, 1989.)

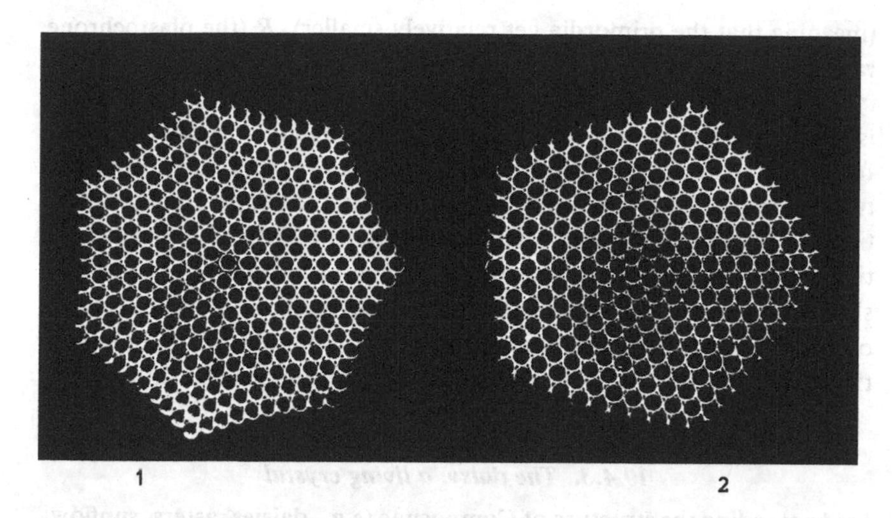

Figure 10.13. Lattice of floating bubbles used to explain discontinuous transitions in phyllotactic patterns. Wedge disclinations alter the normal sixfold symmetry of the lattice. (1) Adding a wedge of bubbles produces sevenfold symmetry. (2) Removing a wedge from the hexagonal raft produces fivefold symmetry. (From Harris, 1977.)

The systemic homology with crystals and bubble rafts, or between soap bubbles and developing primordia, also used by other workers (e.g., van der Linden, 1990; Stevens, 1974), is of course very relevant. The theory of soap-bubble coalescence has been used by Needham et al. (in press) to improve a secondary feature of Bursill and XuDong's (1988) model reported in Section 3.5.1, namely, by replacing the disks by shapes assumed by bubbles in aggregates. But it is not necessary here to resort to an homology and to exotic terms to throw light on the phenomenon of discontinuous transitions. Here is an explanation which uses known results in phyllotaxis, found in Chapter 4 on differential growth, and which supports the homology with bubble rafts.

In Section 4.6 we analyzed the problem of the relative sizes of apices and primordia, which are important factors of pattern generation. It is postulated here that a discontinuous transition arises as the result of possible local abrupt changes in these relative sizes. Consider for example Church's definition of bulk ratio B (Section 4.6.1), or Richards's definition of area ratio A (Section 4.6.2) precisely concerned with the relative sizes of primordia and apex. It follows from Equation (4.25) for B, or from Equation (4.27) with $1/A$, that when $1/A$ or B suddenly decreases (meaning that the primordia get relatively smaller), R (the plastochrone ratio) suddenly decreases. Looking then at Equation (4.12), the allometry-type model, we can see that m or n or both must increase, where (m, n) is the opposed parastichy pair. Let us thus assume that there is an interference of some kind in the phyllotactic mechanism {e.g., an abrupt disturbance in the local conditions of nutrition, in the temperature (see Section 5.2.2), or in the plastochrone [see Equation (4.29)]}, resulting in drastic changes in the relative sizes of apices and primordia. The equations show that $m+n$ will change, that is, new spaces on the apex will be created or spaces will be lost, resulting in the addition or subtraction of parastichies in the system, thus bringing about discontinuous transitions.

10.4.3. The daisy: a living crystal

Understanding the structure of Compositae (e.g., daisies, asters, sunflowers) is considered by crystallographers as a problem of crystallography. The daisy is the ideal crystal with cylindrical symmetry. A parastichy is called a reticular spiral. The three families of parastichies determined by triple contact points $(n, m, m+n)$ in phyllotaxis, called **grains** in crystallography, correspond to the lattice planes of conventional crystallography. Rivier explains crystallographically the structure of the daisy, using as a genetic spiral, Vogel's cyclotron parabolic spiral defined by $r = a\sqrt{n}$ and $\theta = 2\pi nd$, $n = 1, 2, 3, \ldots$, where $0 < d < 1/2$ is the divergence angle. Voronoi polygons (Section 9.1.2) are placed around each point n on the spiral lattice, to constitute floret n, a natural unit cell of the structure. The cyclotron spiral, also used by Ridley (1982a) to reconstruct the sunflower, gives uniform floret density. When the florets of the daisy have not six neighbors but five or seven, we observe defect lines called **grain boundaries** (see Fig. 9.4).

The direction of growth of a crystal sometimes changes. In phyllotactic patterns such a change corresponds to a transition. These phyllotactic transitions, discussed in Section 8.3, are given in Harris and Erickson (1980) as examples of the defects known in crystallography as dislocations and disclinations (Frank, 1951; Nabarro, 1967). The defect lines in Figure 9.4 representing a capitulum, act like crystal dislocations. An edge dislocation in the atomic or molecular networks of a crystal arise when a new line of atoms or molecules is inserted between two others, or when two lines combine to form one. This can be compared to Church's addition or substraction of a parastichy in a family of parastichies to explain transitions (Section 8.3.2). Edge dislocations are the means to insert additional material, so as to produce phyllotactic transitions. Now, a grain contains hexagonal florets exclusively; it consists of a circular annulus. For Rivier, grain boundaries (the defect lines, arrays of dislocations) are quasicrystals. A daisy has one-dimensional quasicrystals.

Crystallographic requirements deliver noble numbers for possible values of the divergence angle in botanical structures. Structures based on noble numbers have maximal uniformity, local homogeneity, self-similarity, and structural stability. Self-similarity (see Section 9.1) corresponds to the operation of inflation–deflation familiar in quasicrystals. The structure of the daisy is thus dominated by self-similarity and maximal structural stability. The unijugate normal patterns in Table 6.1 are ordered according to decreasing stability $S(T)$ (see Proposition 1a of Chapter 6), and the Fibonacci pattern of the daisy has maximal stability in the sense used in the interpretative model. This means that the structure of the daisy is very robust.

10.4.4. Minimal energy costs with regular transitions

Grain boundaries (Section 10.4.3) are quasicrystalline arrays of dislocations where transitions occurs. These transitions are necessary to accommodate more florets of the same size on a longer circular perimeter. Rivier (1990) associates energy costs to what he calls regular and singular transitions (Rivier et al., 1984; Rothen & Koch, 1989a,b). These are two kinds of continuous transitions. Consider the grain $(n, m, m+n)$, $n<m$, representing three families of parastichies and two consecutive contact parastichy pairs. A regular transition is obtained if by the decrease of the plastochrone ratio the next grain is $(m, m+n, 2m+n)$, the family of n spirals having been replaced by a family of $2m+n$ spirals. Another kind of continuous transition can occur, if the family of m spirals is replaced by a family of $2n+m$ spirals, to give thc grain $(n, m+n, 2n+m)$. This is a singular transition.

Let us discuss these transitions in terms of the theorem of Chapter 2. The theorem associates intervals for the divergence with the grain $(n, m, m+n)$. That is one interval with (n, m), let us say $[p, q]$, and one with $(m+n, m)$, let us say $[s, q]$, with $p<s$. The interval $[p, q]$ on which the structure is locked is then subdivided by s in two intervals $[p, s]$ and $[s, q]$. At a regular transition the interval obtained from the pair $(m+n, 2m+n)$ is a subinterval of $[s, q]$. At a singular transition the interval obtained from the pair $(2n+m, m+n)$ is a subinterval of $[p, s]$. The regular transition corresponds to what is called the left extension of (n, m), that

is $(m+n, m)$. The singular transition corresponds to a right extension of (n, m), that is $(n, m+n)$.

Regular transitions correspond to the rhythmic alternation of the neighbors of the vertical axis that we already noticed in previous figures (in agreement with the Bravais–Bravais assumptions; Section 9.1.1) and in the contraction algorithm of Section 2.3. Regular transitions are possible with noble numbers only. In this case, the Fibonacci-type relation from one grain to the next is precisely the relation between the denominators of consecutive principal convergents of the continued fraction of the divergence angle. For a singular transition, the Fibonacci-type relation in each consecutive grain is made possible by the presence of intermediate convergents in the continued fraction (Coxeter, 1972). Singular transitions can occur with non-noble numbers only.

The question then arises: On which side of the bifurcation point s will the divergence move as the plastochrone ratio decreases? For Rivier (1990) each alternative at a bifurcation costs some energy. A capitulum being considered as a close-packing of deformable florets (soft material like bubble rafts, magnetic bubbles, colloidal crystals, Bénard convection cells (Rivier et al., 1984; Occelli, 1985; Rivier, 1990), flux lines in superconductors (Levitov, 1991a,b) in cylindrical symmetry, Rivier introduces a measure of the elastic strain needed to confine the structure crystallographically. This is the elastic energy necessary to deform a floret in order to fit it in the structure. A grain boundary is the place where shear and strain dissipates in a capitulum made with hexagonal cells in the grains. Using the principle of economy, Rivier explains why nature prefers regular transitions: their energy costs are minimal; phyllotaxis is not the result of an infinitely accurate biochemical protractor not likely capable of leaving fossil records, but the result of a structurally stable partition of space.

Now we have seen in Chapter 6 that a principle of economy allows us to have the pattern (m, n) that will minimize energy costs under certain constraints. Singular transitions correspond to rhythms with period $w > 2$. There are more than two production rules in the L-systems generating the corresponding hierarchies. In Proposition 5 of Chapter 6, we have seen that the principle of minimal entropy production gives a rhythm with period $w = 2$, that is, Fibonacci hierarchies showing consecutive regular transitions needing only two production rules in the L-system. Phyllotaxis is a very simple example of a structural archetype produced by a sequential process.

11
Morphogenetical parallelism and autoevolutionism

11.1. General comparative morphology

The *systemic viewpoint* I introduced in phyllotaxis drove me to stress the fact of the presence, in other areas of research, of patterns similar to those found in phyllotaxis. Showing parallelism between structures with various substrata is what general comparative morphology is about. It proceeds from a synthetic and homological approach, and it attributes more importance to resemblances than to differences, as is the case in what is known in biomathematics as relational biology.

We have seen in Chapter 10 that the phenomenon of morphogenetical parallelism allows us to extend the method of phyllotaxis to other fields, such as molecular biology, for the study of microorganisms, viruses, and polypeptide chains. On the other hand, the methods used in other fields such as crystallography can be applied to phyllotaxis precisely because of structural parallelism. The principles of physics that pertain to crystallization are the same regardless whether one is dealing with protein molecules in viral capsids, with atoms in a crystal lattice, or with primordia in a phyllotactic pattern. The physical environment must be structurally the same in every case.

In physics Maxwell has shown that the action of a magnet on steel and the passage of a light beam through the air, two phenomena that seem to be worlds apart, have in fact a lot in common. Meyen (1973) used the word "nomothetics" to describe this type of research in which one looks for general principles, common structures, and universality among various sets of organisms or objects. According to Meyen morphological laws cannot be reduced to history, function, and adaptation. There exists some general structural principles, manifested within various groups of organisms. Urmantsev's laws of correspondence (see Meyen, 1973) have great heuristic value and can bring understanding by showing the parallelism and convergences between structures and functions in one field and those in other areas.

The *phylogenetic viewpoint* I stressed in phyllotaxis drove me to look in the plant kingdom for ancestors of the phyllotactic patterns. Chapter 3 presented Church's and Zimmermann's theories as important keys for a better understanding of phyllotaxis. But looking at the nice application of morphogenetical parallelism that relates crystal growth to phyllotactic pattern generation (Chapter 10) leads us to move beyond the beginning of plant evolution with ancestral land plants and algae. Darwinism and neo-Darwinism generally consider that biological patterns are the outcome of genetic activities. But there are no genes in minerals. In any attempt to deal with major issues of phyllotaxis, the critique of genocentrism generally aimed at neo-Darwinism is of particular significance. Using the tenets of autoevolutionism, which appears as a key for understanding phyllotactic patterns, a step further can apparently be made regarding their origins. One of these tenets is that we simply have to go back to the evolutions of minerals and chemicals to integrate biological evolution into its natural context. In autoevolutionism, morphogenetical parallelism is called homology, isofunctionalism, and isomorphism. The latter term is the one used in systems research.

According to Meyen (1973), "from the point of view of the spatial position of elements, the flower having juxtaposed sepals and petals will be closer to carbohydrate molecules with the same arrangement of hydrogen atoms, than to other flowers bearing alternating sepals and petals." This point of view has important consequences on the search for insights into the problem of phyllotaxis. Trying to explain the same structure observable in many areas by referring to one context only, botanical or not, seems to be an impossible task. But

> for the overwhelming majority of biologists, the comparison of flowers with molecules, and shells of foraminifers with Roman pottery, seems either senseless or improper (or both). The overcoming of this psychological barrier, and the spreading of the application of system laws to structural investigations, will be a revolution in biology comparable to the introduction into biology of statistical thought (Meyen, 1973, p. 248).

The phenomenon of morphogenetical parallelism is certainly an important key for extending the frontiers of our knowledge on phyllotaxis.

The aim of comparative morphology is to discover homologies in the organic world, and more often only in a specific part of it. We are interested here in isomorphisms with phyllotactic patterns, and these do not stop at the artificial frontier between inanimate and animate worlds. Beyond genes there are fundamental mechanisms, such as branching processes (introduced in Chapter 3) and gnomonic growth (defined below), which reveal deeper isofunctionalisms with phyllotactic pattern, as the present chapter shows. This chapter presents more structural similarities suggesting that the origins of phyllotactic patterns are more ancient and lie deeper than it is generally thought. This is perhaps why these origins have been overlooked.

11.2. Isomorphisms with phyllotactic patterns

11.2.1. Minerals, animals, and artifacts

The phyllotactic pattern developed by a specific plant is clearly dependent, as we said, upon the relative sizes of the apical meristem and of the primordia produced by it. Homologously this is also true for patterns of stripes and spots on animals. Given that the black spots on the cylinder-like tail of a jaguar can be generated by a diffusion–reaction equation controlling an inhibitor and an activator (Murray, 1988), then by homology, it is not surprising that many models in phyllotaxis try to reproduce the primordia (the spots) on the cylindrical surface of a plant by using diffusion equations (see Appendix 8).

Stevens (1974) abundantly illustrated the geometrical relationship among the arrangements of scales on fish, tortoise shells, snakeskins [Fig. 11.1(2)], and the seeds of asclepiads [see Fig. 11.1(1)]. He compared these arrangements to the geometry of soap films and bubbles, of turtle shells, and of crystal grains, stating that underneath all these arrangements there is a minimization of work or surfaces of contact, or potential energy, by means of triple junction points. Such triple points are clearly found for example at the junctions of the hexagonal scales on the surface of pineapples (see Fig. 9.2).

The patterns of scales on the anteater *Manis temminckii* [Fig. 11.1(3) and on a spruce cone [Fig. 11.1(4)], were correlated by the geneticist Lima-de-Faria (1988) with the pattern on the shell of the mollusk *Conus milneedwardsi* [Fig. 11.2(2)], and even with a crystal of common salt (NaCl) [Fig. 11.2(1)]. Pine cones are excellent examples of the phenomenon of phyllotaxis, but to understand similar orderly arrangements of scales, whatever be the nature of these scales, it is better not to concentrate on pine cones only.

The word "pattern" refers to arrangements of parts or elements. Of course the word is applicable to many fields, whether they be technical,

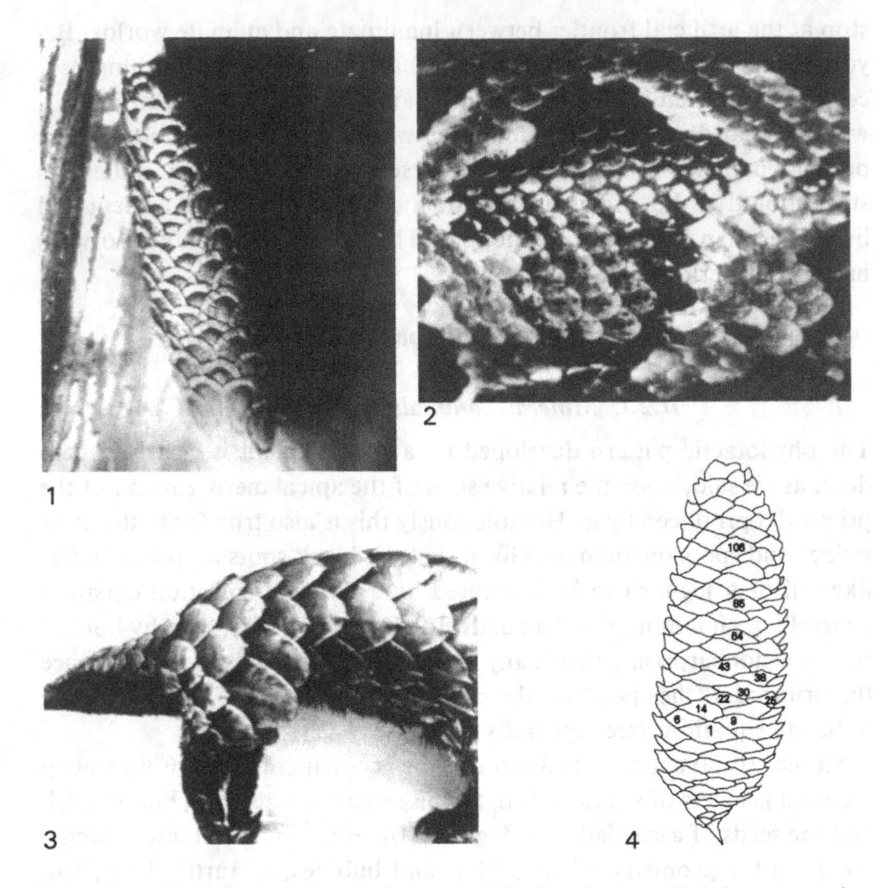

Figure 11.1. (1) Scales on an asclepiad seed. This plant is also called dompte-venin (venom), because formerly it was regarded as an antidote to poison. (2) Scales on a snakeskin. (3) Pangolin. [(1), (2), and (3) from Stevens, 1974] (4) Spruce cone, *Picea excelsa*, having (5, 8) phyllotaxis.

scientific, or artistic. Similarities with phyllotactic patterns are not found in nature only, as Figure 11.3 shows. Philosophers, such as Teilhard de Chardin, defended the idea that man-made designs and forms are extensions of natural ones, and in a sense are still part of nature. Crystal sections and wallpaper designs obey the same structural rules. The presence of the golden number τ in phyllotactic patterns is often said to be mimicked in the works of musicians (e.g., Bartok), architects (e.g., Le Corbusier), and artists (e.g., Dali), in the ancient as well as in the modern world. Creative artists naturally seek inspiration from the Creation. The huge pattern on the cupola of a chapel from the Renaissance resembles the capitulum of a *Chrysanthemum*.

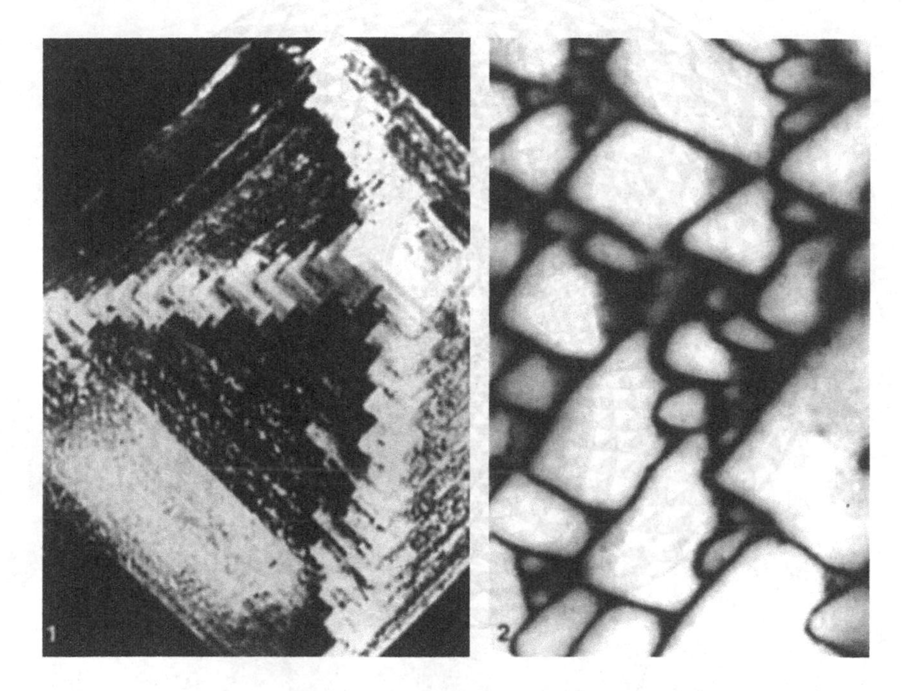

Figure 11.2. (1) Crystal of halite. (2) Surface of mollusk shell. (Both figures from Lima-de-Faria, 1988.)

11.2.2. Colloidal crystals, quasicrystals, and polymers

In growing crystals there exist two main categories of structures: dendrites and fractals. Figure 11.4(1) shows the apical dome of a plant generating primordia in a phyllotactic arrangement. The dome has a structural similarity with the dendrite in Figure 11.4(2) and with the leaf of the common fern in Figure 11.4(3). That phyllotactic patterns are linked to fractals, as we have seen in Chapter 3, is illustrative of a connection between the two crystalline structures.

According to Rothen and Pieranski (1986), colloidal crystals represent a very pedagogical model of real crystals, on a different scale. The tobacco mosaic virus, shown in Figure 10.5(3) and analyzed by the botanist Erickson using phyllotactic methods, is a colloidal crystal. This term covers organic, inorganic as well as biological materials. A colloidal crystal has a size intermediate between microscopic and macroscopic. Geometrically, there is no difference between the virus, a piece of opal, and a ball of latex; they are all colloidal crystals showing compact packing of spheres. Just as in colloidal crystals, phyllotactic arrangements are the results of

1

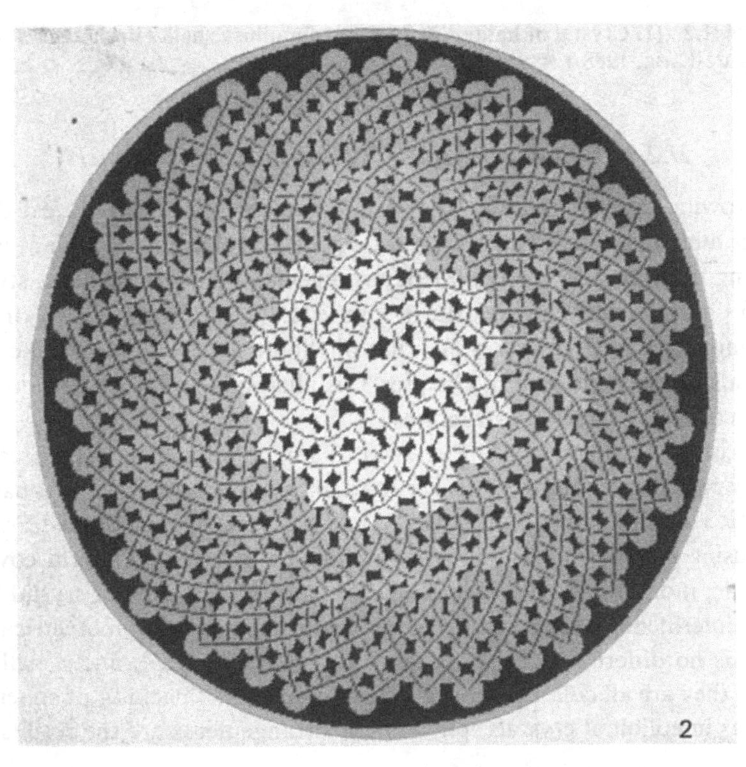
2

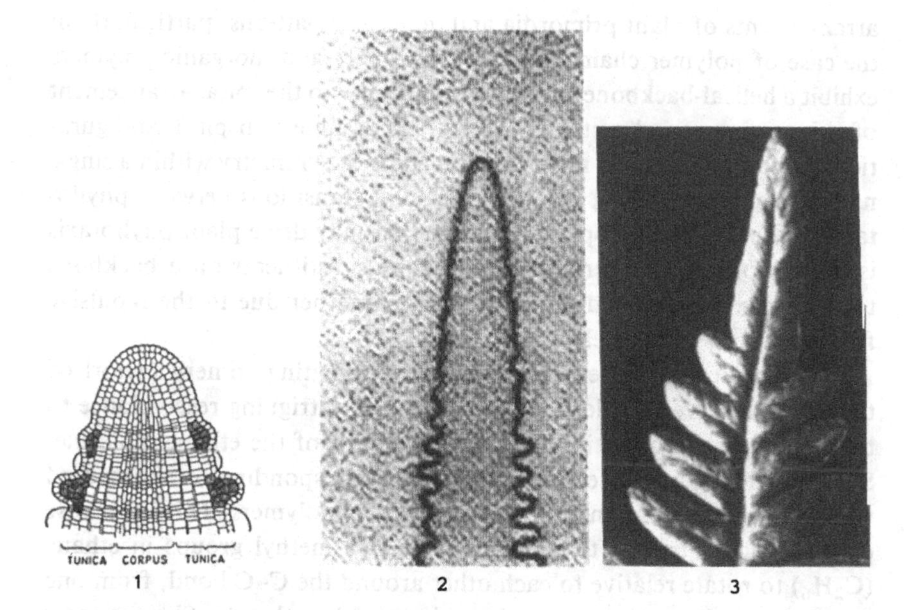

Figure 11.4. (1) Longitudinal section of the parabolic shoot apex of *Hippuris vulgaris* showing the leaf primordia. (From von Denffer et al., 1971.) (2) Crystal of ammonium chloride growing in a dendritic way at a parabolic speed. (From Sawada & Honjo, 1986.) (3) Tip of a leaf of *Pteridium aquilinum*, a widespread fern. (From Novak, 1966.)

packing efficiency of units, and the protoplasm of the apex may be regarded as a mass of fluid colloidal plasma.

Although the crystallographic restriction refers to the absence of pentagonal symmetry in crystals, this type of symmetry can be observed in what are called quasicrystals. Indeed, the microscopic structure of rapidly cooled metal alloys shows fivefold rotational symmetry. It is also impossible to tile a plane using only shapes that have fivefold symmetry. But the two-dimensional analogue of quasicrystals - Penrose tilings - have this symmetry. It is interesting to note that the golden number τ arises in this tiling. The tiling is made of two types of parallelograms, and in an infinite Penrose tiling the ratio between the numbers of units in these two types is τ (Nelson, 1986).

Friedman (1992) compares the generation of phyllotactic patterns with a model of polymer conformation. She shows interesting analogies between

Caption for Figure 11.3
(1) Roman mosaic from the Museo Nationale in Rome. (2) Four colors "Fibonacci" carpet by the botanist R. F. Williams (personal communication). The outer zone shows "(34, 55) phyllotaxis," middle zone "(34, 21)" and inner zone "(13, 21)."

arrangements of plant primordia and molecular patterns, particularly in the case of polymer chains. Some biopolymers and inorganic polymers exhibit a helical-backbone conformation similar to the spiral arrangement of primordia around a stem, while others assume nonspiral configurations. A few may display more than one type of symmetry within a single natural sample, recalling plants that exhibit transitions between phyllotactic patterns. The acropetal influence that may drive plant phyllotaxis is shown to have its chemical counterpart; substituents on a backbone tend to maximize their distance from one another due to the repulsive forces between pairs of electrons.

Friedman insists more particularly on alternating trimery (whorl of three), as observed in oleander, which has an intriguing resemblance to the staggered conformation (alternating whorl) of the ethane molecule. She says that the eclipsed conformation, corresponding to superposed whorls in plants, is an unfavorable structure in polymers. This is because of a rotational barrier that prevents the two methyl groups in ethane (C_2H_6) to rotate relative to each other around the C–C bond, from one staggered conformation to another, via the eclipsed state. She proposes that repulsion between the hydrogen atoms in polymers causes this barrier, as in models of phyllotaxis that postulate a repulsive influence between primordia. Friedman concludes that the factors governing the existence of symmetry in plants and in polymers are similar in many respects.

11.2.3. Properties of space–time

Phyllotaxislike patterns are commonly seen in the global cloud-cover pattern, in particular in hurricane cloud bands, as underlined by Selvam (1990a,b, & personal communication). She developed a cell dynamical system model for atmospheric flows, which predicts the spiral circulation pattern with the golden mean winding number as intrinsic to the quantumlike mechanics of atmospheric flows. Such logarithmic spiral patterns are signatures of deterministic chaos indicating long-range spatiotemporal correlations. Selvam brought to my attention (personal communication, October 1992) the cover picture of the journal *Weather* [vol. 37(12) 1982], which shows "frost flowers" that bear a striking resemblance to real flowers. These "flowers" formed over a gravel surface after a severe overnight frost with daytime temperatures well below the freezing point.

The omnipresent properties of space–time play a role in the structuring and patterning of matter and energy. We learned from Einstein about the control exerted by curved space. The patterns and functions we are concerned with in phyllotaxis may fundamentally come from general

Figure 11.5. The "winter daisy." Time, space, turbulance, centrifugal forces, melting snow on the road, speed, and appropriate atmospheric conditions generated a daisylike structure in the center of the wheels of my car. The V in the center of the figure probably stands for the victory of the intelligence of nature.

properties of space–time. Here is an example of how space, time, and appropriate constraints can create patterns. Figure 11.5 shows the pattern around a central cap (about 8 cm in diameter) of a honeycomb rim of one of my cars. The four caps displayed the same structure. It was generated in three hours by turbulence, centrifugal forces, wind speed, temperature, etc., created by the rotation of the wheels at 120 km/hr over 300 km on a highway covered with a thin layer of melting snow. It offers a striking resemblance to the structure of a daisy – it has a "capitulum," disk "flowers" (protuberances) in the "capitulum," and ray "flowers."

In his equipotential theory Church (1904b) considered that a capitulum was a display of lines of forces. According to him, new centers of lateral growth originated at the points of intersection of parastichies, along paths of distribution of equal growth-potential which may be homologized with lines of force. We have encountered in Section 6.2.2 resonating metal sheets producing lines of forces imitating decussation.

When there is an isomorphism of structures, beyond a variety of chemical composition, there is a common chemical denominator that is not always evident. It may be a radical or an atomic or electronic configuration that is at the basis of the isomorphism. Equally important may be the pressure, the temperature and the gas or liquid in which the materials are allowed to expand. The lesson to be learned is that the tendency to consider such similarities as accidents, curiosities or mere analogies derives mainly from the ignorance of the physical and chemical processes underlying them. (Lima-de-Faria, 1988)

Branching processes and gnomonic growth are among the manifestations of the properties of space–time, and the oldest forms of growth. The next section stresses their importance. From these processes result patterns that have particular functions. Isomorphism means isofunctionalism, just as matter means energy through Einstein's well-known law.

11.3 Isofunctionalism with phyllotaxis

11.3.1. Branching processes

With the search for the origins of phyllotactic patterns, naturally comes the search for the remote mechanisms responsible for the existence of those patterns. The origins of phyllotactic patterns in higher plants must be sought in gnomonic growth and in branching processes, two very pervasive mechanisms present at all levels of evolution. In order to understand the phyllotactic mechanisms in higher plants, we must appreciate that we are dealing with a situation complicated by subsequent development from primitive and simple laws of growth. Branching processes are a common denominator for many of the similarities we started to look at in Chapter 3, where we saw that the skeletons of the phyllotactic patterns are treelike structures I called control hierarchies. The same kind of tree structures are found, for example, in the ramified structures of algae, in the structure of the system of tributaries of rivers, and in chemicals and minerals.

Consider the network of tributaries of the river depicted in Figure 11.6(1). When two tributaries of the same class i, or of different class j and i, $j>i$, meet, the resulting class is $i+1$ or j, respectively; all ends are assigned order number 1 (this is called Strahler's ordering of a branched structure). The parameter m in the figure represents the highest class, that is, the class of the main stream, obtained by applying the above rules. A significant percentage of these networks are Fibonacci networks, where the number of tributaries of class 1, denoted by N_1, is F_k for a certain k. In the example here, $k=6$ given that $N_1=8$ is the sixth Fibonacci number. The graph of this Fibonacci network, Figure 11.6(2) obtained

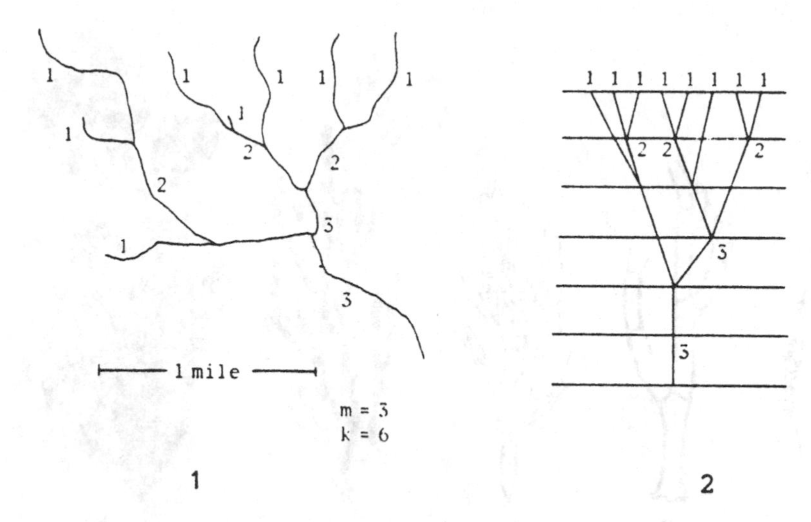

Figure 11.6. (1) Tributaries of a stream in South Carolina. (From Sharp, 1971.) (2) Graph of same stream. (From Sharp, 1972.)

from the Strahler's topological method, is essentially Figure 3.2(2) representing a capitulum of *Cephalaria* or Figure 3.7(2) representing a brown alga. If N_i represents the number of tributaries of class i, it can be proved that $N_{i+1}/N_i \approx \tau^{-2}$, the Fibonacci angle of botany.

While considering Zimmermann's telomes theory in Chapter 3 we met with *Rhynia* and the Psilophytales which are among the oldest known land plants (see Fig. 3.6) that show ramifications. The Psilophytales still resemble quite closely in their gross morphology the dichotomously branched thalli of many of the more highly organized brown algae [Fig. 11.7(4)]. And the latter show affinity with the Chrysomonadales [Fig. 11.7(3)]. The algal ancestor of the Pteridophyta is believed to be found in the Chlorophyceae (green algae) which contain *Ulotrix* and *Cladophora* [Figs. 11.7(1) & 11.7(2)]. That a physical determinant induced the branching process in plants can be imagined when we consider that such a branching tendency already exists at the level of elementary inorganic material [Fig. 11.7(5)].

11.3.2. Gnomonic growth

Gnomonic growth produces the beautiful logarithmic spiral on the shell of the very old mollusk called *Nautilus pompilius*, by the constant addition of new material at the open end of the shell, according to the same

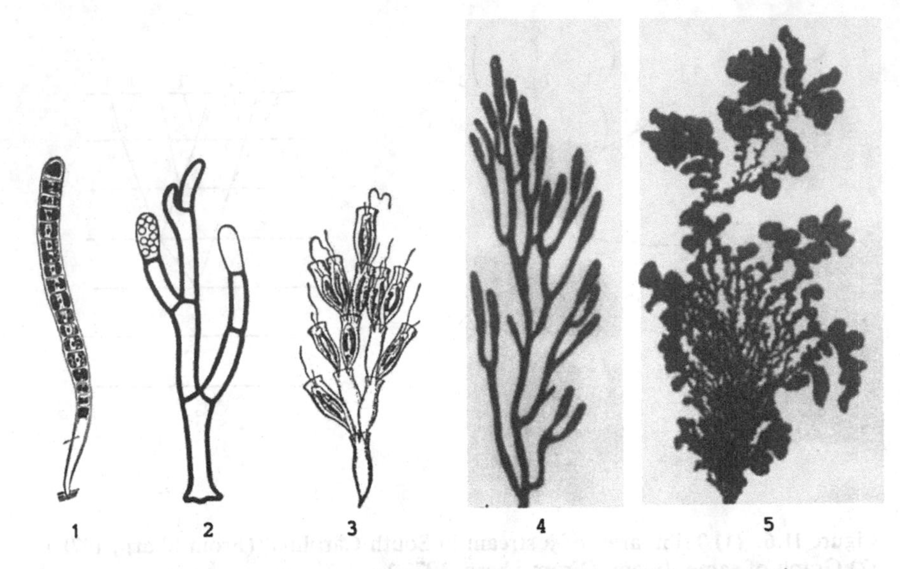

Figure 11.7. (1) Young filament of the green alga *Ulotrix*. (2) Diagrammatic representation of the green alga *Cladophora*. (3) Chrysomonodales *Dinobryon*. (4) *Fucus bifurcatus*. (5) Native copper dendrite. [(1), (2), and (3) from von Denffer et al., 1971; (4) and (5) from Lima-de-Faria, 1988.]

accretionary law. As pointed out by D'Arcy Thompson (1917), we can conceive of no simpler law of growth, the simplest of laws which nature tends to follow, than that observed in the growth of the shell of the *Nautilus*, which widens and lengthens in the same unvarying proportions. The result is that the shell or the organism grows in size but does not change in shape; it shows constant similarity of form. This is precisely what symmetry, the core of mathematics, is literally about - the repetition of the same proportions. This is the way crystals grow; it is called accretionary, or additive, or gnomonic growth. It comes from a single growth gradient, and is present at all levels of evolution, just like the spirality of the shell.

A gnomon can be geometrical or numerical. A geometrical gnomon is an elementary structure that repeats itself, to a near similarity relation, at every step, and which is the building block of the whole. Aristotle defined it as any figure which, being added to any figure whatsoever, leaves the resultant figure similar to itself. As an example of a geometrical gnomon, consider a rectangle with sides of lengths $m-n$ and n, $n<m$. Add a square, the gnomon, with sides of lengths n as illustrated in Figure 11.8 with precise values, and then a square with sides of lengths m in the

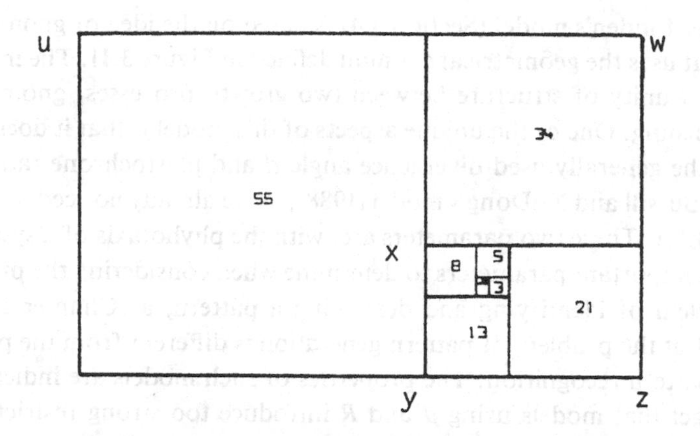

Figure 11.8. Gnomon based on squares with sides of consecutive lengths 1, 1, 2, 3, 5, 8, 13, 21, 34, 55, ... generates quasi-similar rectangles approaching the ratio $\tau/1$.

same direction, and so on along the Fibonacci-type sequence $\langle m-n, n, m, m+n, 2m+n, \ldots\rangle$. It is easily proved that the ratio of the sides of each rectangle is approximately equal to the golden ratio τ, and as we move along the sequence this ratio gets closer to τ. It follows that the rectangles become closer to a golden rectangle whose sides are exactly in the ratio $\tau/1$. By linking the vertices x, y, z, w, u, etc. in Figure 11.8, in the appropriate way, we begin to generate a logarithmic spiral, the essence of constant similarity form.

The isosceles triangle (Δ) with two angles of $\pi/5$ (36°) is the gnomon of the isosceles triangle with two angles of $2\pi/5$ (72°, the golden triangle), and vice versa. For example, looking at Figure 1.8 and at the golden triangle *ONF* we see that

$$\Delta ONF + \Delta NFJ + \Delta FOJ \text{ (a golden triangle)},$$

$$\Delta FOJ + \Delta OJH = \Delta JFH \text{ (a golden triangle)},$$

$$\Delta JFH + \Delta FHA = \Delta HAJ \text{ (a golden triangle)}.$$

The definition of geometrical gnomon includes the case of numbers considered geometrically. A numerical gnomon is a rule or algorithm by which a number in a structure is approximately generated, over and over again. The ancients defined it as what must be added to a quantity to obtain a similar quantity. Thus $2k+1$ is a gnomon of k^2, given that $k^2+2k+1=(k+1)^2$.

Van der Linden's model (Section 3.4) is based on the idea of gnomonic growth. It uses the geometrical gnomon defined in Figure 3.11. The model suggests a unity of structure between two growth processes, gnomonic and branching. One of the unique aspects of this model is that it does not require the generally used divergence angle d and plastochrone ratio R, just like Bursill and XuDong's model (1988), as we already noticed in Subsection 3.5.1. These two parameters are, with the phyllotaxis of a system, the most important parameters to determine when considering the practical problem of identifying and describing a pattern, as Chapter 5 has shown. But the problem of pattern generation is different from the problem of pattern recognition. The properties of such models are indicative of the fact that models using d and R introduce too strong restrictions which prevents the fundamental natural processes, such as gnomonic growth and rhythmic branching, to be fully operational.

Green (1991) reports a mechanism from J. Ungar, called the cut–grow algorithm, which he considers to be a "simple polygonal analog for phyllotaxis . . . having the major features of phyllotaxis." The mechanism is based on the golden triangle. To understand it let us refer again to Figure 1.8. The golden triangle BDE is considered to be the apex viewed from above, equivalent to the top view of a meristem. Its sides have lengths 1, 1, and $1/\tau$. In this triangle a cut is made with DF, such that $r = BF = 1/\tau$, called a "golden cut." This cut demarcates a leaf; ΔDFE (a golden triangle) is considered to be the newly formed leaf or the new appendage. The apex grows in a way that DF increases in size to DA to form the new golden triangle, the new apex, ΔBDA. The new apex is the original one rotated by 144°, which represents the divergence angle. This process is qualified as self-similar, and self-similarity is said to be related to self-stability. If instead of making a golden cut $r = 1/\tau$ of side BE, other values of r are chosen, then there is a drift in size of the apex toward either 0 or infinity, what is called self-stability. Also, if r is equal to $1/\tau$ then the repeated application of the process to any initial triangle will ultimately give the golden triangle; the eigenvalue of the matrix transformation tends to 1.

Goodal (1991) discussed the sequence of shapes constructed by a cut–grow map of triangles, and proposed an eigenshape analysis of the map with vivid graphical depiction of the convergence to the eigenshape. The shape of a triangle is represented by a point in the complex plane and the eigenshapes are the fixed points of a reflection followed by inversion in a circle. The cut–grow map is equivalent to a 3- by 3-matrix acting on a complex vector. In its actual development the map can generate any spiral system with divergence greater than 120°. Distichy and decussation are considered to be special cases of spiral systems. The mechanism does not demonstrate the primacy of the Fibonacci angle or of any one of the accessory sequences.

The τ-model of Chapter 9 uses a numerical gnomon set up by a minimality principle. For every integer m the model determines the integer n such that $(m/n)(1/\tau)$ is closest to 1, and the initial placement of the first m primordia; that is, the system shows phyllotaxis $(m-n, n)$. After the initial configuration is set up at time m, a rhythm of gnomonic growth starts. A dynamical process is replaced by a deterministic one, and the system grows along the Fibonacci-type sequence $\langle m-n, n, m, m+n, 2m+n, \ldots\rangle$. If for example $m = F_k$ and $n = F_{k-1}$, to go from $m/n\tau$ to $(m+n)/m\tau$ we just have to add to $m/n\tau$ the gnomon $(-1)^k/mn\tau$, because of the relation $(F_k)^2 - F_{k-1}F_{k+1} = (-1)^{k-1}$. In the model of Chapter 6 the numbers m and n of nodes in two consecutive levels are such that $(m/n)(1/\tau)$ approaches 1.

11.4. Levels of organization and layers of models

If we are to understand the origins of primordial patterns in plants, we must realize that they are a product of the overall evolution that brought the hierarchical organization of the universe from quarks to atoms, to molecules, to macromolecules, to organites, to cells, to primordia, to populations of plants, to ecosystems, etc. This type of hierarchy or multilevel system, called **compositional hierarchy**, concerned with nested entities, is one of the objects of study of general systems theory (e.g., in Mesarovic, Macko & Takahara, 1970; Pattee, 1974; Salthe, 1985; Auger, 1983, 1989).

Phyllotactic patterns can be studied by considering the hierarchy of composition of which they are a part. Here is an elementary description of adjacent levels of particular interest in that hierarchy:

Level 1: The set of cells in the cytoplasmic (protoplasmic) mass or phragmoblasteme, whose activities bring up the primordia to the organogenic region of the apex (see Fig. 11.9).

Level 2: The set of primordia of a plant, whose growth activities generate the primordial patterns.

Level 3: The set of primordial patterns in plants whose functions serve and derive from the internal physiology and external environment.

Level 4: The populations of plants in the ecosystems.

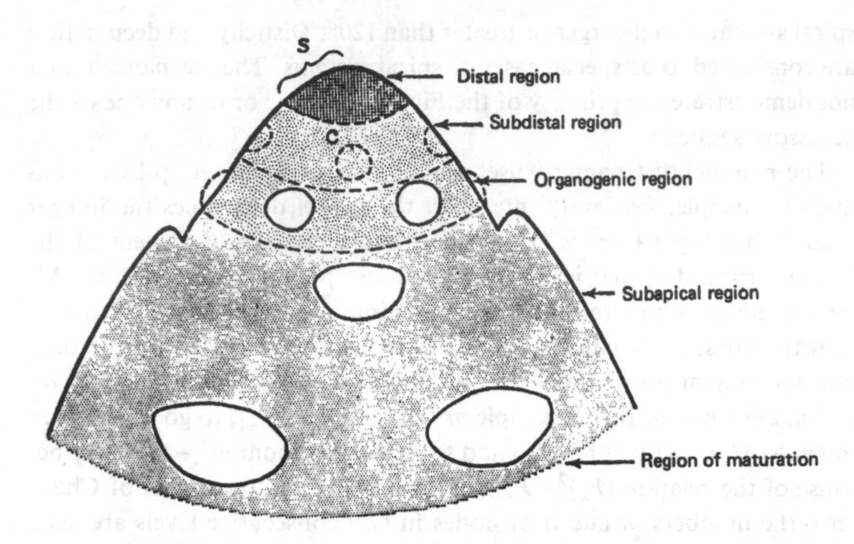

Figure 11.9. Plant apical meristem. Primordia are initiated in the subdistal region and are observed in the organogenic region. As the distal region moves up by growth, the primordia increase in size in the subapical region. (After Wardlaw, 1968b.)

The hierarchies used to represent phyllotactic patterns (Chapter 3) are **control hierarchies** dealing with the coupling or *interface between levels 2 and 3 of the compositional hierarchy* in which the phenomenon of phyllotaxis is embedded.

Atoms make molecules, and primordia make patterns. It is generally stated that the primordial patterns are meant to provide maximum light to each leaf, floret, etc., to maximize radiant energy collection by the plant, to facilitate transpiration and transport of substances in plants. Phyllotactic patterns are obviously nature's solutions to optimality problems. Higher plants have evolved under pressure from the pollinators. A sunflower capitulum for example is a society of flowers forming spiral patterns, which have been shown to be particularly attractive for pollinators (see Leppik, 1970). Consequently, they can pollinate more flowers in less time, and thus they maximize the survival probabilities of the species. The existence of primordial patterns in plants has consequences on the upper levels of the compositional hierarchy; trying to understand these consequences involves delving into the functional aspects of phyllotaxis.

It is clear that the aggregate of entities forming patterns have properties that are not the sum of the properties of the entities and that transcend the individual entity. Even if we know all about primordia and organs (the description of the units and their functions in levels 1 and 2 is mainly a botanist's task), we would know nothing about their patterns in level 3 (the description and modeling of level 3 is mainly a mathematician's task). The structures in level 3 are serving functions vital for the species, functions that emerge from the whole set of organs. Phyllotaxis is a problem of coordination or interface, that is, a problem of determination of the interaction between the organs (level 2) and their patterns (level 3) meant to serve the plants having an internal (level 1) and an external (level 4) environment. The evolution of matter that preceded the emergence of life, produced an internal environment, so that lower levels intervene on upper levels. But there is also an intervention of the upper levels on the lower, just as in human organizations a coordinator (president) dictates or influences the motion of the units in the sublevels, and is responsible for their performances. There is an interpenetration of organism and environment (Lewontin, Rose & Kamin, 1984).

In physics the understanding of the infinitely small proceeds from and supposes the understanding of the infinitely large (Hawking, 1988). In phyllotaxis working from upper levels to lower levels implies an a priori, intuitive, inductive understanding of the whole system. With interpretative-like theories we are not bound to be mystified and powerless when facing *the emergence of new properties from the aggregation of units.* Pattee (1974) wrote that "the harnessing of the lower levels by the collective upper levels is the essence of hierarchical control." But this harnessing is included as we have said earlier in the lower levels themselves, given that biological evolution is a prisoner of previous evolutions. In a compositional hierarchy it might be hard to say who harnesses who.

When the phenomenon of phyllotaxis is considered from lower levels to upper levels, the modeler tries to imagine the mechanism at work in the protoplasmic mass or the internal environment. But how is it possible to explain level 3 from level 1 if the aggregate in level 3, called a primordial pattern, has properties absent from levels 1 and 2? The phenomenon of emergence in living systems indicates that concentrating on lower levels is clearly not sufficient. A better understanding of phyllotaxis comes when the two layers of models (from lower to upper, and from upper to lower levels) are integrated to rebuild the whole system. The model of Chapter 6 proceeds from upper levels to lower levels, but is based on control hierarchies originating in lower levels. The model is dictated by the external as well as by the internal environment. The whole universe is active inside the isolated phenomenon of phyllotaxis. The commonplace saying that the whole is in the part is still true, and the statement is in agreement with this view on evolution called autoevolutionism, a view that is radically different from neo-Darwinism.

11.5. A universal framework for the study of phyllotaxis

11.5.1. Tenets of autoevolutionism

The homologies between forms and functions in plants, animals, and minerals, which I put forward previously to underline similarities with phyllotactic patterns, may appear strange to many. It is only that these homologies are far removed from their trail of thoughts generally springing from Darwinism and neo-Darwinism. For autoevolutionism the emergence of biological patterns is totally conditioned and canalized by the autonomous evolution of elementary particles (from quarks and leptons), of chemical elements (from hydrogen), and of minerals (seven crystallographic systems), which established *the physical, chemical, and mineral imprints,* respectively, on biological evolution. Biological forms are already present in minerals and are derived from them by means of similar atomic and molecular constructions.

Autoevolution is a word introduced by Lima-de-Faria (1988) to describe the transformation phenomenon that is inherent to the construction of matter and energy, and that consequently produced and canalized the transformation of biological processes. The interpretation of this phenomenon is called **autoevolutionism**. "The direct consequence of autoevolution has been the emergence of forms and functions that were all derived and framed by the initial properties of matter and energy. It has had as an obligatory consequence the emergence of isomorphism and isofunctionalism." One of the central themes in Lima-de-Faria's stimulating book is that every biological pattern and every biological function has its predecessors in the inorganic world. Though Lima-de-Faria is a geneticist, he believes that forms and functions in biology have not been created by genes and chromosomes. As an echo to that statement we have seen (e.g., in Section 8.3.1) that phyllotaxis is not directly controlled by genetic factors. In neo-Darwinism the gene generally appears to be omnipotent and what existed before it is considered irrelevant to evolution.

Pattee (1969a) "believes that all the molecules in the living cell obey precisely the laws of normal physics and chemistry." Morowitz (1968) assumes "that biology is a manifestation of the laws of physics and chemistry operating in the appropriate system under the appropriate constraints." In the context of autoevolutionism, these apparently reductionistic views would mean that living organisms are moulded by the evolutions that preceded the emergence of the cells, and that nothing is essentially new at the biological level, each level introducing novelties only by new combinations of earlier components. In Darwinism, on the contrary, biological evolution would consist of totally new forms not present previously.

A tenet of autoevolutionism is that "self-assembly is automatic and hierarchic, it occurs at all levels from that of elementary particles to that of organisms, it is the visible consequence of autoevolution, it is a phenomenon with a strict physico-chemical basis disclosing order and hierarchy in self-organization" (Lima-de-Faria, 1988). Cells and organisms are apparently not machines since they are built by self-assembly processes in which every molecule and every cell recognizes the others and is directly related to the whole structure. In relation to that tenet, Morowitz (1968) shows that the tendency to organize is a very general property of a certain class of physical systems and is not specifically dependent on living processes. Systems have a spontaneous tendency to self-organize in a hierarchical manner. In Darwinism, randomness and selection, rather than atomic order and hierarchy, would lead to organization.

The central conclusion emerging from autoevolutionism is that in nature no order, no form, and no function is created or lost; it is simply transformed by combination. Order can only arise from order. The developments of molecular biology show that genes rather act as entities introduced by evolution to fix a successful tendency of pregenetic evolutions. In autoevolutionism, genes introduced controlled repetition of a given structure, extra order by exact location in an organelle, tremendous speed, and fixation of alternatives. If one considers that the emergence of the Fibonacci sequence in plants is a consequence of previous evolutions, then we must explain why the sequence is overwhelmingly present in botany, more than in other sectors of inorganic and organic nature.

11.5.2. Autoevolutionism and neo-Darwinism

It is not the aim of Part III of this book to enter into the details, the difficulties, the controversies, and the philosophical issues raised by neo-Darwinism. The interested reader can consult, for example, the works of Sober (1984), Depew and Weber (1986), and Ho and Fox (1988). Autoevolutionism is simply considered here as a complement to Meyen's nomothetics, and as a complementary evolutionary framework able to provide an insight into the problems of phyllotaxis.

More generally speaking, in my opinion autoevolutionism avoids the conceptual traps that create false problems. One is the trap set forth by the strong dichotomy between the doctrines of structuralists and functionalists in biology. (For an introduction to structuralist and functionalist philosophies in biology, see for example Lambert & Hughes, 1989 and Sattler, 1990). In autoevolutionism form means function, process, and vice versa. For Sattler (1992) each form is a process combination. There is no dichotomy anymore, only two aspects to be considered. Autoevolutionism is a view of evolution by which it is possible not to fall into problems generated by reductionist and organismic theories. Pattee (1971) presents the dichotomy as follows: "we observe living matter evolving hierarchies of collective order, and non-living matter evolving a collective disorder." In phyllotaxis, the new context established by the relationships developing in collections of primordia is not found in the detailed structures of the primordia, yet plants obey physicochemical laws given the inherence of physical and chemical evolutions in organic matter whose order must proceed from the order in nonliving matter. Phyllotactic patterns cannot emerge from chaos in the usual sense of turbulence, disorder or war between elements, but can emerge from chaos in the sense of a reality awaiting manifestation.

Pattee (1974) recorded that "the theory of evolution (we understand neo-Darwinism) lacks an independent measure of fitness (selection), without which it has very little predictive value." For autoevolutionists neo-Darwinism has no method for predicting the emergence of new species. Moreover, in Darwinism organisms are at the mercy of the environment by means of selection, while in autoevolutionism every level of life's organization is integrated with and counteracts the environment, and this counteraction increases with evolution.

Autoevolutionism avoids the problems generated by the concepts of selection and randomness, by the fact that there are as many Darwinistic interpretations as there are authors, and by the fact that there is no unified consensus within neo-Darwinism. Some of the criticisms of autoevolutionists toward Darwinism are as follows.

Selection and randomness is a morass of concepts, words used to cover ignorance; they must be banished from evolution. The use of selection at the molecular level is tantamount to introducing an element of mysticism, the term is totally foreign to exact sciences. These concepts have been the opium of the biologist for over 100 years. One of the main reasons why Darwinism is so firmly entrenched in our minds is that it fitted perfectly the Victorian age and it continues to serve the ideals of our industrial society. Interpretations of evolution should be based on the physical agents and on the chemical components of organisms, not on population relationships" (Lima-de-Faria, 1988). "Darwin merely translated the prevailing English political ideas and applied them to nature (Radl, 1930).

Whatever may be the value of these polemical statements, "molecular analysis leads to the inescapable conclusion that life has no beginning; it is a process inherent to the structure of the universe" (Lima-de-Faria, 1988). Life is a phenomenon that started long before the evolution of biological organisms. As a manifestation of life it is clear that phyllotaxis will not reveal its secrets unless we are willing to go beyond the over worn-out lines of thoughts such as diffusion and contact pressure.

12
The challenge redefined

12.1. Early hypotheses

12.1.1. General dissatisfaction with chemical hypotheses

The Introduction to Part II lists the main hypotheses on phyllotaxis. Schwabe (1984) lists about 30 hypotheses. Steeves and Sussex (1989) give an account of the early hypotheses on phyllotaxis, such as Snow and Snow's (1931) first available space hypothesis elaborated from surgical experiments on *Lupinus albus*, Wardlaw's (1949a) physiological field hypothesis developed from experiments on *Dryopteris* (a fern) and originating in Schoute's (1913) work, and Plantefol's (1948, 1950) hypothesis of foliar helices.

Based on some of the assumptions regarding the mechanism at work in the complex apical area, numerous mathematical models were formulated in attempts to reproduce the various arrangements of shoot appendages. Successful reproductions of some types of spiral phyllotaxis were obtained, mostly for the types represented by the main sequence, and sometimes only by the first few terms of this sequence. The simulation generally instructs a computer to continue to reproduce a pattern already initiated by the modeler with a few primordia.

In Steeves and Sussex (1989) the field theory is presented as the most plausible hypothesis to explain the placement of leaf primordia in helical phyllotaxis. The fields are generally visualized in terms of the production of inhibiting substances or in terms of withdrawal of nutrients. Meicenheimer (1980) worked with *Epilobium hirsutum* in order to test his working hypothesis that auxin-induced growth-gradient interactions are the source of phyllotactic control in this species. He concluded that "final acceptance or rejection of the inhibitor hypothesis awaits accumulation of evidence . . . it is recognized that the auxin hypothesis is probably too simplified to correspond to biological reality."

Schwabe and Clewer (1984) stress "that none of the schemes have been successful in explaining whorled patterns as well as spiral patterns, with the same underlying theory and assumption." They reject other chemical field theories and claim that their computer simulation, postulating a chemical inhibitor or morphogen and using a mechanism of diffusion, is capable of explaining both whorled and spiral phyllotaxis in the same system, the transitions between various systems, and even false-whorls (when not all primordia in a whorl are laid down at once). The simulation shows that variation of the parameters allows for transitions among whorled, spiral, decussate, and distichous patterns. But it deals only with the main sequence and with divergence angles in the range of 135° to 145°. It covers neither the accessory nor the lateral sequences. Also the authors have not considered the case of superposed whorls. According to Rutishauser and Sattler (1985), comparative studies by various botanists have shown contradictions of Schwabe and Clewer's generalizations. In particular, one of the assumptions in their model is that new leaf primordia are never initiated below and between existing older primordia. This is not always so.

Chapman and Perry (1987) champion the inhibitor theory. They adopt similar ideas as in Schwabe and Clewer's and in Veen and Lindenmayer's approaches, but with some important differences to overcome the limitations they identified. One of these differences is that leaf positioning is controlled by changing concentrations of a promoter substance. And they also feel that a simulation should develop spiral phyllotaxis from a distichous seedling, instead of assuming a particular growth pattern at the start. Starting with distichy they observed a striking similarity between their predicted values for the divergence angles for consecutive leaves, and the values observed by Williams (1975) on flax seedlings. Also they obtained three groups of divergence angles – around 100°, 135°, and 180°. They are confident that it should be possible to generate whorled patterns.

The patterns described by Sattler and Singh (1978) in the Alismatales, by Leins and Erbar (1985) in *Asarum europaeum*, and the pattern described by Leins and Schwitalla (1984) in *Pereskia*, where the fascicle primordia of the stamens appear half-way between alternation and superposition with the inner perianth members, are reported as contradicting inhibitor models. For Lacroix and Sattler (1988) the mechanical theories cannot explain the superposition of the five stamens and five tepals observed in *Basella rubra*. Moreover, they state that superposition between perianth members and stamens occur in about 13% of all angiosperm families. Even Thornley's extended diffusion model (1975a) fails for superposed whorls. The fact is that we can observe a general scepticism and uneasiness with the diffusion hypotheses. But the case of superposition reported for *Basella* may simply have to do with clear-cut orthostichies in spiral systems; the authors state indeed that there is no break in the 2/5 (144°) phyllotaxis, the divergence being maintained throughout the flower. If this is simply a usual case of spiral phyllotaxis expressed by the Fibonacci sequence, then what about the value of the 13%? Apparently in *Basella* we have false whorls. In the case of *Pereskia* we may have a spiral system with slightly twisted orthostichies, a spiro-whorled system as in Figure 8.2(1).

Not only is the inhibitor diffusion hypothesis not supported by any data, but also there is contradictory data. The diffusion models succeeded in showing that hypothetical mechanisms are possible. But their actual operation cannot be confirmed. In particular the existence, nature, and mode of transport of the inhibitor, whose diffusion is assumed to give rise to the phyllotactic patterns, is unknown. The role of auxin in the determination of the patterns still awaits experimental support after several decades in which these hypotheses have been favored in various circles. These models, built on Hofmeister's rule according to which the primordia should arise as far away from the existing primordia as possible (thus giving spiral patterns and alternating whorls), are contradicted by the mere existence of superposed whorls. Even if the growth, form, and position of young leaves depend on chemical constraints, we cannot identify these constraints and the inhibitors.

12.1.2. General dissatisfaction with physical hypotheses

Plantefol's hypothesis is omnipresent in the French literature where it is said to explain the most diverse phyllotactic arrangements. Camefort (1956) and Loiseau (1969) present a complete report of it. The phenomena of anisoclady and anisophylly apparently support the hypothesis. For Philipson (1949) and Carr (1984) the theory, despite its speculative nature, accounts for many facts of the apical development. But according to Cutter (1959, 1965), it is not supported by any conclusive evidence. Much has been written against this hypothesis, in particular by Wardlaw (1965b) who made an extensive critical analysis of it. The hypothesis has not been applied to complex phyllotactic systems such as the sunflower capitulum. A preliminary obvious remark about the hypothesis is that it would explain phyllotactic patterns using one family of parastichies only (the foliar helices), while it is already so hard to explain these patterns with parastichy pairs. Moreover, it avoids mathematical treatments and fails to solve the problem of origins.

For Adler (1974, 1977a) contact pressure is responsible for the positional accuracy at the apical meristem. But for Schaffner (1938) contact pressure represents "a purely superficial view." Richards (1948) underlines that there is no contact pressure in the arrangements of leaf primordia in ferns showing Fibonacci phyllotaxis. Williams (1975) reached the conclusion that a physical constraint is responsible for phyllotactic patterns, but he is careful not to identify the constraint with contact pressure postulated by Schwendener and Adler. According to Erickson (1983), it is doubtful that Williams's data provide a test of the contact-pressure hypothesis. For Chapman (1988) this hypothesis is open to several physiological objections, while for Schwabe and Clewer (1984) it "rather lacks experimental evidence and does not command sufficient support for further consideration."

Green (1985) underlines the idea that the mechanistic theories of phyllotaxis are first approximations to a complex process of biological pattern formation by plants. From his observation of cell files and of structural reorganization in the epidermis of the apical dome in *Vinca major* and *Ribes viburnifolium*, Green (1985) states that the sites for future leaves can be identified, well before a contact appears between primordia. He concludes that this weighs against the theories based on mutual contact pressure of primordia or on physical constraints determined by the older primordia whose role is a positive, not an inhibitory one, and that this possibly renders the mechanism of diffusion unnecessary.

Hardwick (1984) rejects Snow and Snow's first available space hypothesis, the hypotheses based on diffusion of inhibitors, the contact-pressure hypothesis, Larson's procambial strands hypothesis, and Wolpert's (1971) general theory that cells would respond to positional information. Hardwick suggests a temporal mechanism called the "clockface hypothesis," which combines ideas of mechanical stresses at the cellular level, tensions on the epidermis, circulating waves of cell division, and minimal free energy at the level of cell plate. The inducements he offered (bottles of champagne) to those whose work would bring support to his equation involving the divergence angle do not seem to have produced any results.

Green (1985, 1987, 1991), and Jesuthasan and Green (1989) advocate a major role for the dynamics of the cells and for the epidermis in the spacing mechanism in phyllotaxis. While in most of the models the leaves are treated as either points or circles, in Green's approach the leaf is dealt with as a hoop-reinforced appendage. Using polarized light he described what happens in the microstructure of the apex epidermis as leaf primordia are initiated. The site of leaf initiation is considered to be determined largely by angular discontinuities on the cellulose reinforcement pattern of the epidermis of the shoot apex. As an organ is initiated, a region of the surface (tunica) layer of the apical dome bulges under the pressure coming from the corpus. Green observed a good correlation between distichous, decussate (in *Vinca major*), and trimerous (in *Abelia*) patterns, and the average orientation of the reinforcement patterns of cells and cellulose microfibrils. But primordia do not always form at sites that are precisely predictable, especially in spirals systems for which the approach is underdeveloped. The transitions between the patterns have not yet been described.

Green's reinforcement pattern of cells depends upon interaction of surface forces. But working on sunflower capituli at early stages of their development, with the scanning electron microscope, and using close-ups on receptacles such as the one in Figure 1.7, Hernandez and Palmer (1988) showed the subsurface origin of the disk floret initials. The initials form beneath the surface of the receptacle at precise positions determined by Fibonacci phyllotaxis. Sectioning of the tissues of the receptacle confirms that the first cell divisions creating the initials occur beneath the receptacle epidemis well before the raised surface pattern can be seen. Also, the pattern is seen to be determined before the florets come into contact; that is, contact pressure is a subsequent phenomenon and does not induce the pattern.

Church (1920b) argues that one point is firmly established: the mechanism of phyllotaxis is completely independent of cell segmentation. The parastichies indicate a type of segmentation of the protoplasmic mass wholly independent of cell formation.

> New primordia arise, as seen in longitudinal sections, as waves of lowest elevation, often involving a considerable number of cells; there is no sharp demarcation of such undulations, nor can it be said where to a single cell each exactly begins or ends. The mechanism has no relation to the more obvious cell-framework of the plant apex. . . . The protoplasm of the apex may be preferably regarded as a mass of fluid colloidal plasma, in which the secondary production of denser colloidal cellulose films may have but little effect on the physical condition of the living and fluid mass as a whole. [The mechanism is] quite distinct from cytological details; it is due to some invisible cause beyond the reach of microscopic observation.

As an echo to Church and in contrast to Luck and Luck (1986), Kaplan and Hagemann (1991) also stressed the idea that the explanation of plant pattern in higher plants, independent of cell division, involves principles above cell theory. The cells of plants are merely markers of growth, rather than the source of morphogenesis. How the plant engine works out a pattern does not seem to depend on the response of its outer layer of cells.

12.2. Well-grounded models

12.2.1. Light and water

All models, by the very definition of "model," are formal. Even if some models appear more biologically or physically grounded than others that seem more formal, they may not be in fact. The mechanistic models have varying amounts of appeal given such apparently physical principles as contact pressure and diffusion of an inhibitor. Moreover they can presumably undergo scrutiny in the laboratory. But recent experimental research shows that such scrutiny discredits these models.

Other examples of the so-called physicochemically grounded hypotheses include Larson's procambial strand hypothesis elaborated from *Populus* and Roberts's diffusion–nutritional hypothesis. Hernandez (1988) says that the former does not appear to be applicable to the formation of the sunflower capitulum, and he refutes the latter, given that nutrients are not transported with sufficient efficiency to be able to exert any kind of control over the location of the floret primordia in the sunflower.

Despite the evidence there are still many workers using these old hypotheses, even after they were judged, by Sinnott (1960) for example, as too simple and not logical enough. Church (1920b) speaks about the futility of attempting to reach a final solution by mere observation of the ontogeny of primordia at the shoot apex, stating once again that "all primary structural relations trace back to the sea."

However successful the so-called physically grounded models may be, they do not address the fundamental question of *why* the patterns arise. According to Wardlaw's (1965a) eighth principle of plant organization, the environment must take part in the establishment of the patterns. This means again that the problems of phyllotaxis will not be solved by merely observing the ontogeny of the primordia at the shoot apex. It also means that we need a systemic overview on the phenomenon of phyllotaxis.

One of the leitmotifs found in the literature is the interpretative-like idea that phyllotaxis is meant to maximize light-capturing efficiency. According to Bonnet (1754) and to Leonardo da Vinci (see MacCurdy, 1955), the arrangement of leaves is functional and teleological, in that it is designed by God for the purpose of giving the plant better access to water and light. Leigh (1972) tried to prove mathematically the pretention that the Fibonacci angle maximizes the efficiency to capture light by using a simple treatment with continued fractions. The economy principle in Section 9.1 is another variation on the same theme. Certainly, phyllotaxis has to do with photobiology, though many morphological and physiological factors can circumvent the influence of phyllotaxis. For example, leaf orientation can compensate for any inefficient phyllotaxis that might limit the quality of direct solar radiation reaching the leaves.

A decisive quantitative treatment of the question was given by Niklas (1988). He proposed a computer modeling procedure and two methods taking into account most of the variables that could influence light interception, including a variety of morphological features (shape, area, and number of leaves; petiole length, internodal distances between leaves; leaf opacity; and deflection angle of the leaves from the subtending stem), as well as latitude, season, time of the day, and the divergence angle. By varying the divergence and holding all the other variables constant, he assessed the influence of phyllotaxis on light interception. The simulations indicate that phyllotaxis can significantly influence the quantity of light intercepted by leaf surfaces in rosette plants with stiff, long, and nearly linear leaves, and that in this case the most effective phyllotactic pattern for the sake of collecting light is the common one. Plants with other patterns readily attain efficient light interception by varying leaf shape, orientation, and stem length to compensate for the negative effects of leaf overlap produced by phyllotactic patterns.

An emerging trend, which is in line with the systemic view and the idea of diffusion of substances at the same time (expressed by the well-known Fick's laws), is that *the free energy of the hydrous network is the interface between the genetic, organismal, and the environmental determinisms,* and the center of a system of functional coordination controlling the architecture and the distribution of resources of the plant as a whole.

Monroy Ata (1989) proposes that the arrangement of the resistances to the hydrous flow is at the origin of the emergence of the Fibonacci sequence in the geometry of plants, and that the genetic determinism is not sufficient to explain this emergence. Fibonacci phyllotaxis operates at the epigenetic level. Primordia would be harmonic functions of the hydrous potentials diffusing throughout the plant which would translate the influence of the various environmental tensions (generated for example by gravity, thermal radiation, light, electrical phenomena, wind speed) and would modulate the expressions of the genotype.

> The lack of consistent evidence of hormonal control of plant development is attributed to the overriding influence of water, ... The fundamental effects of water on growth and metabolism suggest that water may have played a major role in the evolution of the various physiological mechanisms that regulate the pattern of plant development (McIntire, 1987).

McIntire (personal communication, March 1990) considers that the distribution of water in the shoot apex, as determined by the pattern of translocation and localized accumulation of osmotically active solutes, may be a major factor in determining the sites of primordium formation.

12.2.2. Lines of force and energy

According to Niklas's simulation (Section 12.2.1), phyllotaxis could operate as a limiting factor in photobiology, provoking compensatory adjustments in other morphological features not directly controlled by patterns of leaf initiation. It is considered here that phyllotaxis is rather the result of a very complex optimization scheme, which started to operate long before the existence of higher plants, and which operates at a much larger scale than postulated by Niklas. It encompasses the ability of plants not only to capture energy (radiant or not), but also to conserve it and to use it for themselves and for the needs of other living organisms.

The question must be addressed in terms of what is achieved by phyllotactic patterns. We are driven to state, as an epistemological and physically grounded assumption, that more comprehensive solutions will come via approaches interpreting the patterns and the morphogenetic processes involved in more systemic terms. What looks like a more formal approach can prove to be more physically grounded and realistic than the popular hypotheses. The author's theory of phyllotaxis (summarized in the Introduction to Part II) stresses the necessity of a global and well-grounded approach that makes use of the results of various sectors of biology, and of comparative morphology throughout nature, in order to dig out the so-called invisible cause mentioned by Church (p. 254).

However inspiring the mechanistic models may be, their links to plant physiology and biochemistry are rather weak, which lowers their heuristic value. But even the physiological evidence does not permit a satisfactory account of phyllotaxis, given that plants live in an environment in which numerous factors interact to make phyllotaxis a functional process. The premises of the interpretative model are more general and more in line with a body of ideas in biochemistry, according to which the study of biology would start with the principles of thermodynamics, which for

Lehninger (1982) constitute the most fundamental point of view from which to study all biological phenomena.

It is mentioned in Section 6.2.2 that Green compares what happens on the tunica to bulging metal plates and to potato chips. It is remarkable that figures from Chai (1990) illustrate decussate phyllotaxis with contour plots, that is, lines of equal out-of-plane deformations of a thin plate of metal anchored at its boundary. A pair of opposite humps appears, and then another small pair appears at 90° to the first, mimicking one cycle of decussate phyllotaxis. Green (1992) advocates the idea that patterns in phyllotaxis result from large-scale coordination and he makes an analogy with buckling processes. In the theory of bucklings (applied mechanics, Szilard, 1974), a stressed surface buckles so that the energy input is accommodated by the surface bending to reach the configuration of minimal strain energy and of maximal entropy. Chai (1990) considered only symmetrical modes of deformations. What kind of asymmetrical pattern can be generated with thin plates? The answer to this question would throw light on spiral phyllotaxis. But the homology between the tunica layer and a metal plate seems to be limited by the fact that the dynamics of the thin plate come from pressures applied to its rim while for the tunica they come from the corpus.

Whatever it may be, all this is reminiscent of Church's statements, given in Sections 3.2.2, 11.2.3, and 12.1.2, on moving waves, undulations, and lines of force. One point is firmly established. Green's homologies illustrate the necessity for a larger framework of thought to attack the problems encountered in phyllotaxis. We have seen (Section 10.4) that botanists dare compare phyllotactic pattern with bubble rafts and crystals, following those who like Erickson (Section 10.2.2) compare them to microorganisms and to patterns of spheres around cylinders. Scientists in other areas than botany compare phyllotactic patterns with patterns in polymers, Inoviruses, and jellyfishes. We will see that they can be compared with patterns of droplets of ferrofluid in magnetic fields. All this is characteristic of a new state of mind on phyllotaxis. Time is at a global approach. In Part II I propose more general concepts and a more logical kind of quantified and axiomatic approach, which is consistent with the necessity to find homologies. These homologies develop their full meaning in the context of autoevolutionism (Section 11.5).

The most fundamental question of phyllotaxis concerns the reason for the initial arrangement of the primordia. Most models start with an initial configuration of primordia based on postulated ontogenetic mechanisms. Given that almost nothing is known about the intimate processes of evolution, about the properties of space–time responsible for it, and about the actual ontogenetic mechanisms inside shoot apices, the answer to the question has been inferred in Chapter 6 in terms of the ultimate effects phyllotactic patterns are considered to achieve as expressions of the rhythmic activities of plants. The methodological paradigm introduces general concepts of rhythm, energy, entropy, stability and complexity, and general principles. We believe that reductionistic and ontogenetic approaches alone are not adequate. Rather, how the plant engine works has to do with its phylogenetical history. The treatment given in Part II is not optimal, but indicative of the interest of functional approaches. By stressing the importance of perturbed patterns (Section 8.6.3), by integrating the question of primordial sites in root systems (see Barlow, 1989b), by giving more importance to the various rhythms operating inside plants – the fundamental feature of the phenomenon of phyllotaxis – by coupling more appropriately the ontogenetic and the evolutionary aspects, and by examining phyllotaxis in its physicochemical pregenetic setting, bolder theories will come to the fore that will enable us to make more predictions and more substantial applications.

12.3. Synergy and systems research

12.3.1. A pyramid of models

In the history of Hinduism we are told that God is like an elephant surrounded by blind men. Each man touches the elephant – the trunk, the tail, the foot, the head, and forehead, and so on – from the point at which he is standing and believes he holds the whole. Thus, we have the wars of religion. There are many investigators in the field of phyllotaxis; each has a particular point of view, and his or her metaphysical view is a secret guide. I believe that each perceives a part of the reality and that a productive approach now is to try to integrate all the known parts which, although to some extent antagonistic and contradictory, are complementary to each other.

I agree with Rutishauser and Sattler (1985) that by a commitment to only one model we risk overlooking phyllotactic phenomena that tend to confirm alternative models. According to their philosophy of perspectivism, complementary models (theories, interpretations, descriptions, etc.) represent different perspectives of the same phenomenon. To complement that point of view, I propose to view the modeling work pyramidally. Perspectivism seems to refer to a horizontal dimension among models or theories viewed as complementary. The idea of a pyramid refers to both horizontal and vertical dimensions of modeling in phyllotaxis, and is consistent with the idea of layers of models explained in Section 11.4.

We have seen in Chapter 9 for example, convergences among the results found in various models of phyllotaxis, where the approaches to the problems are very dissimilar. Some models have concepts, structures, and results that appear to be more general than those of other models. Over and above the traditional distinctions between developmental and evolutionary biology, between causal and descriptive phyllotaxis, and between biological and mathematical phyllotaxis, theories of phyllotaxis must welcome any attempt to deal with the multiple aspects of the phenomenon. The systemic coherency afforded by general concepts can be a ground for supermodels that generalize and unify the approaches to phyllotaxis. The key words in this area are evolution and system, phylogenism and holism, rhythm and cycles, energy and entropy, branching processes and gnomonic growth, functionality and teleonomy, and homology and morphogenetical distance. Theoretically, more general models would include previous models, and every attempt would have its place in the pyramid. The pyramid must be based on reliable data and well-grounded descriptions. The top of the pyramid, if attainable, represents the best solution that embodies a comprehensive biomathematical model concerned with general principles incorporating ideas of energy, entropy, and rhythm, and having all the explanatory power the other models have. This monograph proposes a Copernican style revolution – the apical dome is not the center of the world – that will correct the errors of perspective of the various approaches to phyllotaxis and the errors of alignment of the various stones in the pyramid.

12.3.2. Biological and mathematical phyllotaxis

Phyllotaxis has both biological and mathematical facets. Expressing himself for biomathematics in general, Rosen (1977) emphasized "that no resolution, or even proper formulation, of these problems can be forthcoming unless the mathematician and the physicist can bring their skills, techniques and insights to bear upon them." This is particularly true for phyllotaxis, and it is in line with d'Arcy Thompson's thoughts that "problems of forms are in the first instance mathematical problems, and problems of growth are essentially physical problems."

Mathematics not only deals with descriptive phyllotaxis, but also with causal phyllotaxis, via hypothetico-predictive constructs that can be confirmed or disconfirmed by the biologist. The place of mathematics in phyllotaxis, too often considered only as possibly useful for a posteriori descriptions and as not being able to reach the causes and generating principles, must be generally redefined. For example, the question "why is maximal uniformity so important in plant growth?" (Section 9.1) does not belong only to the realm of biology but to that of mathematics as well. Phyllotaxis is the simple (but not elementary) mathematical solution of an optimality problem caused by complex environmental constraints that can be modeled as a whole.

On the other hand, botany very often appears to be primarily based on data collecting. Moreover, this data collection seems rarely to be organized around the idea of testing models. Thus there is a lack of integrated data, despite the abundance of data in general (Chapter 7). Biological descriptive research can point to the large variety of patterns in plants, but this is not sufficient for scientific understanding. There comes a time when we must go beyond detailed descriptions. Otherwise these descriptions will eventually appear useless and unscientific. They will even generate false problems, if no framework is shown in which the apparently disconnected data can be organized and interpreted scientifically. We need research that can provide more coherent frameworks in which to organize the data, in order to make testable predictions. The fact is that

> every natural science has necessarily an experimental aspect and a theoretical aspect. The absence of one of the two to the exclusive profit of the other can only lead either to an accumulation of dissimilar and disconnected observations, having no scientific meaning, or to more or less gratuitous speculations that risk to have no link with the observable realities (P. Delattre, 1987).

12.3.3. Systemic phyllotaxis

Phyllotaxis is a scientific discipline which is not adequately described by the word used to represent it. Indeed, the word was coined 160 years ago simply for the study of the arrangement of leaves on stems. Even in that restricted sense, the knowledge of phyllotaxis has important consequences in ecology and anatomy. "It is a prerequisite for understanding the internal vascular system in the shoots of plants, because leaf arrangement closely determines the course and interconnection of vascular bundles, and is a factor in the photosynthetic strategy of a plant, since the primary arrangement of leaves can determine maximum light interception" (Tomlinson & Wheat, 1979). Phyllotaxis includes the study of vegetative shoots and shoot meristems, the initiation of leaf primordia, the development of leaves and buds, and the inception of floral appendages. It includes the study of many types of growth centers – such as stamen and tepal primordia, hairs, axillary buds, stipules, enations, bristles, leaf pinnae, and all other intermediate structures.

The primordial patterns must be studied not only ontogenetically, but also phyletically, functionally, and homologically. The phyletic aspect of the problem refers to the problem of origins. The functional aspect (interpretative and teleonomic approaches) concerns the upper levels of the compositional hierarchy in which phyllotactic patterns are embedded. In botany phyllotaxis is concerned with the factors determining the position of shoot primordia. But this problem is a part of a more general problem concerning arrangements of organs in general. We have seen indeed that there are arrangements of primordia in *Hydra* and *Botryllus*. Tentacles, canals, and zooids in some jellyfishes conform exactly to (2, 3) and (3, 5) phyllotaxis. Furthermore, phyllotaxis-like patterns arise at other levels of organization, such as in microorganisms showing 2(2, 3) and 2(3, 5) patterns, and proteins with (7, 11) phyllotaxis and 99.5° divergence. Crystallographers use terms such as parastichy (spiral) and rising phyllotaxis to describe their lattices, and their own terms such as cylindrical crystal, dislocation and disclination, scale-invariance, inflation and deflation, and quasi-periodic lattices are beginning to invade the field of phyllotaxis. Because of morphogenetical parallelism, the methods of phyllotaxis can be exported to other fields (e.g., proteins, Section 10.3).

A more appropriate name for the discipline, beside Systemic Phyllotaxis, stressing the presence of homologies, structural similarities, correspondences among various structures in various areas, and the complementarity of the efforts coming from various sectors of science, may be Primordial Pattern Morphogenesis (PPM), the initialism also standing for Pyramidal Pattern Modeling. The name gives to phyllotaxis its full dimension; situates it in a general environment; and unifies mathematical, physicochemical, and biological phyllotaxis, which all three have to do with descriptive and causal aspects. This term also underlines the universal synergy aimed at the solution of the problems recast in mathematical systems theory. Future research in the field will have to give strong emphasis to systems research, as will now be underlined again.

12.3.4. Phyllotaxis, magnetic fields, and superconductors

The application of methods from other fields of research to phyllotaxis is bound to deeply and irreversibly affect the way to look at the phenomenon. A striking illustration is provided by Douady and Couder (1992)

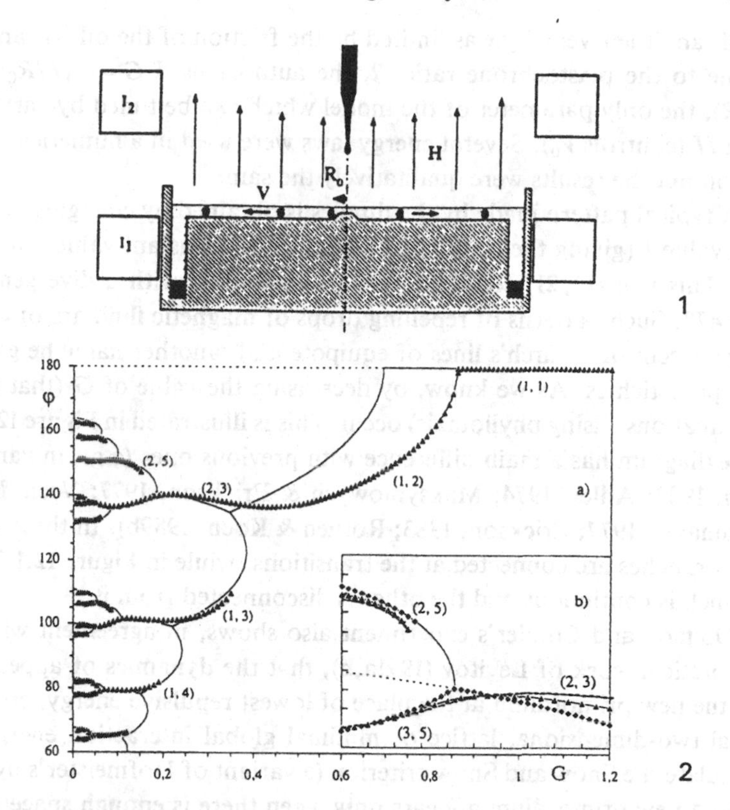

Figure 12.1. (1) Sketch of experimental apparatus in which dipoles under the influence of a magnetic field H created by two coils fall with periodicity T at a given radius R_0 from the center of a dish having in its center a small truncated cone that simulates the apical dome. Due to the interaction with previous drops, the new drop falls from this cone swiftly in the direction of the point of minimal energy. The drops ultimately fall into a ditch at the periphery. (2) Steady divergences obtained as a function of G for two energy profiles (represented by triangles and squares). Inset is a detail of a transition. (Both figures from Douady & Couder, 1992.)

who demonstrated that phyllotactic patterns can result from purely physical mechanisms. They devised an experiment in which phyllotactic patterns were reproduced dynamically by drops of ferrofluid. These drops were of equal volume and fell with constant periodicity T (the plastochrone) at a given radius R_0 from the center of a dish filled with silicone oil and placed in a vertical magnetic field H created by two coils [Fig. 12.1(1)]. Each drop was polarized by the field and formed small magnetic dipoles which generated a repulsive energy $E(d)$ proportional to $1/d^4$ where d is the distance (realizing Hofmeister's condition). These dipoles were moved horizontally at speed V_0 by a radial gradient of the magnetic field, and their velocity was limited by the friction of the oil. As an analogue to the plastochrone ratio R, the authors used $G = V_0T/R_0$ ($G = \ln R$), the only parameter of the model which can be tuned by varying T and H (controls V_0). Several energy laws were used in a numerical simulation but the results were qualitatively the same.

A typical pattern made by the dipoles is obtained by changing G from the value 1 (giving the distichous mode) to the constant value 0.01 after $6T$. This is a (13, 21) conspicuous parastichy pair with a divergence of 137.47°. Such patterns of repelling drops of magnetic fluid are of course reminiscent of Church's lines of equipotential, another name he gave to the parastichies. As we know, by decreasing the value of G (that is R), bifurcations (rising phyllotaxis) occur. This is illustrated in Figure 12.1(2). The diagram has a main difference with previous ones (e.g., in van Iterson, 1907; Adler, 1974; Maksymowych & Erickson, 1977; Veen & Lindenmayer, 1977; Erickson, 1983; Rothen & Koch, 1989b). In these works the branches are connected at the transitions, while in Figure 12.1(2) one branch is continuous and the other is disconnected from it.

Douady and Couder's experiment also shows, in agreement with the theoretical work of Levitov (1991a,b), that the dynamics of appearance of the new primordium at the place of lowest repulsive energy, creates a final two-dimensional lattice of minimal global interaction energy. To simulate the Snow and Snow criterion (a variant of Hofmeister's hypothesis: a new primordium appears only when there is enough space for its formation), they vary the value of T. That is, they allowed a particle to fall only when the minimal energy falls below a given threshold. In addition to the spiral systems the simulation was then able to reproduce multimery, including distichy and decussation, and abnormal cases.

This experiment shows that a hypothesis for the physiological process of the interaction of the primordia is not necessary, and that the various patterns do not seem to be directly encoded in genes. The existence of the patterns seems once again to transcend the decrees of botany, as the phenomenon of phyllotaxis is available without referring to a strictly botanical substratum. This is also the conclusion to which Levitov (1991b) arrives. His study of the "phyllotaxis" of flux lattices in layered superconductors concerns a physical system quite unrelated to botany, yet it gives rise to structures very similar to those known in phyllotaxis. Under variation of a magnetic field a lattice undergoes transformations represented by a diagram identical to Figure 12.1(2). The discontinuities are called quasi-bifurcations, and the branching structure is referred to as a Cayley tree with branching number $N=3$ and underlying symmetry $SL(2, Z) \otimes Z_2$. The evolution of the system under compression can be viewed as a directed walk on this tree. A directed walk can be coded by a sequence of 1 and 0 corresponding to right or left turns at branching points. Special preference is given to the sequences 101010... and 010101... . This is reminiscent of the Bravais–Bravais assumption of alternation giving noble numbers (see Proposition 4 in Appendix 4). Levitov proves the identity of the tree of all possible lattices with the hierarchy of Farey numbers. These numbers are introduced in Appendix 4 to deal with visible pairs of points in cylindrical lattices.

These studies bear upon the future of experimental, biochemical, and physiological research in phyllotaxis. What has been referred to in the introduction of this monograph as the mystery of phyllotaxis – that is, the occurrence of consecutive Fibonacci numbers in conspicuous parastichy pairs – happens to be a property of all lattices built of soft objects subjected to repulsive forces and submitted to strong deformations. It follows that the models based on diffusion of phytohormones, which have always been favored in botanical circles, must be reassessed, if not discarded. And if Fibonacci patterns resulting from the evolution of inorganic matter are more frequent in the plant kingdom than in biology and nature in general, we must look for the causes of this peculiarity. Plants appear to be the perfect media for the perennial expression of simple laws of nature. It might be at the level of the functional and teleological aspects of growth that the resources of botanical research are to be concentrated. Why is it that the set of lattices obtained in phyllotaxis (the set **J** of divergence angles according to the interpretative model) forms a subset of the set of lattices obtained for example by Levitov (1991a,b)?

12.4. The whole is in the part

12.4.1. Spirality and branching everywhere

This book deals with *morphogenesis,* the study of the origins and development of forms and patterns. The search for origins in phyllotaxis is concerned with the search for the remote mechanisms responsible for the patterns observed. Researchers have always tried to reproduce the amazing and complex patterns observed at the ontogenetic level of highly developed plants, but phylogeny reveals simple hidden structures and fundamental mechanisms at work in phyllotaxis. These mechanisms are the phenomena that must be modeled.

It has been postulated that behind the untold diversity of plants hides a limited number of basic structures and patterns. Among them are the spirals and the ramifications, and the dendrites and the fractals, which abound in nature (see the many illustrations of Lima-de-Faria, 1988; Stevens, 1974; Cook, 1974) and in the spectrum of human activities including anthropology and semiotics. Spirals are visible in the structure of galaxies as well as in colonies of *Heliodoma* (invertebrate), in the unfolding leaves of ferns, and in the fronds of *Cycas revoluta* (a primitive seed-bearing plant), as well as on many shells. Related curves and helices are found in proteins. These curves combine to form elaborate patterns in sunflowers and pine cones, on the trunks of palm trees made by leaf scars, and with primordia in transverse sections of buds. Ramifications are also very important in phyllotaxis as we have seen. Ramifications occur in minerals, such as pyrolusite which forms branching crystal dendrites in sedimentary rocks, as well as in fungi and in Purkinje cells, in the structure of the tributaries of rivers as well as in colonies of *Aglaophenia* (invertebrate), and in lightning, etc. The basic structures arising in phyllotaxis come from universal natural processes of growth such as rhythmic gnomonic growth and rhythmic branching, both of which are present at all levels of evolution.

From *Ulotrix* [Fig. 11.7(1)] to *Fucus spiralis* [Fig. 11.7(4)], to the giant sunflower (Figs. 3.12 & 3.13), a fractal is developing in time through organic matter, as it already did through inorganic matter [Fig. 11.7(5)], from a single branch to a fully ramified structure. Complex phyllotactic patterns are the end result of a process by which nature reproduces at the biological level simple characteristics of chemico-physico-mineral laws. This follows from the properties of light, gravity, temperature; and from those properties of the carbon atom, and of the minerals inside the tissues, etc.

12.4.2. Prebiotic and modern creations

Part III of the book argues for the general idea that the patterns and functions we are concerned with in phyllotaxis do not come from genes: they are properties of space, time, matter, and energy. This conclusion is in agreement with autoevolutionism according to which genes only fix a structural aspect present at the levels of evolution that preceded biological evolution. Genes do not create symmetries; they only choose between those that exist, for example, in minerals and quasicrystals, and those that are inherent to the cytoplasmic construction. One of the tenets of autoevolutionism is that no order, no form, and no function is created or lost; it is only transformed by combination.

Since phyllotaxis is not a closed system, it must not be treated, as it is generally done, as a self-contained phenomenon. The phenomenon cannot be separated, as it has always been, from the rest of the world and from evolution. The part is controlled by the totality of the systems. Part III is a sketch of the universal framework in which the analysis and modeling of phyllotaxis must take place. Given the actual state of development of physics (Hawking, 1988) we can say that phyllotaxis is the tip of an iceberg resulting from the "Big Bang." The former phenomenon will be fully understood when the processes initiated by the latter are clarified. Mathematical modeling can shed light on these processes by elaborating deductive mechanisms able to reconstruct the fascinating patterns and by devising inductive models able to initiate the patterns. "Our experience hitherto justifies us in believing that nature is the realization of the simplest conceivable mathematical ideas" (A. Einstein).

Hermant's (1946, 1947) relatively old work regarding "transverse sections of buds" (see Fig. 3.15) should be re-examined with the help of modern techniques. Computer generation of phyllotactic patterns is a subject that has been dealt with successfully by many authors. Others have reproduced with a remarkable realism colorful shrubs, trees, and flowers (e.g., Prusinkiewicz and Hanan, 1989), using rules of production such as L-systems. These reproductions may be of some help in the determination of the causes and mechanisms that induce the natural objects; by necessity they cast each time a new look at nature. This possibility provided by the development of computer graphic techniques raises the problem of the meaning of these reproductions, with respect to the understanding of the fundamental causes in phyllotaxis and of the processes of creation itself.

An exact replica of a sunflower capitulum generated on the screen of a computer and the capitulum itself are both natural objects, one mimicking the other just like a spotlight mimicks the sun. The difference is at the level of their radiating media. They are essentially alike and substantially different. In order to throw more light on the fascinating phenomenon of phyllotaxis, we will have to discover new properties of the primordial substance from which everything is made and of light itself with its various radiations modulating the substance and inducing forms, patterns, colors, and harmony. In the meantime, the best we can probably do is to build more general models that take into consideration the effects those patterns supposedly achieve and their functions in the environment (such as mutual services with pollinators, production of food, manifestation of beauty).

12.4.3. A multidisciplinary problem

It is too often said that biology is not well developed when compared for example to physics which has been mathematized. But Sections 11.4 and

11.5 show that physics in particular will have to undergo more progress if we are to be able to solve the problems generated at the more complex biological levels. One of the reasons with respect to phyllotaxis is that the phenomenon has a physicochemical basis. The levels of investigation of the problem are not only botanical. Also the time has finally ended when the study of phyllotaxis can be made without any reference to fields of research such as paleobotany, evolutionary biology, population genetics, population ecology, physiology, molecular biology, systems analysis, and cybernetics. Apparently we should add to these disciplines epistemology, ontology, and metaphysics, in the following sense.

Whether we like it or not, most research is secretly guided by one's own metaphysics (Bohm, 1969). For example, Morowitz's (1968) preconceived philosophical ideas, the thesis of his book, is that "the flow of energy through a system acts to organize that system." Salthe (1985) "believes that the more complete hierarchy theory is one with a metaphysical commitment to individual entities (things as opposed to processes) as primary phenomena." For Heisenberg, quantum theory does not deal with nature but with our knowledge of nature, or what is usually called epistemology. Einstein "want[s] to know God's thoughts. The rest are details." The primacy of metaphysics in connection with biology is beginning to be more commonly recognized. Popper (1968) argues that "we do not know, we can only guess, and our guesses are guided by the unscientific, the metaphysical faith in regularities." Phyllotaxis is a part of the integrated whole, including the observer. The whole has repercussions on the part which must be dealt with in a systemic way.

Consequently, what would be needed to cope with the challenge of origin beside the possibilities that can be opened by an autoevolutionary theory of compositional hierarchies, is a bio-philosophy which considers that nature is far richer than usually thought. This knowledge may have nothing to do with scientific research as it stands today. A more systemic ontological orientation may do more for the advancement of science in general and the understanding of biological phenomena in particular than detailed analyses of structures and functions. Then shall we be able to better understand the origin of function and control in collections of primordia. In that direction Pattee (1974) wrote that

> the physical basis of hierarchical control does not lie at any one level of physical theory. This means that we cannot understand the nature of biological compositional hierarchies [Section 11.4] simply by a finer look at molecular structure, nor by solution of detailed equations of motion, nor by application of non-equilibrium statistical thermodynamics.

Maybe more stress should be put on teleomatic, teleonomic, and teleological arguments, better called end-resulting, end-directed, and goal-seeking arguments (or goal intendedness, functionality, finality, goal directedness, and purposiveness), in spite of the aversion scientists may have to these words (as reported by Wicken, 1988; O'Grady & Brooks, 1988; Sattler, 1988). As someone already said, the fact is that teleology (the science of ends) is a concept without which (almost) no biologist can live; yet he is ashamed to show himself in public with it.

Epilogue

Parts I, II and III of the monograph can be summarized in the following way: (1) what the patterns are (description), (2) what they are for (functionality), and (3) what they have been (origins). Part III pledges a broader approach to phyllotaxis than students are generally accustomed to. The object of the monograph was to present a universal theory of phyllotaxis. Despite its flaws in the general setting of morphogenesis, it is the first such theory.

Part III introduced two important phenomena: the phenomenon of emergence and the phenomenon of homology. It has been recognized that the whole system called a primordial pattern has properties that are not exhibited by the constituting units, the primordia. Emergence means that the whole is more than the sum of its parts. A famous example is sodium chloride (NaCl), which has (emergent) properties quite different from those of sodium and chlorine taken separately. Another example is an impressionistic painting, which is more than the sum of its dots of paint. Of course, as the systems become increasingly complex the results of emergence become more and more striking. Sattler (1986) emphasized that some biologists continue to argue that the most important levels of organization are the cellular and molecular. He discusses the inadequacy of this view in the context of the phenomenon of emergence [the plant forms cells, not cells the plant, as Barlow (1982) points out]. Although a knowledge of the elementary units is a prerequisite, it is not sufficient for understanding phyllotaxis. Mechanistic hypotheses, working from the lower levels of the compositional hierarchy of phyllotaxis, have produced not only what they probably had to but also a general dissatisfaction.

The biomathematical theory of phyllotaxis presented here brings together all the resources available, in particular those of general comparative morphology, which deals with the comparison of structures, forms, and functions. Homology is the central concept of comparative biology. Sattler (1986) emphasizes that many modern biologists tend to find this field either uninteresting or outdated. He hopes that the discussion he proposed has shown that this is not the case. For him comparative studies are important to provide a balanced overall perspective. Meyen's strong position for the importance of general comparative morphology can also be recalled. The theory of phyllotaxis presented here supports the idea that the worlds of crystals, of proteins, and of microorganisms, to name just a few, and the processes of branching and of spiral formation in the universe, can shed light on the study of the general structures and processes involved in phyllotactic pattern generation. We have seen that many researchers outside of botany and mathematics claim that their study of symmetry properties is *generalized* phyllotaxis. With the recent discovery of quasicrystals, phyllotaxis has become a playground for crystallographers, who believe that phyllotaxis belongs to crystallography – as exotic crystallography. Phyllotaxis is a *universal* field of research, and Part III was intended to show it.

Through 160 years of development phyllotaxis has earned the privilege of being called a discipline. It is in fact a field of research ramifying within the core of botany and of mathematics, and situated at a crossroads of several disciplines. The bibliography given in Part IV, which shows the extent to which phyllotaxis has been studied, corroborates this statement. Part III has shown that the subject surprisingly ramifies other areas of science as well, and that the acquisition of knowledge in phyllotaxis does not depend any more solely on the botanist's endeavors. The problem was therefore raised about what constitutes the specificity of the botanical aspect of the problem of phyllotaxis. The structural problem seems to be a purely mathematical problem concerned with general laws of the universe, whereas the problem of botany is concerned mainly with questions of viability of the structures in a particular substratum.

In spite of the tremendous progress accomplished in the study of phyllotaxis during the last 20 years, there is a particular need for practical applications, and there will always be a need for holistic minds able to grasp inductively and creatively the multiple facets of phyllotactic problems, whether they be physiological, ecological, morphological, or mathematical, so as to produce more coherent corpuses where past developments will find their legitimate places. Phyllotaxis is at a crossroads of its development where the dialogues between mathematicians, botanists, and comparative morphologists, need to be strengthened, so as to produce a synergy aiming at better and more universal solutions for the problems encountered.

As a final remark, may future students find in this book a tool that will inspire them to write another page of this never-ending quest for scientific knowledge of phyllotaxis. It is hoped that the in-depth treatment of phyllotaxis given here, especially the state-of-the-art treatment given to pattern recognition in Part I (in Chapter 5 in particular) and the systemic approach to and the insight into the origins of phyllotactic patterns given in Part II and III, will be an incentive for those who want to pursue the study of this most beautiful subject. This monograph presents the state of the art and general guidelines for new programs for investigation in phyllotaxis, and gives an approach to pattern formation which can be transferred to other areas.

Part IV
Complements

Introduction

Part IV contains four types of sections:

Appendixes on gradual learning
  Appendix 1 – Glossary of biomathematical terms
  Appendix 2 – Answers to problems in Chapters 1 to 4
  Appendix 3 – Review questions
Appendixes on special topics
Bibliography
Author Index and Subject Index

The special topics in Appendices 4 to 9 contain material that could not be fully accommodated in the main text. Appendix 4 entitled "General Properties of Phyllotactic Lattices" is the result of my introduction of Farey sequences to this area of botany. Farey sequences is a subject close to continued fractions, but it is in some respects a stronger tool, able to give more general mathematical results. The results presented spring from my generalization of the Hardy and Wright (1945) notion of visible point in a lattice, that is, the notion of visible pair of points in a lattice. The main result states that in a regular cylindrical lattice with divergence angle d, the pair (m, n) is visible if and only if the points m and n are such that $|m(nd)-n(md)|=1$, where (x) means the integer nearest to x. Based on the results, four algorithms can be devised to determine the visible parastichy pairs from d, or an interval for d from a pair: the computational, and contraction algorithms discussed in Chapter 2 and a diophantine together with a graphical algorithms presented in Appendix 4.

Appendix 5 is an analysis of Williams and Brittain's (1984) model, and shows relations to the author's allometry-type model of Chapter 4. The numerical values of the parameters obtained from the former model are seen to be comparable with the values obtained from the Pattern Determination Table in the latter model.

Appendix 6 contains an in-depth discussion initiated in Chapter 5 on Fujita's normal curves, which raise some questions with respect to the standards of data collecting.

Appendix 7 deals with a mathematical by-product of the treatment of phyllotaxis with control hierarchies. It brings forward new results in the study of growth functions of L-systems, obtained by bridging these systems with the Perron–Frobenius spectral theory in this area of mathematics called discrete functional analysis. This appendix presents a table which can be used in the study of filamentous organisms. As mathematical by-products, the results in Appendices 4 and 7 are an indirect testimony to the richness of the approach to phyllotaxis implemented here.

Appendix 8 deals with a diffusion–reaction theory that expresses the very essence of the general approach to pattern generation in phyllotaxis with diffusion–reaction equations.

Appendix 9 shows that the variation of the parameters (m, n), d, r, g (the girth of the cylinder) and γ, in the cylindrical representation, controlled by the allometry-type model of Chapter 4, produces hyperbolic transformations of the plane.

The Bibliography is intended to be an exhaustive compilation of the publications on phyllotaxis. It contains a selection of about 920 titles. It does not of course list all the literature on related fields. Some of the articles and books listed here are discussed in Section 4 of the Prologue.

Appendix 1
Glossary

This Glossary sometimes refers to the main text, but the reader is encouraged to consult the indexes in order to locate text references to the terms defined here.

Accessory sequence: Any one of the sequences of normal phyllotaxis with $t > 2$, such as the Lucas sequence where $t = 3$. The term "accessory" is relative to the main sequence, which is the Fibonacci sequence where $t = 2$.

Acropetal: Toward the apex.

Angiosperms: Seed plants in which the seed is enclosed in a fruit. In the gymnosperms (conifers) the seeds are naked (gymno) on scales, which are arranged on cones.

Angle of intersection γ (of a parastichy pair): The angle made by two parastichies in a parastichy pair.

Anomalous phyllotaxis: A system shows anomalous phyllotaxis when the visible opposed parastichy pairs of the system are made with consecutive terms of the sequence of anomalous phyllotaxis (obtained from the model of Chapter 6), that is the sequence $\langle 2, 2t+1, 2t+3, 4t+4, \ldots \rangle$, where $t \geq 2$ is an integer. Putting $t = 2, 3, 4, \ldots$ gives respectively the first lateral sequence $\langle 2, 5, 7, 12, 19, \ldots \rangle$, the second lateral sequence $\langle 2, 7, 9, 16, 25, \ldots \rangle$, the third lateral sequence $\langle 2, 9, 11, 20, 31, \ldots \rangle, \ldots$.

Apex (shoot apex): The center of a bud where leaf or reproductive primordia are formed. The center of morphogenesis in vascular plants, giving rise to stem, leaves, and flowers. Leaves originate as leaf primordia on the flanks of the shoot apex.

Apical dome (distal zone, bare apex): The tip of the shoot apex, more or less hemispherical, or flat, where there is no primordium.

Autoevolution: The transformation phenomenon, inherent to the construction of matter and energy, which produced and canalized the transformation of biological processes. The interpretation of this phenomenon is called "autoevolutionism" (see Section 11.5).

Axillary bud: A bud situated in the angle between the upper side of the leaf and the stem.

Basipetal: Toward the base.

Bract: A leaflike structure below a flower.

Bravais–Bravais assumptions: Assumptions introduced in the 1830s, according to which (1) the divergence angle $d < 1/2$ is an irrational number, and (2) by moving upward along the vertical axis of the cylindrical lattice, the points of the lattice that are neighbors of this axis are seen alternately on each side of the axis and get closer to it [see Figure 2.3(2)]. The axis is the asymptote of two infinite polygonal lines determined by linking consecutive neighbors on each side of the axis.

Bravais–Bravais theorem: On an n-parastichy, in a family containing n parastichies, the numbers on each consecutive primordia differ by n.

Centric representation (spiral lattice): A diagrammatic representation of the phyllotactic patterns, where the primordia (or their centers) are points in a plane, numbered according to the Bravais–Bravais theorem, where each family of parastichies is a set of identical, evenly spaced logarithmic spirals with the same pole, such that the divergence angle and the plastochrone ratio are constant.

Companion hierarchy: A hierarchy generated by a companion (Frobenius) growth matrix, with one double node only. The double node is the only node in level $T = 0$ of the hierarchy (see Section 3.5 & Fig. 3.17).

Conspicuous parastichy pair: A visible opposed parastichy pair for which the angle of intersection of the opposed spirals is closer to 90° than the angle of intersection of any other visible opposed parastichy pair of the system. The value of the plastochrone ratio R determines which visible opposed parastichy pair is conspicuous. Given a conspicuous pair and its angle of intersection γ we have the value of R from the Pattern Determination Table (see the definition below). The conspicuous parastichy pair of a system represents the phyllotaxis (see definition #2) of the system.

Contact parastichy pair: A visible opposed parastichy pair obtained by following the directions determined by opposite sides of the polygonal primordia sharing their sides (contacts). If the primordia are quadrilateral in shape (as in Fig. 1.3), then there is one contact parastichy pair determined by two families of contact parastichies. If the primordia or organs are hexagonal in shape (as it is generally the case for the scales on pineapples), then there are two contact parastichy pairs determined by three families of contact parastichies. Figure 4.2 shows the possibility of three such pairs because of the particular shape of the primordia: (2, 3), (5, 3) which are visible and opposed, and (5, 2) which is not an opposed pair. The pair (5, 8) is not a contact but a conspicuous pair.

Continued fraction: The continued fraction of a number is a particular way to write that number, obtained from the simple usual process of Euclid's division, by recording the consecutive quotients. For example, $27/8 = [3; 2, 1, 2]$, given that $27/8 = 3 + 3/8 = 3 + 1/(8/3)$, $8/3 = 2 + 2/3 = 2 + 1/(3/2)$, $3/2 = 1 + 1/2$. Also, $\pi = [3; 7, 15, 1, 293, \ldots]$, given that $\pi = 3 + 0.141592654\ldots = 3 + 1/(1/0.141592654\ldots)$, $1/0.141592654\ldots = 7 + 0.062513285\ldots = 7 + 1/(1/0.062513285\ldots)$, etc. A semicolon follows the integral part of the number. The continued fraction of $1/\tau$ can be obtained in the following way, knowing that $\tau = 1 + 1/\tau$: $1/\tau = 1/(1 + 1/\tau) = 1/[1 + 1/(1 + 1/\tau)] = \ldots$, that is $1/\tau = [0; 1, 1, 1, \ldots]$. This continued fraction is thus made of 1s only. The continued fraction of a divergence of normal or anomalous phyllotaxis is made of 1s only starting from a certain term or quotient. The continued fraction of $\sqrt{2} - 1$ is $[0; 2, 2, 2, 2, \ldots]$, made of 2's only.

Every positive number a can be written $a = [a_0; a_1, a_2, a_3, \ldots]$, where $a_0, a_1, a_2, a_3, \ldots$ are positive integers. If a is a rational number then the continued fraction terminates; if it is not, then it does not terminate. In any case a continued fraction can be truncated at any term, that is $[a_0; a_1, a_2, \ldots, a_k] = p_k/q_k$, $k = 0, 1, 2, 3, \ldots$, called the kth (principal) **convergent** of the continued fraction of a. The even convergents of a ($k = 0, 2, 4, \ldots$) form an increasing sequence of rational numbers smaller than a, and converging towards a. The odd convergents of a ($k = 1, 3, 5, \ldots$) form a decreasing sequence of rational numbers larger than a, and converging towards a. We have the relations

$$p_0 = a_0, \quad q_0 = 1, \quad p_1 = 1 + a_0 a_1, \quad q_1 = a_1,$$

$$p_k = a_k p_{k-1} + p_{k-2}, \quad q_k = a_k q_{k-1} + q_{k-2},$$

$$p_{k-1} q_k - p_k q_{k-1} = (-1)^k, \quad p_{k-2} q_k - p_k q_{k-2} = (-1)^{k-1} a_k.$$

Using three consecutive convergents p_{k-2}/q_{k-2}, p_{k-1}/q_{k-1}, p_k/q_k, we can generate other fractions by making Farey sums of consecutive numerators and denominators (the **Farey sum** of p/q and r/s is

$(p+r)/(q+s))$, that is

$$\frac{p_{k-2}}{q_{k-2}}, \frac{p_{k-2}+p_{k-1}}{q_{k-2}+q_{k-1}}, \frac{p_{k-2}+2p_{k-1}}{q_{k-2}+2q_{k-1}}, \ldots, \frac{p_{k-2}+a_k p_{k-1}}{q_{k-2}+a_k q_{k-1}} = \frac{p_k}{q_k}.$$

The fractions are in a decreasing or increasing order depending if k is odd or even, and they lie between p_{k-2}/q_{k-2} and p_{k-1}/q_{k-1}. The new fractions obtained in this way are called **intermediate convergents.** When $a_k = 1$, there is no intermediate convergent for this value of k. τ has no intermediate convergent given that $a_k = 1$ for every k. Detailed analysis of the properties of continued fractions can be found for example in Khintchine (1963) or LeVèque (1977).

Bravais and Bravais (1837) were the first to recognize the relevance of continued fractions in botany. Coxeter (1972) was the first to prove that the alternation of the neighbors of the vertical axis in the Bravais-Bravais cylindrical lattice requires the presence of $1/\tau$ in the continued fraction of the irrational divergences. The following authors frequently used continued fractions: Adler (1974, 1977a) in his contact pressure model, Ridley (1982a), and Marzec and Kappraff (1983) in their packing efficiency analysis, and Rothen and Koch (1989a,b) in their study of self-similarity and shape invariance.

Contraction of the visible opposed parastichy pair (m, n): The pair $(m, n-m)$ if $n > m$, and the pair $(m-n, n)$ if $m > n$. The contraction of a visible opposed parastichy pair is a visible opposed parastichy pair.

Cotyledon: A seed leaf containing food material formed in the embryo before the shoot apex is formed.

Crystallography: The science of symmetrical structures, primarily at the level of the arrangement of atoms. Symmetry means that the structural motif of one part of the design is repeated elsewhere. Units plus recurrence rules or operations give overall patterns. Until 1912, crystallography consisted mainly in the classification of crystals of mineral substances into one or other of the 32 crystallographic point groups, according to their idealised external forms. We may take "crystallography" to be the science studying the structures and properties of assemblies made of large numbers of copies of a single type of unit, or of a few types of units. Crystallography has come to mean a science dealing with structures at the atomic level – the way in which the aggregation of atoms combine to give the manifold properties of matter, living as well as inanimate (definitions taken from Mackay, 1986). It is clear that the extended definition of crystallography includes phyllotactic patterns. Compare indeed with the definition of phyllotactic patterns given in Section 3.5 on Jean's hierarchical representation: a phyllotactic pattern is a partially ordered system of interrelated elements (the primordia) interacting in an aggregative fashion. See also Section 11.4 on the emergence of new properties from an aggregation of units giving phyllotactic patterns. "A crystal lacks rhythm from excess of pattern, while a fog is unrhythmic in that it exhibits a patternless confusion of details" (Whitehead, 1925). In phyllotaxis rhythm demands, as we saw, the essential mechanism. In Mackay's proposal for generalized crystallography, classical crystallography is linked to the Fibonacci pattern of the seeds in the sunflower head via fibers. Mackay (1986) says that there is evidence that when fibers are packed in a dense bunch, twisted like a rope, they produce this kind of pattern. "It may be observed as the lines of Retzius in elephant ivory and perhaps in collagen as well as in commercially produced ropes" (Mackay, 1986). From the early 1930s fibers have been examined with the X-ray diffraction methods used for crystals (see Section 10.3.1).

Cylindrical representation (cylindrical lattice or Bravais–Bravais lattice): A diagrammatic representation of the spiral patterns, where the primordia (or their centers) are, in a first step, considered as points at the surface of a cylinder (e.g., points of insertion of leaves on a cylindrical stem) and numbered according to the Bravais–Bravais theorem. The basis of the cylinder is considered to have a perimeter equal to 1. Each family of parastichies is a set of identical, evenly spaced helices winding around the cylinder, such that the divergence angle and the plastochrone ratio are constant. When the cylinder is unfolded into a plane, by cutting its surface along a generator going through leaf 0, the helices become straight lines, and the primordia form a lattice which resembles that made, for example, by well-aligned apple trees in an orchard. The distance between the two representations of leaf 0 on the same level is equal to 1, and the points of the lattice are distributed in a vertical stripe. In this representation the divergence angle is the horizontal distance between any two consecutive primordia along the genetic spiral. Corresponding to the plastochrone ratio R of the centric representation is the rise denoted by r. This is the vertical distance between consecutive primordia along the genetic spiral. This representation was introduced by L. and A. Bravais in the 1830s.

Decussation (decussate): Leaf arrangement in which pairs of leaves at the same level of insertion alternate with pairs inserted at 90° to them in successive nodes on the stem. When viewed from above the pattern resembles a + or an × (Roman symbol for 10, decem), hence the term. This is a whorl of two, forming four orthostichies.

Dicotyledons: One of the two divisions of the angiosperms, consisting of plants having two cotyledons.

Distichy (distichous): Leaf arrangement (general among monocotyledons) in which individual organs are inserted alternatively in two rows 180° apart with respect to the axis of the plant. The leaves are thus distributed on two opposed orthostichies, one leaf at each level. See the term "multimerous system."

Divergence angle *d* (or divergence): The smaller of the two angles at the center of a transverse section of a growing shoot tip (see Figure 1.3) determined by consecutively initiated primordia. This angle may vary depending on the particular primordia on the shoot. In the centric representation, the divergence angle is considered to be constant. It is denoted by d and is smaller than 180°, or equivalently it is a number d between 0 and 1/2. In the vertical stripe of basis 1 (called the "cylindrical representation"), the divergence is the horizontal distance between any two consecutive points n and $n+1$ on the genetic spiral.

Encyclic numbers: Let (m, n) be a visible opposed parastichy pair in a system with divergence angle $d<1/2$. The encyclic numbers u and v are the integers nearest to nd and md, respectively. They satisfy the relation $nu-mv=\pm 1$.

Extension of a visible opposed parastichy pair (m, n): The visible parastichy pair $(m, m+n)$ or the visible pair $(m+n, n)$. One of the two is opposed.

Family of parastichies: Given any parastichy in the centric representation, the corresponding family is the set of parastichies with the same pitch, winding around a common pole in the same direction and going through the centers of all the primordia. On a cylinder with leaves on it, a family of parastichy is a set of helices going through all the points and partitioning the set of points. The helices are parallel equidistant straight lines in the (flat) cylindrical lattice.

Fibonacci angle: The angle of approximately $137\frac{1}{2}°$, whose precise value is $360(\tau^{-2})$, where τ is the golden ratio given by $(\sqrt{5}+1)/2 \approx 1.618$. By extension the Fibonacci angle is the divergence angle equal to τ^{-2} or its approximation 0.3819.

Fibonacci hierarchy: Any control hierarchy of simple and double nodes generated by the Q-matrix (see Section 3.5.2), with at least one double node in level $T=0$, such as the hierarchies in Figures 3.1 and 3.4.

Fibonacci number: Any number in the Fibonacci sequence. The kth number is denoted by F_k, and the numbers are linked by the relation $F_{k+1}=F_k+F_{k-1}$, for $k=2,3,4,5,\ldots$.

Fibonacci phyllotaxis: A system shows Fibonacci phyllotaxis when the secondary numbers in the conspicuous or contact parastichy pair are consecutive terms of the Fibonacci sequence.

Fibonacci sequence (main sequence, or hauptreihe): The sequence of integers $\langle 1,1,2,3,5,8,13,\ldots\rangle$. The kth term is denoted by F_k. It is the main case of normal phyllotaxis with $J=1$ and $t=2$. It is built using the rule $F_{k+1}=F_k+F_{k-1}$ from the initial values $F_1=F_2=1$.

Fibonacci-type sequence: A sequence built from the same recurrence rule as the Fibonacci sequence but starting with other initial terms, such as the sequences of normal and anomalous phyllotaxis.

Genetic spiral (also called generative or ontogenetic spiral): When the primordia arise successively at the shoot apex, the genetic spiral is the hypothetical spiral drawn through all primordia by following the shortest path from one primordium to the next. Of course, this path can be followed clockwise or counterclockwise. We make the convention that the genetic spiral goes from the oldest primordium near the rim of the system in the centric representation, to the youngest near the center of the apex. The oldest being numbered 1, the genetic spiral follows the natural number in succession. In the cylindrical representation the genetic spiral links the origin of coordinates to point 1 by the shortest path, and then the points $2,3,4,\ldots$. The genetic spiral is the 1-parastichy of a spiral system.

Golden ratio (or number): The number $\tau=(\sqrt{5}+1)/2$. See Section 1.3.2.

Growth matrix: Any square irreducible matrix whose entries are 1s and 0s only, such that the sum of the entries in every column is 1 or 2. The term "irreducible" refers to the strong connectedness of the graph generated by the matrix. Growth matrices are used to generate the phyllotactic hierarchies with simple (1) and double (2) (bifurcation) nodes only defined in Chapter 3.

Hierarchy: (1) a **control hierarchy** is a representation of a phyllotactic pattern as a multilevel system of simple and double nodes linked between the levels denoted by $T=0,1,2,3,\ldots$. In the Fibonacci hierarchies the linkage follows the Lestiboudois–Bolle law of phyllotaxis (see Section 3.1). A control hierarchy is an interface between two levels in the compositional hierarchy of phyllotaxis, the level of primordia, and the level of patterns of primordia.

(2) A **compositional hierarchy** is a multilevel system where the levels

go for example from quarks, to atoms, to molecules, to cells, to organs, to organisms, to communities. One of the objectives of systems research is to study the interrelations between the levels.

Hofmeister's axiom (or rule): Hofmeister (1868) postulated that new leaf primordia arise on the shoot apex in those positions that are farthest removed from the lateral margins of the bases of the nearest existing leaves. He considered that new primordia always arose in the largest gaps between existing primordia.

Hyperbolic turn: The jump on the straight line of slope -2 in the log-log grid, defined by the allometry-type model of Chapter 4, to go from (m,n), $m>n$, to $(m,m+n)$ phyllotaxis with the same angle of intersection of γ (see Section 9.2.1 and Appendix 9).

Induction line: A term introduced by Bolle (1939) to describe the pathways of induction of leaves along a stem to form a pattern of simple and double nodes (see Section 3.1).

Inflorescence: The part of the plant consisting of the flowers and structures bearing them.

Initials: The cells at the tip of the meristem that are the progenitors of all the other plant cells.

Jugy (or jugacy): See Multijugate system.

Lateral sequence: Any one of the sequences of anomalous phyllotaxis obtained by giving an integral value to t, such as the first lateral sequence $\langle 2,5,7,12,19,\ldots\rangle$ obtained with $t=2$.

Law of duplication (law of phyllotaxis): Law first formulated by Lestiboudois in 1848 and rediscovered by Bolle in 1939, which controls the distribution of simple and double nodes (duplications) in a control hierarchy (see Section 3.1).

Limit-divergence angle: Any angle given by the theorem of Chapter 2 as the element in an infinite sequence of closed nested intervals whose lengths tend towards 0, corresponding to a sequence of visible opposed parastichy pairs. The limit-divergence angles are generally of the type $360(t+1/\tau)^{-1}$, $t=2,3,4,\ldots$, whose values are close to $137\frac{1}{2}°$, $99\frac{1}{2}°$, $78°,\ldots$ respectively. More generally, they are expressed by (6.4) and (6.6). Table 2.1 gives the sequence of integers corresponding to important limit-divergence angles.

Lucas number: Any number in the Lucas sequence. The kth number is denoted by L_k, and we have the following relation between Lucas numbers: $L_{k+1}=L_k+L_{k-1}$ for $k=2,3,4,5,\ldots$.

Lucas phyllotaxis: A system shows Lucas phyllotaxis when the secondary numbers in the conspicuous or contact parastichy pair are consecutive terms of the Lucas sequence.

Lucas sequence (also called Schoute's accessory sequence or the first accessory sequence): The sequence of integers $\langle 1,3,4,7,11,18,\ldots\rangle$. It is a case of normal phyllotaxis where $t=3$ and $J=1$.

Main sequence: The Fibonacci sequence.

Meristem: The source of new cells in the growing plants, a tissue of undifferentiated dividing cells found at the apices of shoots and roots.

Microfibrils: Aggregation of cellulose chains in cell walls, having crystalline properties that can be examined with polarized light.

Monocotyledons: One of the two divisions of the angiosperms in which the plants have only one cotyledon in the embryo.

Morphogenesis: Origination, creation of form, organs, and structures in plants and animals (usually by growth). Morphogenesis is concerned with the processes that generate the various morphologies. The positioning of leaf primordia at defined locations on the apical meristem (phyllotaxis) is a key element in plant morphogenesis.

Morphology: Science of the shape and form of plants (or animals).

Multijugate system (or pattern): Generally speaking, spiral systems may have one, two, three or more primordia at each level on the stem or on the plant apex, giving rise to the same number of genetic spirals. The term unijugate is used when the system has one genetic spiral, bijugate when it has two, trijugate for the case of three, and multijugate for several genetic spirals. The **jugacy** (or jugy or jugacity) of a system is expressed by the parameter $J=1,2,3,\ldots$ in the sequence of normal phyllotaxis. Section 6.4.2 proposes a unified vocabulary for naming the various multijugate systems. Multijugate systems are subdivided into normal and anomalous systems. The model of Chapters 6 and 8 presents multimerous systems as special cases of multijugate systems. Multijugate systems include all regular phyllotactic systems known to exist, in particular spiral systems, whorled systems (multimerous systems), and spiro-whorled systems (see Fig. 8.2).

Multimerous systems: Multimerous systems have one (distichy, distichous), two (decussation, decussate, or dimery, dimerous), three (tricussation, tricussate, trimery, trimerous, tristichy, tristichous), or several (tetramery, pentamery, hexamery,...) organs at the same node on the stem, or primordia in the same ring around the center of the shoot apex. This generates what is called orthostiches, that is, vertical rows of primordia parallel to the axis of the stem, or rays of primordia in line with the center of the shoot apex. Multimerous systems can be superposed or alternating. To unify the vocabulary, we have dimery, trimery, tetramery, ..., and the adjectives are dimerous, trimerous, tetramerous, ... The widely used term distichy is equivalent to alternating unimery, and the term decussation to alternating dimery. Chapter 8 proposes a model that considers that multimerous patterns are special cases of multijugate patterns. With the exception of distichy, multimerous systems are also called whorled or verticillate systems.

Noble number: Any number whose continued fraction, from a certain term, contains 1s only, that is, does not have intermediate convergents. The limit-divergence angles corresponding to the sequences of normal and anomalous phyllotaxis expressed as numbers smaller than 1/2 [see Eqs. (6.4) and (6.6) without 360] are noble numbers.

Node: The place on the stem at which the leaf is inserted. The points in the various representations of patterns of primordia or organs. The hierarchical representation of phyllotaxis distinguishes between simple and double nodes following the Lestiboudois–Bolle law of duplication (Section 3.1).

Normal phyllotaxis: A system displays normal phyllotaxis when the visible opposed parastichy pairs are made with consecutive terms of the sequence $J\langle 1, t, t+1, 2t+1, 3t+2, \ldots\rangle$, called the sequence of normal phyllotaxis, $t \geq 2$, $J \geq 1$ (J represents the jugy of the system). With $J = 1$ and $t = 2$ we have the Fibonacci or main sequence. With $t = 3, 4, 5, \ldots$, we have respectively the first accessory sequence or the Lucas sequence, the second accessory sequence $\langle 1, 4, 5, 9, 14, \ldots\rangle$, the third accessory sequence $\langle 1, 5, 6, 11, 17, \ldots\rangle, \ldots$. For $J > 1$, Section 6.4.2 gives names to the sequences. Normal phyllotaxis is a case of multijugacy (see Multijugate systems).

***N*-parastichy**: See the term Parastichy.

Opposed parastichy pair: A parastichy pair in which the two families of spirals (in the centric representation) or helices (on the cylinder) or straight lines (in the cylindrical representation) wind in opposite directions with respect to the center of the plant shoot (point C in Fig. 1.6) or the axis of the plant [the Y axis in Fig. 2.3(2)]. For example, in Figure 2.3(2) the pair (8, 5) is opposed, while (13, 5) is not. In Figure 1.6 the pair (3, 2) is opposed. In Figure 1.4 the pair (13, 21) is opposed, while the pair (8, 21) (not drawn) is not.

Orthostichy: In multijugate as well as in multimerous systems an orthostichy is a particular parastichy, a straight line obtained by joining two leaves or primordia that are approximately superposed with respect to the apical tip or with respect to the axis of the plant. Orthostichies are approximately parallel to that axis in the latter case, and they are almost in line with the pole in the former case.

Overtopping: A branching process where the two segments of the doubled branch grow so that the length of one segment exceeds the other and tends to form a main axis.

Parastichy: Any phylloactic spiral seen on plants, or in transverse sections of plant apices. By extension, a spiral, a helix, or a straight line linking points in the centric or cylindrical representation of spiral patterns. The word comes from the Greek "side row," which reflects a contrast with the genetic spiral (main row) which is also a parastichy. The term was coined by Schimper in the 1830s. If the primordia are numbered according to the Bravais–Bravais theorem and if the numbers on two consecutive primordia along a parastichy are a and b $(b > a)$, then the parastichy is defined by the points numbered $a, b, 2b-a, 3b-2a, 4b-3a, \ldots$, where the consecutive numbers differ by $b-a$. If along a parastichy the numbers differ by n then we say that it is a ***n*-parastichy**.

Parastichy pair: A pair of families of parastichies. A parastichy pair can be visible and opposed, opposed but not visible, visible but not opposed, and neither visible nor opposed. In the text we concentrate on visible opposed parastichy pairs, among which are contact and conspicuous parastichy pairs. In Appendix 4 we analyze visible parastichy pairs. The parastichy pairs are denoted by (m, n) where m and n are the numbers of parastichies in each family (the secondary numbers), and the angle of intersection of any two parastichies in the pair is denoted by γ.

Pattern Determination Table: A table (see Table 4.3) giving the values of some phyllotactic parameters controlled by the dynamic model $\ln R = 2\pi p(\gamma)/(m+n)^2$ (Jean's allometry-type model for pattern recognition discussed in Chapter 4), and by an implicit relation between (m, n) and d (Jean's general theorem in Section 2.3.1), which can be used to gain knowledge of any spiral pattern when some other parameters are given by direct observation. Here d is the divergence angle, R the plastochrone ratio and $p(\gamma) = \tau^3[\sqrt{5}\cot\gamma + \sqrt{(5\cot^2\gamma + 4)}]/2\sqrt{5}$ is a function of the angle of intersection γ of opposed spirals in the opposed parastichy pair (m, n). This table can be used for all possible cases of pattern recognition reported in Table 4.2.

Perturbed pattern: Any pattern that is not multimerous or multijugate. It may be a combination of these. The literature offers the terms chaotic, erratic, pathological, scattered, irregular, abnormal, and teratological. The term proposed here has a dynamic connotation; it supposes that an initial regular pattern has been perturbed (see Section 8.6.3).

Phyllotactic fraction: A fractional approximation p/q of the divergence angle, obtained from two primordia a and b that look in line with the center of the apex, that is, two primordia along an orthostichy. The numerator p of the fraction is obtained by counting the number of integral turns around the apex required to go from a to b by following the genetic spiral. The denominator q is the number of primordia met when going from a to b, excluding a. On a cylindrical surface such as a leafy stem, phyllotactic fractions are obtained by working in much the same way with any two leaves that appear to be approximately one above the other on the stem.

Phyllotactic pattern (or system): A construction determined by organs, parts of organs, or primordia of plants. Phyllotactic patterns can be multijugate, mutimerous, perturbed, normal, anomalous, distichous, decussate, whorled, etc. Phyllotactic patterns are defined in Section 3.5 as partially ordered systems of interrelated elements (the primordia) interacting in an aggregative fashion.

Phyllotaxis: (1) (the phenomenon of): The word refers to the arrangement of leaf or floret primordia at the shoot apex (initial arrangement), or on the stem (mature arrangement). The phenomenon of phyllotaxis is the fact that in 92% of the observations (see Table 7.12), the contact or conspicuous parastichy pairs (m, n) in phyllotactic patterns are such that m and n are consecutive terms of the Fibonacci sequence, and as the plastochrone ratio R decreases in the early development of plants the divergence angle d converges rapidly toward the Fibonacci angle. When m and n do not belong to the Fibonacci sequence, they are generally consecutive terms of a Fibonacci-type sequence. This constitutes a part of the challenge of phyllotaxis.

(2) [(m, n) of a system, or conspicuous parastichy pair (m, n)]: In a spiral as well as in a whorled pattern, the phyllotaxis of a plant specimen is expressed by the conspicuous parastichy pair (m, n), $n \leq m$.

(3) (an area of research, restricted definition): The study of phyllotactic patterns, or of the phenomenon of phyllotaxis (from the Greek *phyllon*: leaf; *taxis*: arrangement). Initially, phyllotaxis was the study of leaf arrangements on the plant stem. Phyllotaxis is also the study of the patterns determined by the florets on the capituli of sunflowers and other composites, by the scales on pine cones and pineapple fruits, and more generally by the primordia on apices of plants or in transverse sections of buds. The term also extends to the study of arrangements of bracts, branches, floral parts (anthotaxis), bristles, etc.

(4) (a discipline, extended definition): **Generalized phyllotaxis** or **systemic phyllotaxis** or Primordial Pattern Morphogenesis is the discipline for the study of phyllotactic patterns; the phenomenon of phyllotaxis; the laws that govern the relative arrangements of leaves and other organs or appendages homologous to leaves; the fundamental mechanisms and factors inducing the phyllotactic patterns (spiral, whorled, perturbed); the genesis of these patterns; and their relations to the environment, genetically, ontogenetically, functionally, phyletically, and systemically. The two immediate challenges are those of **pattern recognition** (object of Part I) and of **pattern generation** (object of Part II). The third is the phyletic challenge of the **origins of phyllotactic patterns** (object of Part III). The fourth is the **functional challenge** dealt with interpretative and teleonomic approaches (see Chapter 6, Sections 9.1, 11.4). The phyllotactic arrangements are then considered with respect to their potential usefulness for the plants themselves and for the environment. The resources of botany, evolutionary biology, thermodynamics, physiology, paleobotany – to name a few – are brought to bear on the study of phyllotaxis. In this endeavor mathematics plays a crucial role via the possibilities offered by predictive modeling based on induction and deduction. Comparative morphology is fundamental to these studies in order to give to the phenomenon of phyllotaxis its full dimension and meaning in nature in general where similar patterns are found. Certain authors studying for example crystals, viruses, superconductors, and polymers speak about generalized phyllotaxis because of phyllotaxislike patterns in these areas. Phyllotaxis is thus part of systems research where the phenomena of homology and emergence play a crucial role.

(5) The term is used in the botanical literature as a synonym for phyllotactic fraction.

Phyllotaxis index: An expression derived from the plastochrone ratio for orthogonal ($\gamma = 90°$) Fibonacci systems, defined by Richards by P.I. = $0.379 - 2.3925 \log\log R$. It yields positive and integral values for these systems.

Plant growth substances (plant hormones, phytohormones): Natural components of plants. They are abscissins, auxins, cytokinins, gibberellins, and ethylene.

Plastochrone *P*: The time interval between the emergence of two successive leaf primordia, or pairs of whorls of primordia, at the shoot apex. Also defined as the time required for two successive leaves (or leaves in consecutive whorls) to reach a given length.

Plastochrone ratio R: Ratio of the distances of two successive primordia from the apical center. It is denoted by $R > 1$.

Primordial Pattern Morphogenesis (PPM): A proposed alternative term to represent and replace the term phyllotaxis in its extended meaning [see definition (4) of the term phyllotaxis]. The initialism also stands for Pyramidal Pattern Modeling (see Sections 12.3.1 and 12.3.3).

Primordium: The early stage in the development of a leaf, bract, petal, stamen, floret, sepal, root or other organ, when it is a small protuberance or protrusion. The primordium is the building unit of phyllotactic patterns.

Rise: The internode distance – that is, the vertical distance between two consecutive points on the genetic spiral – in the cylindrical lattice. The rise corresponds to the plastochrone ratio in the centric representation and is related to it by the formula $r = \ln R/2\pi$.

Rising and falling phyllotaxis: The phenomenon observed on plants with spiral phyllotaxis, when the secondary numbers in a conspicuous parastichy pair is increasing or decreasing, going from (m, n) to $(m, m+n)$ or the other way around. Considering an angle of intersection γ for each parastichy pair, the transition from one pair to the other by rising phyllotaxis produces a hyperbolic turn of the cylindrical lattice (see Appendix 9).

Secondary divergence: The abscissa of a point n in the cylindrical lattice (represented by x_n, see Sections 2.5 and 4.2) or in the spiral lattice (represented by Δ_n, see Section 1.6).

Secondary number (or parastichy number): Any one of the two integers defining a parastichy pair.

Snow and Snow's first available space hypothesis: A variant of Hofmeister's rule, according to which a new primordium appears when the available space on the shoot apex has reached a minimal size or a minimal distance from the tip of the apical dome.

Telome: The branchlets of the oldest known land plants representing, according to Zimmermann, the kind of primitive structure that produced by evolution the plants as we know them today (see Section 3.2.1).

Transverse section: Microscopic section cut through plant material across the tissue (e.g., stems) at 90° to the long axis.

Tunica: The outermost layer of dividing cells in the shoot apex of angiosperms (above the corpus), consisting of the epidermis and of one of two layers of subepidermal cells. The site of the first visible sign of a leaf primordium initiated below it.

Venation: Pattern of veins in a leaf.

Visible opposed parastichy pair: An opposed parastichy pair that is a visible parastichy pair. The value of the divergence angle d determines the visible opposed pairs, and a visible opposed pair determines an interval of values for the divergence (the theorem of Chapter 2). In Figure 2.3(2) the pairs (8, 5) and (8, 13) are visible opposed parastichy pairs, but (18, 5) and (18, 13) are visible but not opposed.

Visible parastichy pair [or visible pair of points (m, n)]: In a cylindrical lattice, the parastichy pair (m, n) is said to be visible if (1) the points m and n are visible, and (2) the ray from 0 scanning clockwise or counterclockwise the first N points of the infinite vertical strip with horizontal base of length 1 between the values -0.5 and $+0.5$, where $\max\{m, n\} \leq N < m+n$, meets m and n consecutively. Otherwise stated, the pair (m, n) is visible when the triangle in the strip, with vertices 0, m, and n does not contain on its sides or inside it another point of the lattice. For example, in Figure 2.3(2) the pairs (8, 5), (18, 5) and (18, 13) are visible, and the pairs (10, 13) and (8, 26) are not. In the centric representation, a visible parastichy pair is such that all the points of intersections of two parastichies are centers of primordia.

Visible point: In a cylindrical lattice, a point n is said to be visible if the ray from 0 to n contains no other point of the lattice between 0 and n. For example, in Figure 2.3(2) the points 5, 8, 13, and 18 are visible points, while the points 10 and 26 are not.

Whorled (or verticillate pattern): A leaf arrangement whereby two or more leaves are positioned in a ring at virtually the same level around the plant stem. See the term "Multimerous system."

Appendix 2
Answers to problems

A2.1. Chapter 1

Problem 1.1

In pineapples we may find families of 3, 5, 8, 13, and 21 spirals, so that the visible opposed pairs are (3, 5), (8, 5), (8, 13), (21, 13). Similar observations can be made on cones, except that the consecutive Fibonacci numbers generally obtained are smaller, as we have seen in the text.

Problem 1.2

1. Figure 1.3(1) shows the visible opposed spiral pair (7, 11).

2. Given a system (7, 11), in the family containing 7 spirals, the numbers on adjacent primordia on a spiral must differ by 7, while the numbers on two consecutive primordia on any of the 11 spirals must differ by 11. So the thing to do is to choose a primordium near the rim of the specimen, let us say the one we marked with a dot in its center, and to put the number 1 on it. This primordium is at the junction of two opposed contact spirals, one in each family. The primordium with two dots in it is primordium #8 and the one with three dots is #12. Then we have everything we need to put a number on every other primordium, by applying the Bravais–Bravais theorem. For example, the primordium adjacent to #1, #8, and #12 is #19; the one on the right of #12 is #5; the one above #5 is #16; the one on the left of #16 is #23; and on the right of #16 is #9; on the right of #9 is #2, etc. As a general rule, the smaller the primordium, the larger is its number, given that it is a younger primordium.

3. With a number on each primordium, it becomes possible to draw the genetic spiral simply by linking the centers of the primordia following the consecutive numbers 1, 2, 3, 4, 5, ..., along the shortest path between any two of them. The genetic spiral winds counterclockwise.

Problem 1.4

1. $F_{16} = 987$, $F_{21} = 10{,}946$.
2. The Lucas sequence is $\langle 1, 3, 4, 7, 11, 18, 29, 47, 76, 123, \ldots \rangle$.
3. Performing the operations suggested, we have $x/1 = (1+x)/x$. From this equation we obtain $x^2 - x - 1 = 0$, whose positive solution is $x = \tau$.

Problem 1.5

The first part of the problem has shown that when k is large, $F_k/F_{k+1} \approx 1/\tau$. This is in agreement with Equation (1.4), given that $F_k/F_{k+1} = 1/(F_{k+1}/F_k)$. Taking the square of each member in the equality gives

$$\lim(F_k/F_{k+1})^2 = 1/\tau^2,$$

meaning that when k is very large the equality to $1/\tau^2$ holds.

Problem 1.6

The first equality follows from the relations: $1 - 1/(F_{k+1}/F_k) \approx 1 - 1/\tau$ for large k, and $1 - 1/\tau = (\tau - 1)/\tau = 1/\tau^2$ (use the fact that $\tau + 1 = \tau^2$). Then, using the relations $F_{k-1}/F_{k+1} = (F_{k-1}/F_k) \times (F_k/F_{k+1})$, at the limit each of the last two ratios is equal to $1/\tau$ and the second equality follows.

Problem 1.7

We have $(m, n) = (3, 2)$, $R = C1/C2 \approx 1.5$, $d =$ angle $2C1$ or $3/8$, $\Delta_2 =$ angle $2C4 \approx \pi/2$, $\Delta_3 =$ angle $1C4 \approx -\pi/4$, $\phi_2 =$ angle $CUV \approx 63°$, $\phi_3 =$ angle $CUS \approx -33°$, $\gamma \approx 96°$. We have $u = v = 1$. Using a calculator, we can verify the other relations. Accurate values of the parameters will deliver the equalities.

Problem 1.8

Given, for example, the Fibonacci pair (5, 3), we have $u = 2$ and $v = 1$. For the Lucas pair (7, 4), we also have $u = 2$ and $v = 1$.

Problem 1.9

1. Using the relations in Equation (1.5), the first multiplied by n, the second multiplied by m, and subtracting, yields the result, given that $mv - nu = \pm 1$.

2. Extracting the value of $\cot \phi_m$ from Equation (1.9) and putting it in the left-hand member of the equality to be proved, and then using part 1 of the problem gives $\pm 2\pi \Delta_n \cot \phi_n / n$ as a left member. Comparing this last expression with Equation (1.8) delivers the result.

3. Multiply Equations (1.7) and (1.8) and bring everything to the left side of the equation obtained. Given that $\gamma = \phi_n - \phi_m$ or $\phi_m - \phi_n$, use the trigonometric relation $\cot(a-b) = (\cot a \cot b + 1)/(\cot b - \cot a)$ to obtain $mn \ln^2 R \pm \Delta_n \Delta_m \cot \gamma (\cot \phi_m - \cot \phi_n) + \Delta_n \Delta_m = 0$. Then use the relation obtained in part 2 of this problem to conclude.

Problem 1.10

Notice the isosceles triangles having two angles of $2\pi/5$. Properties of similar triangles give the first equalities. The law of sines applied to any one of the isosceles triangles gives the last equality.

A2.2. Chapter 2

Problem 2.2

1. The sequence of nested intervals obtained and containing the angle $(3+1/\tau)^{-1}$ is: $[1/4, 1/3]$, $[1/4, 2/7]$, $[3/11, 2/7]$, $[3/11, 5/18], \ldots$, where the denominators are consecutive terms of the Lucas sequence.

2. The pair (L_k, L_{k+1}) made of consecutive Lucas numbers for $k = 1, 2, 3, \ldots$, is visible and opposed if and only if the divergence angle is in the intervals whose end points are F_{k-1}/L_k and F_k/L_{k+1}.

3. To prove that $(3+1/\tau)^{-1}$ is in the intervals in part 2, it is sufficient to prove that $\lim F_k/L_{k+1} = (3+1/\tau)^{-1}$. We have that

$$F_k/L_{k+1} = (1 + F_{k+2}/F_k)^{-1},$$

and at the limit this last expression has the value $(1+\tau^2)^{-1}$. Prove that $\tau^2 = 2 + 1/\tau$, and the result follows.

Problem 2.3

Let $d = 105°$, that is 7/24. The fractions in A up to $k = 20$ are:

$$\frac{0}{1}, \frac{1}{2}, \frac{1}{3}, \frac{1}{4}, \frac{1}{5}, \frac{2}{7}, \frac{3}{10}, \frac{3}{11}, \frac{4}{13}, \frac{4}{15}, \frac{5}{16}, \frac{5}{17}, \frac{5}{18}, \frac{6}{19}.$$

For ordering the fractions we insert Farey sums between 0/1 and 1/2, so as to use all the fractions consecutively:

$$\frac{0}{1} \mid \frac{1}{2},$$

$$\frac{0}{1} \mid \frac{1}{3}\ \frac{1}{2},$$

$$\frac{0}{1}\ \frac{1}{4} \mid \frac{1}{3}\ \frac{1}{2},$$

$$\frac{0}{1}\ \frac{1}{5}\ \frac{1}{4} \mid \frac{1}{3}\ \frac{1}{2},$$

$$\frac{0}{1}\ \frac{1}{5}\ \frac{1}{4}\ \frac{2}{7} \mid \frac{1}{3}\ \frac{1}{2},$$

$$\vdots$$

$$\frac{0}{1}\ \frac{1}{5}\ \frac{1}{4}\ \frac{4}{15}\ \frac{3}{11}\ \frac{5}{18}\ \frac{2}{7} \mid \frac{5}{17}\ \frac{3}{10}\ \frac{4}{13}\ \frac{5}{16}\ \frac{6}{19}\ \frac{1}{3}\ \frac{1}{2}.$$

The visible opposed pairs are (1, 2), (1, 3), (4, 3), (7, 3), (7, 10), and (7, 17), made with consecutive denominators on each side of the vertical representing d in each row.

Problem 2.4

Hint: Here is a general method for solving diophantine equations. Let us apply the method for the resolution of the equation $17x+23y=500$. Solving the equation for x gives: $x=(500-23y)/17=[(29\times 17+7)-(17y+6y)]/17=29-y+(7-6y)/17$. Given that x and y are integers, 17 must divide $7-6y$. Put $(7-6y)/17=t$, t being an integer. It follows that

$$x=29-y+t,$$
$$7-6y=17t,$$
$$6y=7-17t,$$
$$y=(6+1-12t-5t)/6,$$
$$y=1-2t+(1-5t)/6.$$

As previously, 6 must divide $1-5t$. So put $1-5t=6s$, where s is an integer. We have

$$5t=1-6s,$$
$$t=-s+(1-s)/5.$$

Put again $1-s=5p$, and the solution will finally appear, given that we finally obtained a coefficient (of s) which is equal to 1:

$$s=1-5p,$$
$$t=-s+p=6p-1,$$
$$y=1-2(6p-1)+(1-5p)=4-17p,$$
$$x=29-(4-17p)+(6p-1)=24+23p.$$

For example, if we put $p=0$, then we have the particular solution $x=24$ and $y=4$. We can verify that $17\times 24+23\times 4=500$.

Now by using the result in Problem 2.2 we have: the pair $(L_{k+1}, L_k)=(47, 29)$ for $k=7$ is a visible opposed parastichy pair if and only if the divergence angle is in the bounds F_{k-1}/L_k and F_k/L_{k+1}. Putting $k=7$ in the two fractions gives 8/29 and 13/47; that is, d is in the interval [8/29, 13/47].

By the **diophantine algorithm** (Appendix 4) we must solve the equation $47v-29u=\pm 1$. We have:

$$29u=47v\pm 1, \qquad u=v+(18v\pm 1)/29.$$

We put $18v\pm 1=29t$, so that

$$u=v+t, \qquad v=t+(11t\pm 1)/18.$$

We put again $11t\pm 1=18s$, so that

$$v=t+s, \qquad t=s+(7s\pm 1)/11.$$

We put again $7s\pm 1=11p$, so that

$$t=s+p, \qquad s=p+(4p\pm 1)/7.$$

We put $4p\pm 1=7q$, so that

$$s=p+q, \qquad p=q+(3q\pm 1)/4.$$

We put $(3q\pm 1)/4=r$, so that

$$p=q+r, \qquad q=r+(r\pm 1)/3.$$

Now that the coefficient of r in the expression $(r\pm 1)/3$ is equal to 1, the solution arises: we put $(r\pm 1)/3=f$, to obtain:

$$r=3f\pm 1, \qquad q=r+f=4f\pm 1,$$

$$p=(4f\pm 1)+(3f\pm 1)=7f\pm 2,$$

$$s=11f\pm 3, \qquad t=18f\pm 5,$$

$$v=29f\pm 8, \qquad u=47f\pm 13.$$

Putting $f=0$ gives $v=8$ and $u=13$ (they must be positive integers). Putting $f=1$ gives $v=21$ (v must be smaller than 29) and $u=34$ (u must be smaller than 47). Values of $f\geq 2$ are not admissible in the present context, given that u and v must be smaller than m and n, respectively. The value $f=0$ gives the required interval for d which is [8/29, 13/47], given that $f=1$ produces a divergence angle greater than $\frac{1}{2}$.

By the **contraction algorithm** we have:

(47, 29)	$\leftrightarrow$	[8/29, 13/47]. $\uparrow$
$\downarrow$ (18, 29)	$\leftrightarrow$	[8/29, 5/18] $\uparrow$
$\downarrow$ (18, 11)	$\leftrightarrow$	[3/11, 5/18] $\uparrow$
$\downarrow$ (7, 11)	$\leftrightarrow$	[3/11, 2/7] $\uparrow$
$\downarrow$ (7, 4)	$\leftrightarrow$	[1/4, 2/7] $\uparrow$

$\downarrow$ (3, 4), so that $t=3$ and d is in [1/4, 1/3]

Problem 2.5

For the sequence of normal phyllotaxis, denoting the consecutive terms by $N_{t,k}=F_k t+F_{k-1}$, [so that $N_{t,1}=t$ ($F_{k-1}=0$), and $N_{t,2}=t+1$, the first two terms of the sequence], the problem is tantamount to looking for the following limit as k increases: $\lim F_k/N_{t,k}$. This limit is equal to $1/\lim(t+F_{k-1}/F_k)$ which is obviously equal to $1/(t+\tau^{-1})$.

A2.3. Chapter 3

Problem 3.1

1. It is easily seen that the consecutive levels of the tree contain respectively 1, 2, 3, 5, 8, 13, ... branches.

2.

t	Organisms	Total
1	m	1
2	my	2
3	mym	3
4	mymmy	5
5	mymmymym	8
6	mymmymymmymmy	13

3. See Figure A2.1 (top of page 296).

Problem 3.2

The sequence of numbers is always the same, whether it be cell division or phyllotactic diagrams. For example, in the fifth cycle of cell division the sequence is 8-3-6-1-4-7-2-5. This sequence can be observed in level 4 of the hierarchy of Figure 3.1, and is also obtained by projecting radially the first 8 points in the middle of Figure 1.4 on the circumference of the disk.

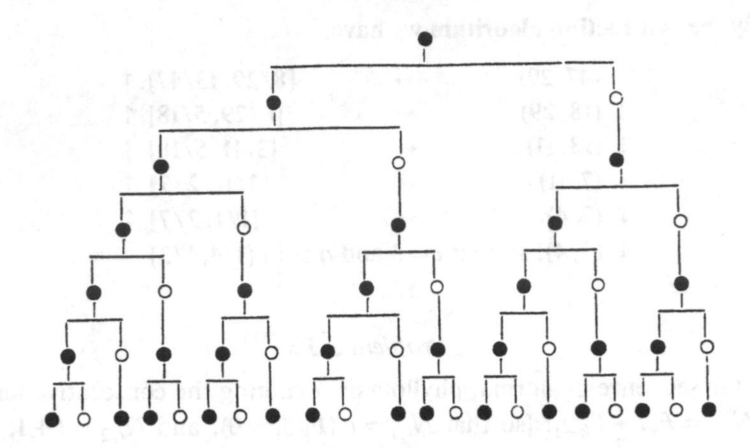

Figure A2.1. Diagram used to explain the role of Fibonacci numbers in the ontogenesis of tissues and organs in plants. (From Berdyshev, 1972.)

Problem 3.3

We must read the sequence of 1s and 2s by turning around the cone. The 2s correspond to two successive junctions $j^{\#}$, and the 1s to one junction j^*. Notice for example that when the scales γ_1, γ_2 and γ_3 are placed, by turning clockwise around the cone, we have two junctions $j^{\#}$ followed by one junction j^*, followed by two junctions $j^{\#}$.

A2.4. Chapter 4

Problem 4.1

Considering the case $k=2$, for example, we have $[\tau^2-(-\tau)^2]/\sqrt{5}=[\tau+1-1/(\tau+1)]/\sqrt{5}=(\tau^2+2\tau)/\tau^2\sqrt{5}=(\tau+2)/\tau\sqrt{5}=1=F_2$. Given that the term τ^{-k} in Binet's formula approaches 0 very rapidly with the increase of k, we have the desired approximation for F_k.

Problem 4.2

1. Putting the value $\gamma=90°$ in Equations (1.10) and (4.4) respectively gives the first two relations. Though Church worked with the Fibonacci sequence, Church's formula is true in general. It can be checked, using Equation (4.12) and the approximate values of m and n given in Equations (4.8) and (4.10) respectively. For example, when $m=F_k$ and $n=F_{k-1}$, we have that $m\approx\tau^k/\sqrt{5}$ and $n\approx\tau^{k-1}/\sqrt{5}$, so that $\sqrt{5}(m+n)^2\approx$

$(m^2+n^2)\tau^3$, given that $\tau^2+1=\tau+2=\sqrt{5}\tau$. Putting this value of $(m+n)^2$ in Equation (4.31) gives Equation (4.32).

2. and **3.** Here m and n are consecutive terms of common sequences.

$m+n$	Jean R Equation (4.31)	Church R Equation (4.32)	Richards R Equation (4.30)
3	3.75302	3.51359	3.79666
4	2.10421	1.87446	2.19950
5	1.60982	1.62145	1.60969
6	1.39186	1.36911	1.39589
7	1.27497	1.28573	1.27580
8	1.20440	1.20298	1.20451
10	1.12640	1.12843	1.12638
11	1.10337	1.10149	1.10357
13	1.07297	1.07315	1.07296
16	1.04750	1.04728	1.04762
18	1.03742	1.03765	1.03741
21	1.02736	1.02733	1.02736
26	1.01776	1.01781	1.01776
29	1.01425	1.01422	1.01426
34	1.01035	1.01035	1.01035
42	1.00677	1.00676	1.00677
47	1.00540	1.00541	1.00540
55	1.00394	1.00394	1.00394

Problem 4.3

1. The table gives $\gamma \approx 106°$, given by the row where $m+n=8$, so that the pair is (3, 5). The other visible opposed parastichy pairs have intersection angles that differ more from 90°.

2. The table gives $\gamma \approx 74°$, given by the row where $m+n=11$, so that the pair is (7, 4).

3. The table gives $\gamma \approx 94°$, given by the row where $m+n=9$, so that the pair is (4, 5).

Problem 4.4

Pair	(1, 2)	(3, 2)	(3, 5)	(8, 5)	(8, 13)	(21, 13)
P.I.	0.955	2.016	2.995	4.002	5.012	6.036

Problem 4.5

1. The value of the divergence angle indicates that we must look in the rows of the Pattern Determination Table where $m+n=3, 5, 8, 13, \ldots$.

The visible opposed parastichy pairs are thus (2, 1), (2, 3), (5, 3), (5, 8), ... The pair (5, 8) is conspicuous, and the angle of intersection of its opposed spirals is approximately 86°.

2. Putting the mean value of R in Equation (4.16) gives P.I. = 3.92 or almost 4. This means, using Problem 4.4, and given that the divergence of the system is close to τ^{-2} ($137\frac{1}{2}°$), that we have an almost orthogonal Fibonacci system. Consulting the table in the answer to Problem 4.4, we see that the pair (8, 5) is the one.

Appendix 3
Questions

1. Define the divergence angle, the limit-divergence angle, and the plastochrone ratio.
2. Write out the first ten terms of the Fibonacci sequence and of the Lucas sequence, and give the divergence angle corresponding respectively to those sequences.
3. Using the relation $\tau = \lim F_k/F_{k-1}$, derive an elementary relation between the observed angle of $137\frac{1}{2}°$ and the visible opposed pair (F_k, F_{k-1}).
4. Prove that the classical relation $mn \ln^2 R - 2\pi \ln R \cot\gamma + \Delta_m \Delta_n = 0$ holds, using the relations $\Delta_m \Delta_n(\cot\phi_m - \cot\phi_n) = \pm 2\pi \ln R$, $m \ln R = \Delta_m \cot\phi_m$ and $n \ln R = \Delta_n \cot\phi_n$.
5. State the theorem relating the visible opposed pairs and the divergence angle $d = \tau^{-2}$, and give an intuitive presentation of it.
6. What is the difference between a contact parastichy pair and a conspicuous parastichy pair? What is the difference between a family of parastichies and a parastichy pair? Do you know of special types of visible opposed parastichy pairs?
7. Using the Pattern Determination Table and considering the divergence angle $(4+1/\tau)^{-1}$, write visible opposed parastichy pairs and determine the conspicuous pair if $R = 1.7$. *Hint:* Consider the sequence corresponding to the angle. Answer: $\gamma \approx 84°$.
8. Derive, from the model given by Equations (4.12) and (4.13), the formula for orthogonal ($\gamma = 90°$) parastichy pairs.
9. Using the Pattern Determination Table, what is the value of γ closest to 90°, and what is the observed conspicuous pair for the system governed by the divergence angle $(3+1/\tau)^{-1}$ if a plastochrone ratio of 1.40 has been measured?

10. If the visible opposed pair (13, 21) is observed, determine a closed interval of values (in degrees) in which we are likely to find the divergence angle. *Hint:* Use Adler's theorem.
11. Determine a closed interval in which we are most likely to find the divergence angle if the system shows the visible opposed pair (7, 4). *Hint:* See Problem 2.2.
12. With systems showing the visible opposed spiral pairs (3, 5), (7, 11) and (4, 5), corresponding to sequences studied earlier, and the corresponding divergence angles respectively, determine the values of u and v defined in the text and verify that the relation $mv - nu = \pm 1$ holds in each case.
13. If the measured value of the angle of intersection of a contact parastichy pair seen on a plant is 86°, in a system governed by the sequence $\langle 1, 4, 5, 9, 14, 23, \ldots \rangle$, give four possible values of the plastochrone ratio.
14. What are the values that the phyllotaxis index can take in the case of orthogonal Fibonacci systems?
15. What are the contact parastichy pairs likely to be observed on pine cones, sunflowers, and pineapples?
16. State a theorem relating the visible opposed pairs for a Lucas system to closed intervals for the divergence angle.
17. Given one of the three most common limit-divergence angles for spiral systems (or observed divergence angles that cluster around one of these angles), write a step-by-step procedure to determine the conspicuous parastichy pair of the system using the Pattern Determination Table, given the plastochrone ratio.
18. State Binet's formula and give an approximation for F_k.
19. What is the centric representation of phyllotaxis?
20. How would you define the Pattern Determination Table?
21. What is a genetic spiral? Give other names for it. What is its relation to parastichies?
22. Use Figure 1.6 to describe some of the parameters involved in phyllotactic pattern recognition and description.
23. How would you describe an ontogenetic theory of phyllotaxis as opposed to a phylogenetic theory?
24. Do you know of biological supports for the hierarchical representation of phyllotaxis?
25. What are multimerous patterns? What are their relations to multijugate patterns according to the interpretative model of phyllotaxis? Do you know of particular multimerous patterns?

26. In which way does the phenomenon of phyllotaxis transcend the strict botanical substratum?
27. What do you know about the origins of phyllotaxis and about theories that can be brought forward to explain these origins?
28. How would you describe the phenomenon of phyllotaxis? What do you mean by "the phyllotaxis of a system is (5, 8)"?
29. What is an apical dome, a meristem, and a primordium? What is the Fibonacci angle, an orthostichy, and a distichy?
30. What is the object of the field of research known as phyllotaxis, in the widest acceptance of the term?
31. What is the main result of the interpretative model of phyllotaxis? Compare it with the main result of Adler's contact-pressure model.
32. What are the main predictions of the interpretative model?
33. According to the interpretative model, what is the relation between superposed whorls and normal spiral systems?
34. How would you describe the importance of comparative morphology in dealing with the origins of phyllotactic patterns?
35. What do you know about the crystallographic approach to phyllotaxis? State some of the results obtained by crystallographers dealing with phyllotaxis.
36. Write a comment on the problem of transitions between phyllotactic patterns. What is Jean's explanation of discontinuous transitions? What is a singular transition in Rivier's approach?
37. Give a report on the priority order of phyllotactic patterns according to four models: (1) Jean's interpretative and (2) τ-models, (3) XuDong and Bursill's model of packing of disks, and (4) Marzec's entropy model.
38. Who invented the bulk ratio, the phyllotaxis index, the Pattern Determination Table?
39. What do you know about allometry in phyllotaxis? State a formula relating r, γ, (m, n). What does it represent in the log-log grid?
40. Report on the general dissatisfaction emerging from old trends in phyllotaxis.
41. What is Barabé and Vieth's explanation of the phyllotaxis of *Dipsacus*? What is the alternative interpretation proposed by Jean?
42. State a relation between Jean's Figure 8.2 expressing the interpretative model, and Richter and Schranner's phase diagram for the diffusion of an inhibitor.
43. State the Rashevsky–Rosen principle of optimal design, and the principle of minimal entropy production used in the interpretative model.

44. Chapter 8 stresses the idea that spiral patterns are primitive. State Church's arguments in favor of this hypothesis. Can you find arguments in favor of the statement that whorled patterns are primitive?
45. What are the names given to the various sequences expressing multijugate pattern?
46. State three applications of the theorem of Chapter 2.
47. In a historical perspective, what can we think about the existence of so many visible opposed parastichy pairs on a single specimen?
48. The floret and seed patterns in the sunflower have been the object of intense studies, for example, botanically by Palmer and mathematically by Bursill. What do you know about such studies?
49. Branching and gnomonic growth: how do you see their relevance in phyllotaxis?
50. What can be the role of mathematics in the effort to understand the phenomenon of phyllotaxis?
51. State the Bravais–Bravais approximation formula, theorem, and assumptions. What do you call the Bravais–Bravais lattice?
52. Draw a normalized cylindrical lattice with a given divergence angle, and consider the region for which the secondary divergences fall in $[-0.5, 0.5]$, and in $[0, 1]$. In each case determine (non) visible pairs that are (not) opposed.
53. Using phyllotactic methods it is possible to predict the relative arrangement of subunits in polypeptide chains where the first accessory sequence is concerned. What are these predictions? Show that they are confirmed with the subunits in the filamentous bacteriophage called *Inovirus* where the third accessory sequence is involved.
54. Explain the parameters in the systemic model of Chapter 6. Calculate the costs for (5, 6) and for (5, 9) phyllotaxis ($E_b \approx 4.27$ and 4.50 respectively).
55. State a necessary and sufficient condition for a parastichy pair to be visible. Define the notion of visible pair in terms of the fundamental triangular region of a lattice, and as a generalization of the notion of visible point of a lattice.
56. In which way is spiromonostichy a particular case of normal systems according to the systemic model? What is the interpretation given to the pattern found in *Costus*?
57. Explain the linear relation between min E_b and the plastochrone (see Proposition 8, Chapter 6). Compare the value of E_b obtained directly from Equation (6.2) for $P_r = 31$ [giving (11, 20) according to Table

6.1], and the value of E_b obtained from the linear relation. *Answer:* Using Equation (6.2): $S(1) = 9/11$, $S(2) = 20/31$, $X(1) = 1$, $X(2) = 2^{31}$, $E_b = 31 \log 2 + \log 31 + \log 11 - \log 9 - \log 20 \approx 9.61$; using the linear relation: $E_b \approx 31 \log 2 + 2 \log \tau \approx 9.75$.

58. What is the mathematical relation deduced from the allometry-type model of phyllotaxis, between (m, n) and Church's bulk ratio, or Richards's area ratio, or the plastochrone, respectively?
59. What is a systemic approach to phyllotaxis? Make a statement regarding the importance of phyllotaxis in plant morphogenesis. Summarize the theory of phyllotaxis presented in this monograph.
60. If you had to write a thesis on phyllotaxis, what aspect of the subject would be most attractive to you, and what references in the Bibliography of Part IV would you like to scrutinize first?

Appendix 4
General properties of phyllotactic lattices

A4.1. Phyllotaxis and Farey sequences

This appendix is a self-contained presentation of the basic concepts of phyllotaxis in a more general setting which has led to the general form of the fundamental theorem of phyllotaxis. The theory presented here (Jean, 1988a) concerns lattices from a general point of view, but it can be particularized to give results known in phyllotaxis. The discussion below will allow the reader to better grasp the concepts of visible and of opposed parastichy pairs.

Figure A4.1 shows a regular point lattice similar to the ones obtained in the study of phyllotaxis. The lattice is made with a **divergence** $d = 85°$, or $85/360 = 17/72$. This is the abscissa of point 1, a number always taken to be smaller than $\frac{1}{2}$. The coordinates of point $n = 1, 2, 3, \ldots$, are $(nd - k, nr)$ where the **rise** r is the ordinate of point 1, and k is any integer. We concentrate here on the points whose abscissae are between or equal to -0.5 and $+0.5$, included in an infinite vertical strip denoted by S (the past analyses, e.g., Coxeter, 1972; Adler, 1977a; Marzec & Kappraff, 1983, are performed in the region of the lattice of points having abscissae between 0 and 1). In this sector of the lattice the coordinates of point n are $(nd - (nd), nr)$, where (nd) is the integer nearest to nd. If nd is half an odd integer, (nd) is allowed to be either $[nd]$ or $[nd]+1$, where $[nd]$ is the integral part of nd. The value $nd - (nd)$ is the **secondary divergence** of point n. For any given integer N, the ordered set of fractions $(kd)/k < 1/2$, $k = 1, 2, 3, \ldots, N$, in their lowest terms, will be denoted by A. The closer point x is to the vertical axis, the closer the value $(xd)/x$ is to d. The fraction is known as a **phyllotactic fraction**, an approximation of d. Rolling the vertical strip S to form a cylinder, (xd) represents the number of turns on

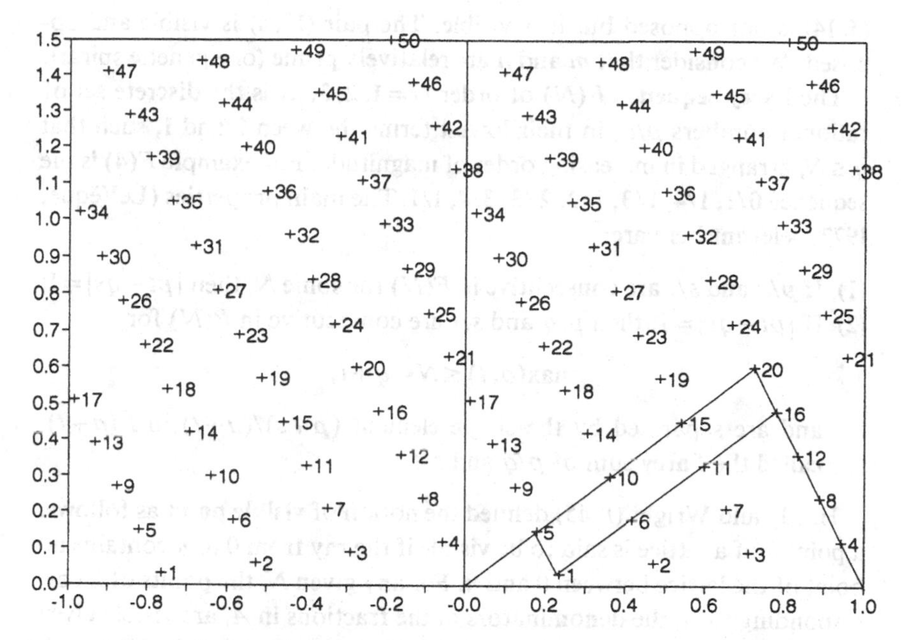

Figure A4.1. Cylindrical lattice built with the value $d = 17/72$. The coordinates of point n in S (see the text) are $(nd-(nd), nr)$.

the helix around the cylinder to reach x from 0 along the **genetic spiral.** Two points x and y on each side of the vertical are such that d is between $(xd)/x$ and $(yd)/y$.

Joining by a straight line segment any two points x and y of the lattice, such that no point of the lattice is between them, determines a **family of parastichy,** that is, $m = |x-y|$ parallel, equidistant straight lines partitioning the points of the lattice and linking them from m to n. Any two such families containing respectively m and n lines is a **parastichy pair** (m, n). When the lines in a family have a positive slope while those in the other family have a negative slope, we say that the pair (m, n) is **opposed.** The pair (m, n) is **visible** when there is a point of the lattice at every intersection of two lines in the pair.

For example, in Figure A4.1 the pair $(5, 7)$ is opposed given that 7 and 5 appear on opposite sides of the vertical axis in the strip S, one with abscissa between $-\frac{1}{2}$ and 0, the other with abscissa between 0 and $\frac{1}{2}$. The pair is not visible since the line going through the points 7, 14, 21, 28, ... does not meet the line going through the points 5, 10, 15, 20, ..., at a point of the lattice. The pair $(5, 13)$ is neither opposed nor visible. The pair

(5, 14) is not opposed but it is visible. The pair (17, 4) is visible and opposed. We consider that m and n are relatively prime (one genetic spiral).

The **Farey sequence** $F(N)$ of order $N = 1, 2, 3, \ldots$ is the discrete set of rational numbers p/q, in their lowest terms, between 0 and 1, such that $q \leq N$, arranged in increasing order of magnitude. For example $F(4)$ is the sequence 0/1, 1/4, 1/3, 1/2, 2/3, 3/4, 1/1. The main properties (LeVèque, 1977) relevant here are:

(1) If p/q and s/t are consecutive in $F(N)$ for some N, then $|pt - qs| = 1$;
(2) If $|pt - qs| = 1$, then p/q and s/t are consecutive in $F(N)$ for

$$\max(q, t) \leq N < q + t,$$

and are separated by the single element $(p+s)/(q+t)$ in $F(q+t)$, called the Farey sum of p/q and s/t.

Hardy and Wright (1945) defined the notion of **visible point** as follows: a point n of a lattice is said to be visible if the ray from 0 to n contains no point of the lattice between 0 and n. For any given N, the points of S corresponding to A, the denominators of the fractions in A, are visible given that the members of A are in their lowest terms. The proof of Proposition A4.1 suggests an equivalent definition for the concept of visible pair that generalizes the notion of visible point. The pair (m, n), of points or of families of parastichies, is **visible** if and only if m and n are visible, and a ray from 0 scanning the first N points of the strip S (clockwise or counterclockwise), where $\max(m, n) \leq N < m + n$, meets m and n consecutively. Otherwise stated the pair (m, n) is visible when the triangle in S whose vertices are 0, m, and n does not contain on its sides or inside it any other point of the lattice.

The original characterization of the visible pair (m, n) given in Proposition A4.1 below involves the relation $|m(nd) - n(md)| = 1$. These are Equations (1.6), or (4.3), arising in the centric and in the cylindrical representations of phyllotaxis. If the pair is also opposed, the numbers (nd) and (md) are the **encyclic numbers**. This relation has been discussed by many botanists (e.g., Richards, 1948; Thomas, 1975; Thornley, 1975b; Erickson, 1983). Given the similar relationship in the theory of Farey sequences (part 1 above), it becomes clear that these sequences allow us to delve into the spatial relationships among leaf primordia represented by lattices of points. On the other hand, the concepts of visible and of opposed parastichy pairs arising from botany are not dealt with in the theory of Farey sequences. That explains why we have the distinctive results brought in this sector of number theory.

A4.2. Visible parastichy pairs

Proposition A4.1. *In the regular lattice with divergence d, the parastichy pair (m, n) is visible if and only if the points m and n are such that* $|m(nd)-n(md)|=1$.

In some cases (nd) or (md) will have two possible values. But as long as for one of these values the relation $|m(nd)-n(md)|=1$ holds, the pair (m,n) will be visible. For the pair (m,n), m and n may be relatively prime; this does not mean that (m,n) is visible, as we already know from the pair (7, 5) in Figure A4.1. For the pair (7, 5) there exists a pair of integers $u=3$ and $v=2$ such that $|mv-nu|=1$. Also we have another pair of integers $(7d)=2$ and $(5d)=1$, for which $7(5d)-5(7d)$ differs from ± 1, and such that the divergence is between $(5d)/5$ and $(7d)/7$ as expected (given that the pair is opposed). The divergence is not between $v/n=2/5$ and $u/m=3/7$. Proposition A4.2, the **fundamental theorem of phyllotaxis**, shows that when the pair (m,n) is visible and opposed, the same pair of integers yields the required relations.

Proposition A4.2. *The parastichy pair (m, n) is visible and opposed, if and only if there exist unique integers u and v such that* $0 \le u < m$, $0 \le v < n$, $|mv-nu|=1$, *and such that the divergence* $d<1/2$ *lies at or is between* u/m *and* v/n.

Generally speaking, neglecting the restrictions on u and v, m and n being relatively prime and the diophantine equation $mv-nu=\pm 1$ having a solution, it has infinitely many solutions. The restrictions $0 \le u < m$ and $0 \le v < n$ show that there is a finite number of solutions. More precisely u/m and v/n being consecutive in $F(N)$ for $\max(m,n) \le N < m+n$, if d is to be between two such consecutive fractions, the choice for u and v becomes unique, as stated in Proposition A4.2.

A proof of Proposition A4.2 can be based on elementary properties of similar triangles. Consider indeed in Figure A4.1 the visible opposed pair (5, 4) ($m=5$, $n=4$). The figure shows three triangles whose vertices 0 and 1 on their respective bases are on, or near, the horizontal axis, and whose other vertices are respectively the points 5, 20, and 16. Those triangles are approximately similar. The first two yield the relations $u/m \le d \le v/n$ and $mv-nu=1$, where $v=1$ (step from 0 to 5) and $u=1$ (step from 1 to 5). In this case note that d is smaller than $\frac{1}{2}$ as requested. Working with the last two triangles delivers the relations $mv-nu=-1$ and $v/n \le d \le u/m$, where $v=3$ (steps from 1 to 16) and $u=4$. But this case does not correspond to a divergence smaller than $\frac{1}{2}$.

If the pair (m, n) is visible, then the pairs $(m+n, n)$ and $(m, m+n)$ are known as the **extensions** of (m, n) (Adler, 1974). The extensions may not be visible given that the point $m+n$ may fall out of the strip S, when forming the fundamental parallelogram with 0, m, and n. For example, in Figure A4.1 the pair (5, 19) is visible but its extensions (5, 24) and (24, 19) are not. The same cannot be said about the **contraction** of (m, n) as Proposition A4.3.1 shows. The contraction of a pair (m, n) is the pair $(m-n, n)$ if $m>n$, and the pair $(m, n-m)$ if $m<n$. Moreover a pair (m, n) may be opposed, but its contraction may not be, as Figure A4.1 shows for the contraction (4, 15) of (19, 15), contrary to what happens when the opposed pair is also visible, as Proposition A4.3.2 shows. That a contraction of a visible opposed pair is visible and opposed is already known from Adler (1977a). Proposition A4.3.3 brings a precision to Adler's (1977a) analysis where it is said that an extension of a visible opposed pair is not necessarily visible. What Adler implicitly means is that the extension is not necessarily visible and opposed. Indeed this author only considered visible pairs that were opposed. Propositions A4.1 and A4.3 are motivated by the possibility of having visible pairs that are not opposed.

Proposition A4.3.

(1) *If (m, n), $m>n$, is visible, its contraction is visible, and*

$$(md)-(nd)=((m-n)d);$$

(2) *If (m, n), $m>n$, is visible and opposed, its contraction is visible and opposed, and*

$$mD_n+nD_m=1, \qquad D_m+D_n=D_{m-n},$$

where $D_x=|xd-(xd)|$, $x=n, m, m-n$;

(3) *If (m, n) is visible and opposed, both extensions are visible, at least one of them is opposed, and*

$$D_{m+n}=|D_m-D_n|.$$

Consider all the points m, n of the lattice such that (m, n) is a visible opposed pair. On each side of the vertical axis through 0, join the points with two infinite polygonal lines [see Fig. 2.3(2), or join the large black spots in Fig. 10.9]. When d is irrational, there is no point of the lattice between the two lines, except 0. The relations in Proposition A4.3 show that the polygonal lines are asymptotic to the vertical axis. The points of the lattice on these lines are known as the **neighbors** of the vertical axis. They correspond to the points of a square lattice on the lines asymptotic to the line $y = dx$ in Klein's well-known geometric representation of the continued fraction for d. With the **Bravais–Bravais assumptions**, requiring the alternation of the neighbors, we have Proposition A4.4.

Proposition A4.4. *If d is an irrational number smaller than $\frac{1}{2}$, if the pair (m, n), $m > n$, is visible and opposed, and if from n and m the neighbors of the vertical axis indefinitely alternate on each side of it, then*

$$d = [\tau(md) + (nd)]/(\tau m + n).$$

Proposition A4.5. *In any of the rows of fractions built according to the computational algorithm of Section 2.3.2, for any d, the consecutive visible pairs (m_i, m_{i+1}), $i = 1, 2, \ldots, k$ (the consecutive denominators in the row) are such that*

$$\sum_{i=1}^{k-1} (m_i m_{i+1})^{-1} = \frac{1}{x},$$

where x is the first point of the lattice, after point 1, on the other side of the vertical axis with respect to point 1.

A4.3. Examples and algorithms

Putting $m = t$ and $n = 1$ in Proposition A4.4, $(nd) = 0$, $(md) = 1$, $d = 1/(t + \tau - 1)$, we obtain the divergence of **normal phyllotaxis**. If $t = 2$, the system shows the Fibonacci sequence. If $n = 2$ and $m = 2t + 1$, then we have $(nd) = 1$, $D_n = 2d - 1$, $D_m = t - d(2t + 1)$ (Proposition A4.3), and $(md) = t$. Then d is the divergence of **anomalous phyllotaxis**. Proposition A4.4 gives a particular form for the set of **noble numbers** (Section 9.1.1). These numbers do not have intermediate convergents from a certain term, meaning geometrically that the infinite polygonal lines mentioned earlier do not have, from a certain vertex, a point of the lattice on the segments between their vertices.

The proof of Proposition A4.1 delivers simple algorithms to determine all the visible pairs, not just the visible opposed pairs, from the subsequence A of $F(N)$ determined by d. Proposition A4.2 delivers a simple algorithm to determine an interval for d from a visible opposed pair. Section 2.3.2 presents two of these algorithms. Here are two more.

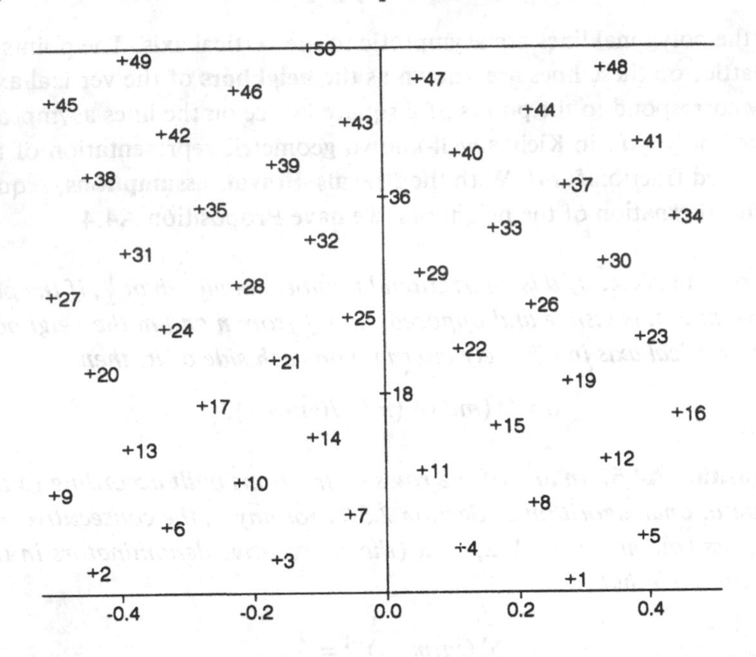

Figure A4.2. A rational value for the divergence gives two sequences of visible opposed pairs, namely (1, 2), (1, 3), (4, 3), (4, 7), (11, 7), followed by (11, 18) and consecutive left extensions (29, 18), (47, 18), ..., or followed by (18, 7) and consecutive right extensions (18, 25), (18, 43), ... The points 9, 27, 45, ... are considered to be on the right side of the strip S, with a secondary divergence equal to $+\frac{1}{2}$.

Graphical algorithm (for the determination of the visible parastichy pairs, from a computer drawing of the strip S corresponding to d and an arbitrary r). We scan the first N points of S with the ray from 0, taking note of the consecutive points met. The visible pairs (m, n), $m, n \leq N$, are obtained by taking any two consecutive points. The contractions of these pairs are also visible (Proposition A4.3). But it is not necessary to use the contractions; we can proceed with the ray from 0 by scanning S for $N=2, 3, 4, \ldots$, or by using the notion of the fundamental triangle.

For $N=29$ and $d=100°$ or 5/18, scanning Figure A4.2 with a ray from 0 gives the following consecutive points:

1 5 9 4 27 23 19 15 26 11 29 18 25 7 24 17 10 13 3 2.

The visible pairs up to $m, n=29$ are (1, 5), (5, 9), (9, 4), ..., (10, 13), (13, 3), (3, 2) and all the contractions of these pairs. The pairs (29, 18) and (18, 25) (denominators on each side of 5/18) are visible and opposed, together with their consecutive contractions which are (11, 18), (18, 7), (11, 7), (4, 7), (4, 3), (1, 3), (1, 2). The figure shows that among these contractions the pair (4, 3) is conspicuous (Figure A4.2 is an elaborated version of Figure 10.8 from Crick). Point 9 is such that $(9d)/9$ is 2/9 and 1/3. Given that Proposition A4.1 is satisfied for $(9d)=2$, the pairs (5, 9) and (9, 4) are visible. The case $(9d)=3$ gives that point 9 is not even visible from the origin. Points 9, 27, 45, ... are considered to be on the right side of S. When d is $1/(3+\tau^{-1})$, an irrational value between 8/29 and 5/18, the pairs (18, 25) and (18, 7) are no longer opposed, and we have a single sequence of visible opposed pairs (see Figure 10.9).

Diophantine algorithm (for the determination of an interval for the divergence angle, from a given visible opposed parastichy pair (m, n)). The algorithm amounts to solving the diophantine equation $|nu-mv|=\pm 1$ given in the fundamental theorem of phyllotaxis. For example, if the visible opposed pair is (29, 18), it is seen that the possible values for u and v are $u=29p\pm 8$, and $v=18p\pm 5$, where p is any integer. The hint given in the answer to Problem 2.4 recalls the usual method for solving diophantine equations. If we are to have $0<u<29$ and $0<v<18$, then we must put $p=1$ (giving $u=21$ and $v=13$) or $p=0$ (giving $u=8$ and $v=5$). But only the value $p=0$ gives a divergence smaller than $\frac{1}{2}$, so that the interval for d is, according to the theorem, [8/29, 5/18]. We could have obtained this result much faster, from the particular form of the theorem deduced in Problem 2.2 (see Problem 2.4). But the diophantine algorithm can be applied in all the cases.

To illustrate Proposition A4.5, consider, for example, in Section 2.3.2 the fourth row of fractions for $d=\tau^{-2}$ given by the computational algorithm. We have $k=4$, and the visible pairs in that row are (1, 3), (3, 5), (5, 7), (7, 2) giving $(1/3)+(1/15)+(1/35)+(1/14)=\frac{1}{2}$ [$x=2$ when d is larger than 90 degrees, and $x=n$ when d is between $180/n$ and $180/(n-1)$ degrees].

Appendix 5
The Williams–Brittain model

A5.1. The mechanism

Section 4.2.2 contains a paragraph introducing the relations between the allometry-type model and the Williams–Brittain model (1984). Here are some details on the latter model and examples of these relations. Both models involve the parastichy pair (m, n), the divergence d, the plastochrone ratio R, and, a parameter generally neglected, the angle of intersection γ of the parastichy pair.

The Williams–Brittain model is formulated in the centric representation. Circular primordia arise one by one according to a mechanism based on **Hofmeister's rule** and **Snow and Snow's hypothesis**. According to the rule a primordium arises in the largest available space, as far as possible from the existing primordia. According to the hypothesis, the primordium arises when the available space reaches a minimal size, and is at a minimal distance from the tip C of the apex.

Primordium X has polar coordinates ∇_X and r_{CX}, where ∇_X does not change during the growth of the system so that the primordium moves radially. The parameter r_{CX}, the distance from the center of X to C, changes continually as a function of the constant plastochrone ratio R, and such that the ratio $q = r_{PX}/r_{CX}$ is constant, r_{PX} being the radius of the primordium. The primordium arises at plastochrone T_X.

The center of the first primordium is always placed at $\nabla_1 = 0°$ and $r_{C1} = 1$. The second primordium necessarily arises at $\nabla_2 = 180°$, and given that it has all the space available at its birth, its center is on the circle of radius 1 centered at C, so that $r_{C2} = 1$ and $r_{P2} = q$. During that time, primordium 1 moved radially to $r_{C1} = R$, and $r_{P1} = Rq$. The system is $T_2 = 2$ plastochrones old. The third primordium will arise in one of the two free spaces between the first two; the choice is arbitrary. The values of r_{C3}, r_{P3}, and

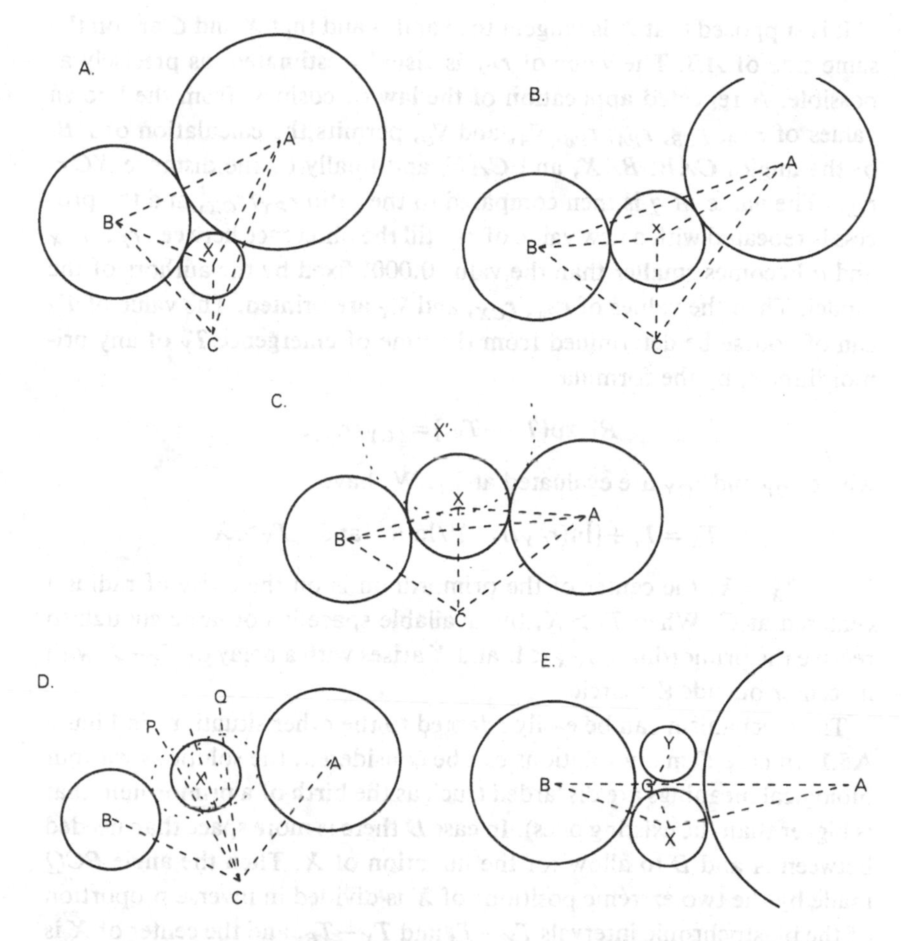

Figure A5.1. Five situations can arise with respect to the placement of primordium *X* between the existing primordia *A* and *B*. (From Williams & Brittain, 1984.)

∇_3 must be determined together with the value of T_3 at which the available space will be large enough for the insertion of this primordium in the existing structure.

Here is the general mechanism allowing the determination of the parameters ∇_X, r_{PX}, r_{CX}, and T_X of primordium X between primordia A and B. The primordium must arise in the largest available space. To determine it with precision, the mechanism will be repeated for each one of the available spaces, and the primordium will be placed where the value of T_X is minimal. Five cases can be considered as Figure A5.1 shows. The most important case only, that is case B, will be analyzed.

It is supposed that X is tangent to A and B and that X and C are on the same side of AB. The value of r_{PX} is visually estimated, as precisely as possible. A repeated application of the law of cosines, from the known values of r_{CA}, r_{CB}, r_{PA}, r_{PB}, ∇_A, and ∇_B, permits the calculation of AB, of the angles CAB, BAX, and CAX, and finally of the distance $XC = r_{CX}$. The value of q is then compared to the ratio r_{PX}/r_{CX}, and the process is repeated with a new value of r_{PX} till the difference between r_{PX}/r_{CX} and q becomes smaller than the value 0.0001 fixed by the authors of the model. Then the values of r_{PX}, r_{CX}, and ∇_X are printed. The value of T_X can of course be determined from the time of emergence T_Y of any primordium Y, by the formula

$$R[\exp(T_X - T_Y)] = r_{CY}/r_{CX},$$

where r_{CY} and r_{CX} are evaluated at T_X. We have

$$T_X = T_Y + [\ln(r_{CY}/r_{CX})]/\ln R \quad \text{and} \quad T_X \geq X.$$

When $T_X = X$, the center of the primordium is on the circle of radius 1 centered at C. When $T_X > X$, the available space is not large enough to receive the primordium, $r_{CX} < 1$, and X arises with a delay of $T_X - X$ with its center outside the circle.

The mechanism can be easily adapted to the other situations in Figure A5.1. In case C many solutions can be considered; the solutions without biological meanings are discarded (such as the birth of a primordium that is bigger than the existing ones). In case D there is more space than needed between A and B to allow for the initiation of X. Then the angle PCQ made by the two extreme positions of X is divided in inverse proportion of the plastochronic intervals $T_X - T_A$ and $T_X - T_B$, and the center of X is consequently placed inside the angle PCQ.

The authors show how this mechanism can be used to generate a (3, 2) pattern from the positions of primordia 1 and 2 defined above, the values $R = 1.49049$, and $q = 0.6946$, by calculating dynamically the position of each consecutive primordia and by using the various algorithms sketched in Figure A5.1, depending on the value of X to be inserted in the structure. With primordium $X = 24$, the system (3, 2) of tangent circles of increasing size is well established, $\gamma = 102.34°$ and $d = 141.28°$.

A5.2. The results

Table A5.1 shows the main results of the modeling work, that is, contact spiral systems (m, n) obtained for various values of R and q. The table

Table A5.1. *Systems developed by Williams and Brittain (1984) from values of the phyllotaxis index P.I. ranging from 1 to 4*

R	q	d	(m,n)	γ
3.54892	0.89236	142.65°	(1,2)	90.27°
3.54892	0.93933	87.63°	(2,2)	63.59°
2.58311	0.83468	136.75°	(1,2)	106.82°
2.18766	0.82484	131.69°	(2,3)	61.74°
2.18766	0.85035	94.64°	(2,2)	90.11°
1.90744	0.80000	136.76°	(2,3)	73.82°
1.90744	0.80000	91.30°	(2,2)	101.14°
1.79771	0.7758	138.07°	(2,3)	79.49°
1.79771	0.7758	90.70°	(2,2)	106.50°
1.70352	0.7528	139.09°	(2,3)	84.82°
1.70352	0.7528	90.01°	(2,2)	111.71°
1.62226	0.7316	139.99°	(2,3)	91.03°
1.62226	0.7316	91.17°	(2,2)	116.73°
1.55181	0.7121	140.55°	(2,3)	96.64°
1.49049	0.6946	141.28°	(2,3)	102.34°
1.43690	0.6790	141.55°	(2,3)	107.66°
1.43690	0.6790	98.46°	(3,4)	
1.38989	0.6653	140.85°	(2,3)	111.91°
1.38989	0.6653	99.14°	(3,4)	73.55°
1.34853	0.6398	101.30°	(3,4)	79.36°
1.31204	0.6066	100.51°	(3,4)	85.44°
1.25111	0.5473	101.45°	(3,4)	97.17°
1.20298	0.5022	101.76°	(3,4)	107.69°
1.20298	0.5022	56.77°	(3,3)	124.02°
1.14850	0.4546	44.53°	(4,4)	109.61°
1.10931	0.4033	79.41°	(4,5)	111.95°
1.07296	0.3236	65.18°	(5,6)	111.44°

Note: R is the given plastochrone ratio [see Equation (4.21)]; parameter q defined in the text; d, (m,n), and γ are the divergence angle, the contact parastichy pair, and the angle of intersection of the opposed parastichies.

gives the values of the divergence d in degrees, and of the angle of intersection γ of the opposed spirals.

The only Fibonacci systems that the authors succeeded to reproduce are (2,1) and (3,2). Apart from a few unstable whorled systems (2,2), (3,3) and (4,4), all the systems obtained are of the type $(m+1,m)$. The authors conclude that the systems having high secondary numbers do not arise *de novo* but from the evolution of systems having small secondary numbers. They succeeded in going from (2,1) to (8,5) and from (4,3) to

(11, 7) by slightly reducing R at the emergence of each primordium. They show that for $R=2.18766$, the decrease of q allows the system to pass from (2, 2) to (3, 2) with a divergence angle moving towards the Fibonacci angle, and with a value of γ increasing towards 64.6°. They finally propose considerations on the stability of distichous, spirodistichous, and spiromonostichous systems that would result from surrounding the apical dome by the margins of each new primordium before the initiation of the next primordium (complexicaules leaves).

Table 4.3 (the Pattern Determination Table) and Table A5.1 can be used to verify the assertions in Section 4.2.2. For example, with $R=1.349$, if $m+n=7$, then $\gamma=79.36°$, and Table 4.3 gives, by interpolation, an approximately equal value of γ. For $\gamma=73.82°$ and $m+n=5$, $R=1.91$ in one case and interpolation gives an approximately equal value in Table 4.3. The slight differences between the values for R or γ deduced from one model or the other come from the fact that the allometry-type model assumes the limit-divergence angles while the Williams–Brittain model does not (101.3° instead of 99.5° in the first case; 136.76° instead of 137.5° in the second case). Note for example with the systems (2, 3) in Table A5.1 that as R decreases, γ increases, a phenomenon that is clearly observable in Table 4.3. For γ approximately fixed in Table A5.1 (e.g., 111°), as R decreases, $m+n$ increases, an obvious phenomenon in Table 4.3. The Williams–Brittain model is limited to a few types of spiral systems. Finally, note that the decussate systems (2, 2) with their corresponding values of R and γ in Table A5.1 are compatible with Table 4.3. As for Williams's (1975) data considered in Section 8.3.2, we can say that whorled systems behave like multijugate systems.

Appendix 6
Interpretation of Fujita's frequency diagrams in phyllotaxis

In Section 5.6.1 we started a discussion on Fujita's diagrams. It was stated that many possible sources of errors leading to these diagrams can be identified. They are, for example, Fujita's use of contact parastichies, and the way he determined them from transverse sections and from direct observations, the presence of triple contact points, the number of species he considered, the levels at which the transverse sections were made, the way he measured the divergence angles, the relative ease with which the parastichies can be observed, and the determination of the axis of the plants (see in Section 5.6.2 the discussion on Fig. 5.2). Let us look into these statements.

The absence of distinctions between contact, conspicuous, and visible opposed parastichy pairs, and the presence of a large quantity of such pairs in the same system (see Section 2.2) weighed heavily on the subject of phyllotaxis since its very beginning (see Section 2.1). The lack of a clear understanding of them has indeed created since the 1830s a long history of confusion and obstruction, inducing serious errors in the conclusions of the investigators. These clarifying notions were introduced by Adler (1974, 1977a) after Fujita's papers were published.

The notion of contact parastichies, used by Fujita (1938, 1939) to determine which pattern was under observation, is intuitively good to some extent. On a theoretical basis, we are not concerned with contact parastichy pairs, but with visible opposed parastichy pairs. A contact pair generally is visible and opposed, but we are not particularly interested in the form of the primordia, giving the contacts, which can be misleading. The fact that many parastichy pairs can be drawn with the same points results in various forms for the primordia (i.e., various sets of contact parastichies with the same divergence angle). The form of the primordium gives the value of the angle of intersection of the contact parastichy pair, but we are primarily interested in the value that is closest to 90°. This latter value may have nothing to do with the former value. It follows that we must use the rigorous concept of conspicuous parastichy pair in order to decide on a regular basis if the system is (1, 2) or (3, 2).

A (3, 2) contact parastichy pair can be in many cases a (2, 1), a (3, 2), or a (5, 3) conspicuous parastichy pair. When three primordia are equally distant from a given primordium in what is known as triple points, two conspicuous parastichy pairs are observable at the same time. It follows that a system identified by Fujita as (3, 2) on the basis of contact parastichies can represent at the same time the systems (2, 1) and (3, 2), or the systems (3, 2) and (5, 3). As a (2, 1) or (5, 3) system, this otherwise (3, 2) system has probably not been tabulated by Fujita.

Why is it that Fujita's normal curves at 137.5° show secondary peaks at 132°? Fujita (1938, 1939) does not show statistical significance tests for his frequency curves. Considering his curve 1 (1939) for example, we can see that 8 angles at 132° are recorded, on a total of 121 angles distributed over 20 categories of angles (from 128° to 147°). The expected mean in each category being 6, two measurements above is hardly a significant peak, especially when it is realized how difficult it is to measure accurately to 1°. Taking all of Fujita's measurements for (2, 1) systems, the percentage is not significantly higher at 132° (5.6%) given that the values at 130°, 131°, 133°, 134°, and 135° are 3%, 3.8%, 4.2%, 3.4%, and 3.8%, respectively. The secondary peaks at 132°, 140° or 142°, come only from natural variations based on relatively very small numbers of measurements. Indeed these peaks drop when more species are considered. For the (3, 2) systems where 2,385 angles were measured on 13 species, these peaks are lower than for the (2, 1) systems where 266 angles were measured on three species. In particular, the peak at 132° is 50% higher in the (2, 1) systems than in the (3, 2) systems, and the peak at 140° or 142° is 100% higher in the (2, 1) systems.

The peak at 142° for (3, 2) systems can be dissolved in a variety of ways. For example in the contact pressure model of phyllotaxis, 142° is the value of the divergence angle when phyllotaxis is at the same time (3, 2) and (3, 5). This peak would thus correspond to the point where a (5, 3) system replaces the (3, 2) system registered by Fujita. In that case a (5, 3) system should have been taken into consideration so that the local maximum at 142° drops. In some systems the trough at 135° may be more important than the peak at 132°. Indeed in the contact pressure model, 135° is the value of the divergence angle when the system shows (8, 5) and (3, 5) phyllotaxis at the same time. If Fujita had tabulated the system as (5, 3) instead of (8, 5), the trough could have been filled. Many scenarios can be imagined that fill the gullies and level the hills in Fujita's graphs. Apparently the meshes in the net (protocol) of Fujita are not tight enough to grasp the subtle behavior of the potential data.

Fujita (1964) made the following observations on the (3, 2) system of five specimens of *Cuscuta japonica*. With the birth of primordium number 5, the (3, 2) system is apparent, the angle between the first two is 180°, and between primordia number 4 and 5 it is (for each of the specimens) 137°, 135°, 144°, 147°, and 127°, respectively. Between primordia number 2 and 3, and 3 and 4 the angles lie in the range [110°, 147°]. What about the species reported in Fujita (1939)? Of course I do not mean that Fujita selected his results, but which among the numerous angles measured on the same specimen did he retain and on which basis?

Fujita's normal curves seem to be irreconcilable with the contact-pressure model of phyllotaxis (Adler, 1974, 1977a). Under contact pressure the divergence angles for (3, 5) systems should be found in the range [135.95°, 142.1°]. For this system Fujita observed however 943 angles (out of 2,201 observations) in the ranges [128°, 135°] and [143°, 147°]. It is experimentally known that for higher Fibonacci systems such as (13, 21) and (21, 34), the divergence angles observed are practically equal to the Fibonacci angle. According to the fundamental theorem of phyllotaxis, if the pattern is (21, 13) then the divergence is to be found in the range [137.14°, 138.46°]. In Adler's model the interval is narrower. Why is it then that 49 out of the 77 angles measured by Fujita for (21, 13) systems are in the intervals [128°, 136°] and [139°, 147°]? One can say that the observation of angles outside the allowed ranges is understandable, given that plants are not as regular as the lattices representing them. Then how can we explain that almost no (1, 2) system has been observed by Fujita with a divergence in the range [147°, 180°]; it is mainly in the range [128°, 147°], while under contact pressure the angles in the rangc [128.57°, 180°] are equally probable?

Taking for granted Fujita's curves and trying to explain them, Roberts (1984a) states that under his chemical contact-pressure theory (a variation on Adler's themes), the frequency-divergence angle diagram is a rectangle. This means that inside the interval in the above paragraph, [135.95°, 142.1°], the angles are equally probable. But a frequency diagram cannot be at the same time a rectangle and a normal curve. From the fundamental theorem of phyllotaxis, it can be deduced that a system showing phyllotaxis along ⟨3, 8, 11, 19, ...⟩ or ⟨5, 13, 18, 31, ...⟩ will produce limit-divergence angles of 132.2° or 139.53°, respectively. These results are used by Roberts in his trial to explain the secondary peaks at 132°, 140°, and 142°. To the pattern (m, n) corresponds a rectangle with a basis (whose length is the difference of the end points of the interval given by the contact pressure model) larger than the basis for the pattern $(m+n, m)$. Rising phyllotaxis amounts to piling up the rectangles to give a pyramid of rectangles (and a normal curve) enclosing at its top the limit-divergence angle. The process sorts out the angle, while there is no finality or intention from the part of the plant toward the angle. It is thus irrelevant to bring the above sequences to deal with static observations. These sequences are also purely hypothetical (see Chapter 7).

Assuming that the contact-pressure theory is compatible with Fujita's curves raises difficulties. These can be overcome either by rejecting the curves or by admitting limitations in the model which cannot rationalize the curves in the absence of rising phyllotaxis. And before resorting to Roberts's arguments, some analysis must be made of Fujita's protocol. Finally averaging the measurements would eliminate the variations and also sort out one angle or another, thus hiding the possible functional relation between the divergence angle and the plastochrone ratio.

The conditions under which Fujita's data were gathered are obscure; his protocol appears to be very rudimentary. Therefore it is reasonable to question the universality of the Fibonacci angle in the systems (2, 1) and (3, 2). Because of methodological shortcomings it can be said that the foundations of Fujita's diagrams are precarious, and it would be appropriate to adopt a cautious attitude towards them. Data collecting must be based on clearly defined concepts and on scientific protocols.

Appendix 7
L-systems, Perron–Frobenius theory, and the growth of filamentous organisms

A7.1. Preliminaries

We have introduced L-systems in the treatment of phyllotaxis as we have seen in Section 3.5.2. These systems are embedded in the structure of the interpretative model of Chapter 6, where L-systems spring from growth matrices and generate the control hierarchies of phyllotaxis. From this construct resulted a by-product in the theory of growth functions of L-systems, obtained by bridging for the first time the L-systems to Perron–Frobenius spectral theory in discrete functional analysis (Jean, 1979a, 1981b). This allowed us to refine the application of the theory of growth functions of L-systems to filamentous organisms. On the other hand, the technique of growth functions of L-systems proves to be a valuable tool for the treatment of the concepts in the Perron–Frobenius theory.

A general bibliography on L-systems can be found in Rozenberg (1976) or Rozenberg and Salomaa (1986). The growth function $f(t)$, $t=0,1,2,3,\ldots$, introduced in Section 6.1.2, has values in the positive integers (the numbers in the consecutive levels of control hierarchies). A Fibonacci hierarchy can be expressed by $\langle f(0), f(1)\rangle$, and P_r defined in Section 6.2.1 is expressed by $\sum_{k=0}^{w-1} f(k)$. For an introductory presentation on growth functions, see Herman and Vitanyi (1976). Salomaa and Soittola (1978) give an idea of the general framework of the subject.

The **Perron–Frobenius theorem** states that a nonnegative irreducible square matrix C has a positive eigenvalue r equal to the spectral radius (called the Perron root), and a positive eigenvector corresponding to r. Moreover if λ_j, $j=1,2,\ldots,h$, are the eigenvalues of the modulus $r=\lambda_1$ (C is said to be cyclic of index h), then $\lambda_j = re^{2\pi i(j-1)/h}$, and $\lim C^t/r^t$ exists if and only if C is primitive ($h=1$) (Gantmacher, 1959, or Varga, 1962).

The proof of Proposition 1c of Chapter 6 can be carried on with the spectral analysis of Frobenius matrices as follows: the characteristic polynomials of the matrices of order $w=2,3,4,\ldots$, are $\lambda^w-\lambda-1$, and the spectral radii r_w, between 1 and 2, are nearer to 1 as w grows. The matrices are primitive, and $\lim_T S_w(T)=r_w-1$ tends towards 0 monotonically with the increase of w. The function $S_2(T)$ is strictly decreasing towards $\tau^{-1}\approx 0.618$ from $T>2$, while for the other matrices $S_w(T)<1/2$ for all T. The growth matrices for this same hierarchy have the same spectral radius.

Among the new results are conditions for $f(t)$ to be asymptotically equal to br^t, where b is a constant and $f(t)$ is the growth function of the L-system generated by the irreducible matrix C with spectral radius r, and conditions for the convergence of the ratios $f(t+1)/f(t)$, $f(t)/r^t$ (Theorem A7.1), and C^t/r^t (Theorem A7.2). The value of these ratios are given in terms of the coefficients of the eigenvalues of C in the analytical expression for $f(t)$, that is

$$f(t)=\sum_{j=1}^{s} p_j(t)\lambda_j^t, \tag{A7.1}$$

where λ_j $(j=1,2,\ldots,s)$ are the distinct eigenvalues of C (needs not be irreducible), and $p_j(t)$ is a polynomial of degree s_j in t such that s_j+1 is the algebraic multiplicity of λ_j. The coefficients of $p_j(t)$ are determined by $f(i),\ldots,f(n-1)$, where $i=0$ if 0 is not an eigenvalue, or i is the multiplicity of the eigenvalue 0. If $s=n$, the p_j's are constants (Goldberg, 1961).

Unless otherwise stated we consider nonnegative irreducible matrices $C=(c_{ij})$ of order n with integral entries. The reader is referred to Section 3.5.2 (where C is a matrix with 0 and 1 entries only) for the definition of the terms whose meanings are easily extended. For example consider

$$C=\begin{bmatrix}1 & 1\\ 2 & 0\end{bmatrix},$$

the alphabet $A=\{a_1,a_2\}$, and $w_0=a_1$; then $P(a_1)=a_1a_2a_2$, $P(a_2)=a_1$, $f(0)=1$, $f(1)=3$, $f(2)=5$, $f(3)=11$, ... The notion of growth function defined on the positive integers with values in the positive integers, and the notion of growth matrix can be generalized. Here is an example of such a generalization. If

$$C=\begin{bmatrix}1/2 & 1/5\\ 1/3 & 0\end{bmatrix},$$

the alphabet $A=\{a_1,a_2\}$, and if $w_0=a_1$, then $P(w_0)=a_1^{1/2}a_2^{1/3}$, $P(w_1)=a_1^{1/4}a_2^{1/6}a_3^{1/15}$, ..., $f(0)=1$, $f(1)=1/2+1/3$, $f(2)=1/4+1/6+1/15$, ..., and the analysis of this matrix can be made using 30*C*. Theorem A7.1 can be generalized; its proof becomes a proof of Perron–Frobenius theorem for the case where $\lim f(t)/f(t-1)$ exists.

Here are the main results on the asymptotic behavior of the ratios mentioned above. See Jean (1981b) for proofs, for examples where the limits do or do not exist, and for developments on the applications.

A7.2. Theorems and applications

Theorem A7.1. $\lim f(t+1)/f(t)$ *exists if and only if* $\lim f(t)/r^t$ *exists. If C is cyclic of index h, the first limit exists if and only if* $p_2 = p_3 = \cdots = p_h = 0$, *where the* p_j *are the coefficients of the eigenvalues of modulus r (other than r itself) in the Equation* (A7.1). *When C is primitive these limits exist. The first limit is equal to r and the second to* p_1.

Theorem A7.2. *If C is a primitive growth matrix of order n,* $f_j(t)$, $j = 1, 2, \ldots, n$, *the growth function determined by* $m_i = \delta_{ij}$, $i = 1, 2, \ldots, n$, *the number of* a_i *in* w_0, p_{1j} *the coefficient of* r^t *in the analytical expression for* $f_j(t)$, *and if* $x = (x_i)$ *is a positive eigenvector corresponding to r, then*

$$\lim_t C^t/r^t = (p_{11}x, p_{12}x, \ldots, p_{1n}x) \sum x_i,$$

and if C is a companion matrix

$$p_{1j} = \left(\sum_{i=0}^{n-1} r^i\right) \frac{g_j(r)}{g'(r) r^{n-j+1}},$$

where g′ is the derivative of the characteristic polynomial

$$g(\lambda) = \lambda^n - \sum_{i=1}^{n} b_{n-i}\lambda^{n-i} \quad \textit{and} \quad g_j(\lambda) = \sum_{k=j}^{n} b_{n-k}\lambda^{n-k}.$$

Theorem A7.3. *If C is a cyclic matrix of index h* (> 1) *or primitive* (*h* = 1), *then*

$$\lim_t \sum_{i=0}^{h-1} \frac{f(t+i)}{r^{t+i}} = hp_1 \neq 0 \quad \textit{and} \quad \lim_t \left[\sum_{i=1}^{t} f(i)^b\right]^{1/tb} = r$$

for every positive real number b, and $\lim_t f(t+h)/f(t) = r^h$.

In the following theorem the symbols in the alphabet *A* of the L-system are denoted for convenience by $a, b_1, b_2, \ldots, b_{m-1}, c_1, c_2, \ldots, c_{n-1}$, instead of $a_1, a_2, \ldots, a_{m+n-1}$ as in Chapter 3. The matrix *C* which determines this system is of order $m+n-1$, and is cyclic of index *h* equal to the largest common divisor of *m* and *n*.

Table A7.1. *The value of* $\lim_t f(t-m)/f(t-n) = r^{n-m}$, *for* $m, n = 1$ *to* 8

	n							
m	1	2	3	4	5	6	7	8
1	1	1.6180	2.1478	2.6297	3.0794	3.5063	3.9146	4.3079
2	0.6180	1	1.3247	1.6180	1.8918	2.1478	2.3933	2.6297
3	0.4655	0.7548	1	1.2207	1.4251	1.6180	1.8012	1.9774
4	0.3802	0.6180	0.8192	1	1.1673	1.3247	1.4741	1.6180
5	0.3247	0.5285	0.7016	0.8566	1	1.1347	1.2627	1.3850
6	0.2852	0.4655	0.6180	0.7548	0.8812	1	1.1127	1.2207
7	0.2554	0.4178	0.5551	0.6783	0.7919	0.8987	1	1.0969
8	0.2321	0.3802	0.5056	0.6180	0.7219	0.8192	0.9116	1

Note: Those ratios can be compared with those directly observable on branched structures such as filamentous organisms.
Source: From Jean (1981b).

Theorem A7.4.

(1) *Consider the L-system generated by the matrix C, such that A contains* $m+n-1$ *symbols,* $P(a) = b_1c_1$, $P(b_1) = b_2$, $P(b_2) = b_3, \ldots,$ $P(b_{m-1}) = a$, $P(c_1) = c_2$, $P(c_2) = c_3, \ldots, P(c_{n-1}) = a$. *Then the spectral radius r of C is in the interval* [1, 2] *and*

$$r = \lim_t [f(t-m)/f(t-n)]^{1/(n-m)}.$$

(2) *When* $n = 2m$, $r = \tau^{1/m}$, $f(t+m) \approx \tau f(t)$ *for large t, and* $f(t)$ *cannot be approximated by an expression of the type* br^t, *where b is a constant, when* $m > 1$.

Salomaa and Soittola (1978) give bounds for $f(t)$ and $f(t+1)/f(t)$, that is pq^t and q respectively, where $p = f(0)$ and q is the number of symbols in the longest word $P(a_i)$. I provided lower bounds given as functions of the spectral radius r when the matrix C is irreducible (Proposition 3, Jean, 1981b). Salomaa and Soittola, working in the theory of L-systems, do not give conditions for the convergence of the ratios $f(t+1)/f(t)$ and $f(t)/r^t$. They do not use terms such as strongly connected, irreducible, primitive, and cyclic found in the Perron–Frobenius theory.

With respect to applications, Theorems A7.1 to A7.4 are relevant to the theory of Leslie's matrices (have non-zero entries only in the first row and the first subdiagonal) in population biology. For example Theorem A7.2 gives what is known as Leslie's theorem (stated in Cull & Vogt, 1973). The results allow us to carry through the analysis of cell populations with lineage control without aging (initiated by Lindenmayer, 1978b), to rectify wrong conclusions and vagueness with respect to the possibility or not of approximating $f(t)$ by br^t, and to clarify the behavior of the growth functions of filamentous organisms (studied by Lindenmayer, 1978a,b).

Also Luck (1975) gives a table for the quotient $f(t-m)/f(t-n)$ (to three decimal places), for large t [such that the value of $f(t)$ is around 10^6; very slow convergence] and for values of m and n between 1 and 15. The knowledge of these ratios is useful in the study of branching structures of filamentous organisms such as *Anabaena catenula* (blue-green alga, $m=4$, $n=5$), *Chaetomorpha linum* (blue alga, $m=3$, $n=5$), *Callithamnium roseum* (red alga), *Protonema* (moss, $m=1$, $n=5$), and *Athyrium filix-Femina* (fern, $m=1$, $n=3$). Based on Theorem A7.4, Table A7.1 is obtained that allow us to make the application with all the needed accuracy. The table gives the value of $\lim_t f(t-m)/f(t-n) = r^{n-m}$ for the various values of m and n. The value of this limit can be obtained from the equation $\lambda^{m+n-1} - \lambda^{m-1} - \lambda^{n-1} = 0$, whose largest root in absolute value is precisely the Perron root r. A particular case of the second limit in Theorem A7.3 is $\lim_t [F_{t+2}]^{1/t} = \tau$ (proposed in Halberg, 1955).

Appendix 8
The Meinhardt–Gierer theory of pre-pattern formation

Many authors consider that the mechanisms of pre-pattern formation constitute one of the most important aspects of morphogenetic phenomena. For example, Edwards (1982) and Hermant (Section 3.4.2) produced phyllotactic pre-patterns using projective geometry and topology, respectively. Many theories propose that the site of organ formation is determined by a chemical pre-pattern, generated by a diffusion–reaction mechanism. By **pre-pattern** is meant in this appendix the establishment of a concentration gradient of substances from an almost equal initial distribution. In this context the patterns observed in organisms are considered to be the results of the interaction of their constituents such as cells, genes, or molecules. One of the models that has received the utmost attention is the one by Gierer and Meinhardt (see these authors in the Bibliography). The model is based on a principle of formation of patterns by a process of molecular activation and inhibition. Activation is an autocatalytic process with short spatial range, whereas inhibition is considered to be a comparatively long-range interaction in order to ensure stability of the developing structures. We have seen in Section 8.6.2 that Richter and Schranner (1978) used this model. Berding, Harbich and Haken (1983) used it to generate pre-patterns of capituli and leafy stems.

In order to better understand the principle, Meinhardt (1984) proposes the following analogy. A river can be formed from a minor depression in the landscape. Raindrops accumulate, erosion is accelerated, and a larger quantity of water rushes towards the newly created valley. The formation of a pattern thus results, partially, from a phenomenon of autoamplification and autocatalysis. It can inhibit autocatalysis elsewhere, in a vaster environment, given that the accumulation of water in the valley is such that erosion at some distance becomes nonexistent. This is what Meinhardt and Gierer call long-range inhibition.

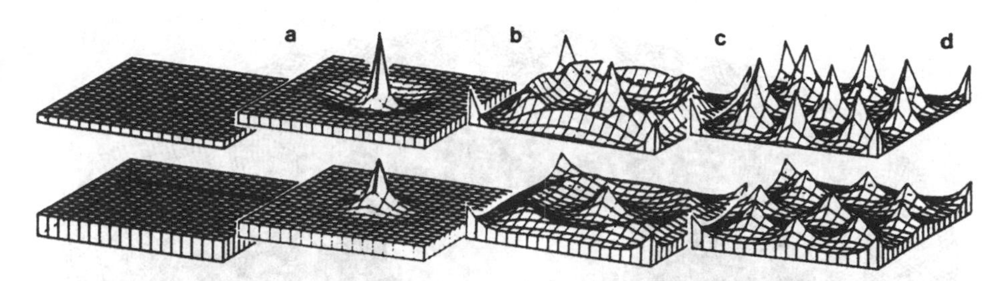

Figure A8.1. Computer simulation of pattern formation in a bidimensional field of cells by the action of an activator–inhibitor. (From Meinhardt, 1984.)

Two types of molecules intervene in the principle, one for short-range autocatalysis, called the activator a, and another which counterbalances its action, called the inhibitor h, whose diffusion is more rapid. The differential equations expressing the interaction between a and h are:

$$\frac{\partial a}{\partial t}=\frac{ca^2}{h}-\mu a+D_a\frac{\partial^2 a}{\partial x^2}+\rho_0, \qquad \frac{\partial h}{\partial t}=ca^2-\nu h+D_h\frac{\partial^2 h}{\partial x^2}. \tag{A8.1}$$

In the first equation for example, where Fick's law of diffusion intervenes (the factor $D_a\partial^2 a/\partial x^2$), the rate of production of the activator, proportional to a^2, is slowed down by the action of the inhibitor ($1/h$), and molecules of activator are decomposed proportionally to the quantity of the molecules that are present ($-\mu a$). The term ρ_0 represents the initial production of activator needed to start the process of autocatalysis.

These equations allow the authors to calculate the variations in the concentrations of a and h on short or long lapses. Figure A8.1 illustrates the mechanism of pattern formation controlled by Equations (A8.1). It is supposed that a cell in the center of the figure has an activator concentration slightly above the mean. This brings a local increase of a, and by the simultaneous production of h, a depression in the concentration of a near the arising peak. Finally a more-or-less periodic pattern arises.

The following considerations intuitively explain why Equations (A8.1) produce stable patterns. To simplify, assuming that h is constant, let us consider that the first equation is $da/dt=a^2-a$. At $a=1$ we have dynamic equilibrium ($da/dt=0$) which is unstable given that for a slightly above 1 the concentration of a will continue to increase because of the exponent. On the other hand, if h is not constant and if $dh/dt=a^2-h$, at equilibrium ($a^2=h$) $da/dt=a^2/h-a$ becomes $da/dt=1-a=0$. This time equilibrium is stable at $a=1$ given that for $a>1$, $da/dt<0$ and the concentration returns to $a=1$.

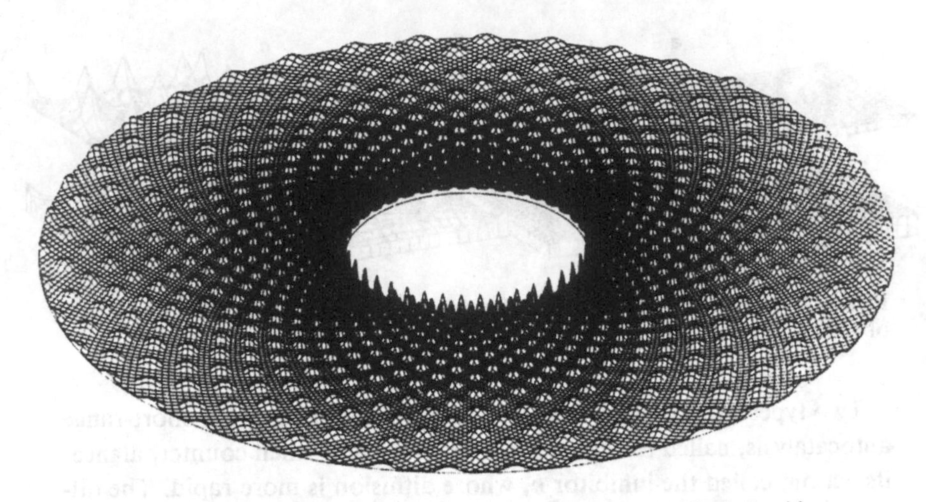

Figure A8.2. Spatial distribution of activating morphogen. Each buckle represents a maximum of morphogen concentration. The number of spirals in each direction is the same (40). (From Berding, Harbich, & Haken, 1983.)

Systems of the Meinhardt–Gierer type and of the Turing type (1952) are similar in that they can generate the same types of patterns (see Bard & Lauder, 1974; Lacalli & Harrison, 1979). However, by extension, Equations (A8.1) generate bands of cells, branching patterns, patterns of leaves, etc. Meinhardt (1984) shows that one can simulate distichy and decussation from a cylindrical arrangement of cells whose growth proceeds by doubling the upper row of cells, and by assuming an inhibition zone at the tip of the stem. As soon as the distance from the last primordium and the center of the apex becomes large enough, a new primordium arises, produced by a new maxima of the activator.

Berding et al. (1983) produced spiral pre-patterns resembling those in the sunflower capitulum (see Fig. A8.2) by means of a mechanism of diffusion–reaction of the Meinhardt–Gierer type, where the coefficients of diffusion are increasing functions of the distance to the pole of the spirals.

Goodwin and Brière (1987) proposed that mechanico-chemical fields based upon cytoskeletal–ionic interactions may be the universal basis of morphogenesis. Consequently, fields based upon Turing's and Meinhardt–Gierer's kinetics of diffusion–reaction may not play the primary role in morphogenesis. There would thus be no such things as morphogens in this sense. This might be the reason why such substances have never been identified. See Harrison (1987) for the status of reaction–diffusion theories in general.

Appendix 9
Hyperbolic transformations of the cylindrical lattice

The model expressed by Equations (4.11) and (4.13), that is,

$$r = p(\gamma)/(m+n)^2,$$

$$p(\gamma) = (\tau^3/2\sqrt{5})(\sqrt{5}\cot\gamma + (5\cot^2\gamma + 4)^{1/2}),$$

where r is the rise and γ is the angle of intersection of the parastichy pair (m, n), shows that the phyllotactic lattice experiences hyperbolic transformations as the three parameters of the model vary.

Taking from Jean (1983a), consider a system for which (m, n) does not change during a period of time while γ varies. In this case, the point $(r, \cot\gamma)$ moves on an hyperbola given by the relation

$$br\cot\gamma = r^2 - a,$$

that is,

$$y^2 - bxy - a = 0,$$

where $a = \tau^6/5(m+n)^4$, $b = (5a)^{1/2}$, $r = y$, and $x = \cot\gamma$. The center of the hyperbola is (0, 0) and its asymptotes are $y = bx$ and $y = 0$. Given that $r > 0$ for the phenomenon concerned, we must consider the right-hand side branch of the hyperbola.

Now, in this development it is considered that the girth g of the cylinder is equal to 1 (see Fig. 4.1), so that as r decreases phyllotaxis rises. Let us consider that g is a variable of the model. This changes only the value of $p(\gamma)$ which becomes

$$p(\gamma, g) = (\tau^3/2\sqrt{5})(\sqrt{5}g\cot\gamma + (5g^2\cot^2\gamma + 4)^{1/2}).$$

Let us hold r constant (equal to 1 for example) and γ constant (the value around 120° at triple contacts or the value 90° for example). Clearly then, if g increases, it means that $p(\gamma, g)$ increases. Consequently, according to the model, $(m+n)^2$ increases. In other words if $m+n$ increases, g increases, and vice-versa.

Considering that $g \neq 1$, it follows that the point $(r, \cot\gamma)$ moves on a segment of an hyperbola given by

$$bgr \cot\gamma = r^2 - a,$$

where a and b are as above. If now $x = g$ and $y = (m+n)^2$ are the variables, instead of r and $\cot\gamma$, the latter equation becomes

$$y^2 - cxy - d = 0,$$

where $c = \tau^3 \cot\gamma / r$, $d = \tau^6 / 5r^2$. If $\gamma \neq 90°$ we have an hyperbola with center (0, 0) whose asymptotes are $y = cx$ and $y = 0$. As g increases, r and γ remaining constant, the lattice experiences an hyperbolic transformation. If $\gamma = 90°$ we have two parallel straight lines $y = \pm\sqrt{d}$. The hyperbolas closer to the y axis in the first quadrant are the analogues of the polygonal lines asymptotic to the Y axis in Figure 2.3(2). If $x = g$ is such that $(m, n) = (F_{k-1}, F_{k-2})$, then $y = (m+n)^2 = ((\tau^k - (-\tau)^{-k})/\sqrt{5})^2 = (4/5)(\sinh(k \ln \tau))^2$ if k is even, and $y = (4/5)(\cosh(k \ln \tau))^2$ if k is odd (where sinh and cosh are hyperbolic functions). Notice that the right-hand side of Equation (5.1) is $\cosh(m \ln R/2)/\cosh(n \ln R/2)$.

In this setting, as g increases, phyllotaxis rises, and the divergence angle d varies in agreement with the fundamental theorem of phyllotaxis (Chapter 2). That is, d moves from one end point to the other of nested intervals of decreasing amplitude, all of them containing the limit-divergence angle of the system.

As phyllotaxis rises the same value of γ gets back. Bodnar (in press) says that an **hyperbolic turn** has been made. Bodnar illustrates the behavior of the points of the lattice in four systems of coordinates, as g increases, causing phyllotaxis to rise. He introduces a few particular formulas derived from hyperbolic geometry to express that behavior. The model of Chapter 4 justifies Bodnar's unusual practice, which consists in varying the girth g of the stem, and integrates his intuitive approach to the hyperbolic turn. From Section 9.2.1 we know that at each hyperbolic turn the distance between two points on the allometric line $r = p(\gamma)(m+n)^{-2}$ (γ fixed) is $\log \tau^{\sqrt{5}}$. Goodal's (1991) eigenshape analysis of the cut-grow mapping for triangles (Section 11.3.2) delivers hyperbolae in the $\lambda - \eta$ phase space, where λ is the "cut" parameter, and η is the dilation factor.

But r and (m, n), or g and (m, n), or r and γ are not the only variables of the system. The relative variations of other pairs of phyllotactic parameters can be considered (e.g., r and g, g and γ). All the parameters of Equation (4.11) vary as a phyllotactic pattern develops. The dynamic of a phyllotactic system is expressed by the equations in this model that control the relative variations of five parameters (including g). The divergence is implicitly controlled by the variation of (m, n) in agreement with the fundamental theorem of phyllotaxis.

Bibliography

Abdulnur, S. F. & Laki, K. (1983). A two-dimensional representation of relative orientations of α-helix residues. *J. Theor. Biol.* **104**: 599–603.

Adler, I. (1974). A model of contact pressure in phyllotaxis. *J. Theor. Biol.* **45**: 1–79.

Adler, I. (1975). A model of space filling in phyllotaxis. *J. Theor. Biol.* **53**: 435–44.

Adler, I. (1977a). The consequences of contact pressure in phyllotaxis. *J. Theor. Biol.* **65**: 29–77.

Adler, I. (1977b). An application of the contact pressure model of phyllotaxis to the close packing of spheres around a cylinder in biological fine structures. *J. Theor. Biol.* **67**: 447–58.

Adler, I. (1978). A simple continued fraction represents a mediant nest of intervals. *Fib. Q.* **16**: 527–9.

Adler, I. (1984). Plant spiral and Fibonacci numbers: A mathematical gold mine. *Mathematical Medley* (Singapore) **12**: 29–41.

Adler, I. (1986). Séparant les aspects biologiques de la phyllotaxie des aspects mathématiques, *54th ACFAS Congress, Montréal,* Unpublished Contributed Paper.

Adler, I. (1987). Some questions of methods in phyllotaxis. *Mathematical Modelling* **8**: 756–61. (*Proc. 1st Int. Symp. on Phyllotaxis, U. of Berkeley,* July 1985; pp. 729–61 of *Proc. 5th Int. Conf. on Math. Modelling in Sci. and Tech.*)

Adler, I. (1990). Symmetry in phyllotaxis. *Symmetry: Culture and Science* **1**(2): 171–81.

Airy, H. (1873). On leaf-arrangement. *Proc. Roy. Soc.* **21**: 176.

Allard, H. A. (1946). Clockwise and counterclockwise spirality observed in the phyllotaxy of tobacco. *J. Agri. Res.* **73**: 237–46.

Allard, H. A. (1951). The ratios of clockwise and counterclockwise spirality observed in the phyllotaxy of some wild plants. *Castanea* **16**: 1–6.

Allsopp, A. (1964). Shoot morphogenesis. *Ann. Rev. Plant Physiol.* **15**: 225–54.

Allsopp, A. (1965). Heteroblastic development in Cormophytes. In *Encyclopedia of Plant Physiology,* XV(1), ed. W. Ruhland. New York: Springer-Verlag, pp. 1172–1221.

Aoki, I. (1987). Entropy budgets of deciduous plant leaves and a theorem of oscillating entropy production. *Bull. Math. Biol.* **49**(4), 449–60.

Aono, M. & Kunii, T. L. (1984). Botanical tree image generation. *IEEE Computer Graphics and Applications* **4**(5): 10–34.
Arber, A. (1970). *The Natural Philosophy of Plant Form*. Darien, Conn.: Hafner Pub. (Facsimile of 1950 edition of Cambridge University Press.)
Auger, P. (1983). Hierarchically organized populations: Interactions between individual, population and ecosystem levels. *Math. Biosci.* **65**: 269–89.
Auger, P. (1989). *Dynamics and Thermodynamics in Hierarchically Organized Systems*. New York: Pergamon Press.
Auger, P. (1990). Self-organization in hierarchically organized systems. *Syst. Res.* **7**: 221–36.
Bailey, K. D. (1984). Equilibrium, entropy, and homeostasis: A multidisciplinary legacy. *Syst. Res.* **1**(1), 25–44.
Bailey, N. T. J. (1964). *The Elements of Stochastic Processes*. New York: Wiley.
Ball, E. (1944). The effects of synthetic growth substances on the shoot apex of *Tropaeolum majus* L. *Am. J. Bot.* **31**: 316–27.
Ball, W. W. (1967). *Mathematical Recreations and Essays*. London: Macmillan.
Bamford, C. H., Brown, L., Elliott, A., Hanby, W. E. & Trotter, I. F. (1954). Alpha- and beta-forms of poly-*l*-alanine. *Nature* **173**: 27.
Barabé, D. (1991a). Chaos in plant morphology. *Acta Biotheor.* **39**: 157–9.
Barabé, D. (1991b). Analyse de la phyllotaxie du genre *Dipsacus* (Dipsacaceae). *Saussurea* **22**: 95–101.
Barabé, D., Brouillet, L. & Bertrand, C. (1991). Symétrie et phyllotaxie: Le cas des bégonias. *Ann. Sci. Nat. Bot. 13ième Sér.* **11**: 33–7.
Barabé, D. & Vieth, J. (1990). La torsion de contrainte et le modèle phyllotaxique de Jean. *Can. J. Bot.* **68**: 677–84.
Bard, J. & Lauder, I. (1974). How well does Turing's theory of morphogenesis work? *J. Theor. Biol.* **45**: 501–31.
Barlow, H. W. B. (1979). Sectorial patterns in leaves on fruit tree shoots produced by radioactive assimilates and solutions. *Ann. Bot.* **43**: 593–602.
Barlow, P. W. (1982). The plant forms cells, not cells the plant: The origin of de Bary's aphorism. *Ann. Bot.* **49**: 269–71.
Barlow, P. W. (1987). The hierarchical organization of plants and the transfer of information during their development. *Postepy Biologii Komorki* **14**: 63–82.
Barlow, P. W. (1989a). Meristems, metamers and modules and the development of shoot and root systems. *Bot. J. Linnean Soc.* **100**: 255–79.
Barlow, P. W. (1989b). From cell to system: Repetitive units of growth in the development of shoots and roots. In *Growth Patterns in Vascular Plants*, ed. M. Iqbal. Portland, Ore.: Dioscorides Press.
Barlow, P. W. & Adam, J. S. (1988). The position and growth of lateral roots on cultured root axes of tomato, *Lycopersicon esculentum* (Solanaceae). *Pl. Syst. Evol.* **158**: 141–54.
Barnsley, M. F. (1988). *Fractals Everywhere*. San Diego: Academic Press.
Batschelet, E. (1974). *Introduction to Mathematics for Life Scientists*. New York: Springer-Verlag.
Battjes, J., Vischer, N. O. B. & Bachmann, K. (1993). Capitulum phyllotaxis and numerical canalization in *Microseris pygmaea*. *Am. J. Bot.* **80**(4): 419–28.
Beard, B. H. (1981). The sunflower crop. *Sci. Am.* **224**: 150–61.
Beck, C. B., Schmid, R. & Rothwell, G. W. (1982). Stelar morphology and the primary vascular system of seed plants. *Bot. Rev.* **48**: 691–815, 913–31.

Bell, C. J. (1981). The testing and validation of models. In *Mathematics and Plant Physiology,* eds. D. A. Rose & D. A. Charles-Edwards. London: Academic Press, pp. 299–309.
Bellman, R. (1970). *Introduction to Matrix Analysis.* New York: McGraw-Hill.
Benoist, R. (1932). La phyllotaxie de *Phyllactis rigida* Pers. *Bull. Soc. Bot. France* **79**: 490–1.
Benoist, R. (1933). La phyllotaxie chez quelques espèces de caryophyllacées et valerianacées. *Bull. Soc. Bot. France* **80**: 367–71, 563–5.
Berding, C. & Haken, H. (1982). Pattern formation in morphogenesis. *J. Math. Biol.* **14**: 133–51.
Berding, C., Harbich, T. & Haken, H. (1983). A pre-pattern formation mechanism for the spiral type patterns of the sunflower head. *J. Theor. Biol.* **104**: 53–70.
Berdyshev, A. P. (1972). On some mathematical regularities of biological processes. *Zh. Obshch. Biol.* **33**: 631–8.
Berg, A. R. & Cutter, E. G. (1969). Leaf initiation rates and volume growth rates in the shoot apex of *Chrysanthenum. Am. J. Bot.* **56**: 153–9.
Berge, C. (1968), *Principes de Combinatoire.* Paris: Dunod.
Berge, C. (1976). *Graphs and Hypergraphs,* 2nd ed. New York: Elsevier-North Holland.
Bernal, J. D. (1965). Molecular structure, biochemical function and evolution. In *Theoretical and Mathematical Biology,* eds. T. H. Waterman & H. J. Morowitz. Boston: Blaisdel, pp. 96–135.
Berrill, N. J. (1961). *Growth, Development, and Pattern.* San Francisco: W. H. Freeman Co.
Bertalanffy, L. von (1973). *Théorie Générale des Systèmes.* Paris: Dunod. Translated from *General Systems Theory.* New York: G. Braziller, 1968.
Bertalanffy, L. von (1974). The model of open systems beyond molecular biology. In *Biology, History and Natural Philosophy,* eds. A. D. Breck & W. Yourgrau. New York: Plenum-Rosetta, pp. 17–30.
Beveridge, W. I. B. (1957). *The Art of Scientific Investigation.* New York: Vintage.
Bicknell-Johnson, M. (1986). Letter to the editor. *Fib. Q.* **24**(4): 309.
Blanco, M., Lorenzo-Andreu, A. & Blanco, J. L. (1968). Modificaciones filotaxicas en mais: Dispersion, espirodistiquia y decusacion. Modificaciones similares en otras partes de la planta. Germination multiple en mais disticos y decusados. *Port. Acta Biol. Ser. A.* **10**: 289–300.
Bodnar, O. I. (1991). Dynamic symmetry. In *Problems of Mechanics and Mathematics.* Lvov: Academia Nauk USSR, pp. 25–90. (In Russian; English version in preparation.)
Bodnar, O. I. (In press). Phyllotaxis geometry.
Bodnar, O. I. (In press). Golden section in nature and art. *Technitcheskaya Estetika* 1.
Bohm, D. (1969). Some remarks on the notion of order. In *Towards a Theoretical Biology,* vol. 2, *Sketches,* ed. C. H. Waddington. Edinburgh: Edinburgh University Press, pp. 18–60.
Boles, M. & Neuman, R. (1990). *Universal Patterns.* Bradford, Mass.: Pythagorean Press.
Bolle, F. (1939). Theorie der Blattstellung, *Verh. Bot. Prov. Brandenb. (Berlin)* **79**: 152–92.
Bolle, F. (1963). Ueber Blattstellungsprobleme und Blutendurchwachsung bei der Nelke. *Ber. Deutsch. Bot. Ges.* **76**: 211–28.

Bonner, J. T. (ed.) (1982). *Evolution and Development*. Berlin: Springer-Verlag.
Bonnet, C. (1754). *Recherches sur l'Usage des Feuilles dans les Plantes*, Neuchâtel. (2nd edition in 1779.)
Bookstein, F. L. (1978). *The Measurement of Biological Shape and Shape Change*. New York: Springer-Verlag.
Bookstein, F. L. (1982). Foundations of morphometrics. *Ann. Rev. Ecol. Syst.* **13**: 451–70.
Brandmuller, J. (1992). Fivefold symmetry in mathematics, physics, chemistry, biology, and beyond. In *Fivefold Symmetry*, ed. I. Hargittai. Singapore: World Scientific, pp. 11–31.
Braun, A. (1831). Vergleichende Untersuchung über die Ordnung der Schuppen an den Tannenzapfen als Einleitung zur Untersuchung der Blattstellung uberhaupt. *Nova Acta Ph. Med. Acad. Cesar Leop. Carol. Nat. Curiosorum* **15**: 195–402.
Braun, A. (1835). Dr. Carl Schimper's Vortrage über die Möglichkeit eines wissenschaftlichen Verständnisses der Blattstellung, nebst Andeutung der hauptsachlichen Blattstellungsgesetze und insbesondere der neuentdeckten Gesetze der Aneinanderreihung von Cyclen verschiedener Masse. *Flora* **18**: 145–91.
Braun, A. (1843). Beitrag zur Feststellung naturlicher Gattungen unter den Sileneen. *Flora* **26**: 349–88.
Braun, M. (1978). *Differential Equations and Their Applications*, 2nd ed. *Applied Mathematical Sciences Series* **15**. New York: Springer-Verlag.
Bravais, L. & Bravais, A. (1837). Essai sur la disposition des feuilles curvisériées. *Ann. Sci. Nat. Bot. Biol. Vég.* **7**: 42–110, 193–221, 291–348; **8**:11–42.
Bravais, L. & Bravais, A. (1839). Essai sur la disposition générale des feuilles rectisériées. *Ann. Sci. Nat. Bot.* **12**: 5–14, 65–77.
Brillouin, L. (1949). Life, thermodynamics and cybernetics. *Am. Sci.* **37**: 554–68.
Brillouin, L. (1959). *La Science et la Théorie de l'Information*. Paris: Masson.
Brooks, D. R. & Wiley, E. O. (1988). *Evolution as Entropy*, 2nd ed. Chicago University Press.
Brousseau, A. (1968). On the trail of the California pine. *Fib. Q.* **6**(1): 69–76.
Brousseau, A. (1969). Fibonacci statistics in conifers. *Fib. Q.* **7**: 525–32.
Bruter, C. (1974a). Dynamique, stabilité et symétrie. *Rev. Bio.-Math.* **45**: 7–8.
Bruter, C. (1974b). *Topologie et Perception*. Paris: Maloine-Drouin.
Buerger, M. (1978). *Elementary Crystallography*. Cambridge, Mass.: MIT Press.
Bursill, L. A. (1990). Quasicrystallography on the spiral of Archimedes. *Int. J. Modern Phys. Letters B* **4**: 2197–216.
Bursill, L. A. & JuLin, P. (1985). Penrose tiling observed in a quasi-crystal. *Nature* **316**: 50–1.
Bursill, L. A., JuLin, P. & XuDong, F. (1987). Spiral lattice concepts. *Modern Phys. Letters B.* **1**(5–6): 195–206.
Bursill, L. A., Needham, A. & Rouse, J. L. (1990). Size and shape of seeds in relation to seed packing in sunflower. *J. Theor. Biol.* **147**(3): 303–28.
Bursill, L. A., Rouse, J. L. & Needham, A. (1991). Sunflower quasicrystallography. In *Spiral Symmetry*, eds. I. Hargittai & C. A. Pickover. Singapore: World Scientific, pp. 295–322.
Bursill, L. A., Ryan, G., XuDong, F., Rouse, J. L., JuLin, P. & Perkins, A. (1989). Basis for synthesis of spiral lattice quasicrystals. *Modern Phys. Letters B* **3**(14): 1071–85.
Bursill, L. A. & XuDong, F. (1988). Close-packing of growing discs. *Modern Phys. Letters B* **2**(11–12): 1245–52.

Bursill, L. A. & XuDong, F. (1989). Scaling laws for spiral aggregation of contact discs. *Modern Phys. Letters B* **3**(17): 1293–1306.
Buvat, R. (1955). Le Méristème apical de la tige. *Ann. Biol.* **31**: 595–656.
Cahen, E. (1914). *Théorie des Nombres,* vol. 1. Paris: Hermann.
Camefort, H. (1950). Anomalies foliaires et variations phyllotaxiques chez les plantules de *Cupressus sempervirens. Rev. Gen. Bot.* **57**: 348–72.
Camefort, H. (1956). Etude de la structure du point végétatif et des variations phyllotaxiques chez quelques gymnospermes. *Ann. Sci. Nat. Bot.* **17**: 1–185.
Candolle, A. C. P. de (1881). *Considérations sur l'étude de la phyllotaxie.* Geneva: Georg.
Candolle, A. C. P. de (1895). Nouvelles considérations sur la phyllotaxie. *Arch. Sci. Phys. Nat.* **33**: 121–47.
Cannell, M. G. R. (1978). Components of conifer shoot growth. In *Proc. Fifth North Am. For. Biol. Workshop,* pp. 313–18.
Cannell, M. G. R. & Bowler, K. C. (1978). Phyllotactic arrangements of needles on elgonating conifer shoots: A computer simulation. *Can. J. Forest Res.* **8**: 138–41.
Carr, D. J. (1984). Positional information in the specification of leaf, flower and branch arrangement. In *Positional Controls in Plant Development,* eds. P. W. Barlow & D. J. Carr. Cambridge: Cambridge University Press, pp. 441–60.
Carton, A. (1948). Études phyllotaxiques sur quelques espèces de *Linum. Rev. Gen. Bot.* **55**: 137–68.
Causton, D. R. (1969). A computer program for fitting the Richards function. *Biometrics* **25**: 401–9.
Causton, D. R. (1977). *A Biologist's Mathematics.* London: Arnold.
Chadefaud, M. (1960). *Traité de Botanique,* vol. 1. Paris: Masson.
Chai, H. (1990). Buckling and post-buckling behavior of elliptical plates. Part II – Results. *J. Appl. Mechanics* **57**: 989–94.
Chapman, J. M. (1988). Leaf patterns in plants: Leaf positioning at the plant shoot apex. *Plants Today* **1**(2): 59–66.
Chapman, J. M. & Perry, R. (1987). A diffusion model of phyllotaxis. *Ann. Bot. (London)* **60**: 377–89.
Charles-Edwards, D. A., Cockshull, K. E., Horridge, J. S. & Thornley, J. H. M. (1979). A model of flowering in *Chrysanthemum. Ann. Bot.* **44**: 557–66.
Charlton, W. A. (1978). Studies in the Alismataceae. VII. Disruption of phyllotactic and organogenetic patterns in pseudostolons of *Echinoderus tenellus* by means of growth active substances. *Can. J. Bot.* **57**: 215–22.
Chauveaud, G. (1921). *La Construction des Plantes Vasculaires Révélée par leur Ontogénie.* Paris: Payot.
Church, A. H. (1901). Note on phyllotaxis. *Ann. Bot.* **15**: 481–90.
Church, A. H. (1904a). The principles of phyllotaxis. *Ann. Bot.* **18**: 227–43.
Church, A. H. (1904b). *On the Relation of Phyllotaxis to Mechanical Laws.* London: Williams & Norgate.
Church, A. H. (1920a). *The Somatic Organization of the Phaeophyceae.* London: Oxford University Press.
Church, A. H. (1920b). *On the Interpretation of Phenomena of Phyllotaxis.* New York: Hafner Pub. Co. (Reprinted in 1968.)
Clark, L. B. & Hersh, A. H. (1939). A study of relative growth, in *Notonecta undulata. Growth* **3**: 347–72.
Cleyet-Michaud, M. (1982). *Le Nombre d'Or,* 4th ed. Que-Sais-Je? No. 1530. Les Presses Universitaires de France.

Clos, M. D. (1861). Cladodes et axes ailés. *Acad. Sci. Inscript. et Belles-Lettres de Toulouse* **6**: 71–101.
Clowes, F. A. L. (1961). *Apical Meristems.* Oxford: Blackwell.
Codaccionni, M. (1955). Étude phyllotaxique d'un lot de 200 plants d'*Helianthus annuus* L. cultivés en serre. *C. R. Acad. Sci. Paris* **241**: 1159–61.
Coen, E. & Meyerowitz, E. M. (1991). The war of whorls: genetic interactions controlling flower development. *Nature* **353**: 31–7.
Cook, T. A. (1974). *The Curves of Life.* New York: Dover. (Reprint of 1914 ed., London: Constable & Co.)
Corner, E. J. H. (1966). *The Natural History of Palms.* Berkeley: University of California Press.
Corner, E. J. H. (1981). *The Life of Plants.* University of Chicago Press.
Cotton, F. A. (1968). *Applications de la Théorie des Groupes à la Chimie.* Paris: Dunod.
Coxeter, H. S. M. (1953). The golden section, phyllotaxis, and Wythoff's game. *Scripta Math.* **19**: 135–43.
Coxeter, H. S. M. (1969). *Introduction to Geometry.* New York: Wiley. (Reprint of 1961 ed.)
Coxeter, H. S. M. (1972). The role of intermediate convergents in Tait's explanation of phyllotaxis. *J. Alj.* **20**: 167–75.
Crafts, A. S. (1943a). Vascular differentiation of the shoot apex of *Sequoia sempervirens. Am. J. Bot.* **30**: 110–21.
Crafts, A. S. (1943b). Vascular differentiation in the shoot apices of ten coniferous species. *Am. J. Bot.* **30**: 382–93.
Crank, J. (1976). *The Mathematics of Diffusion.* Oxford: Clarendon Press.
Crick, F. H. C. (1953). The packing of α-helices: Simple coiled coils. *Acta Cryst.* **6**: 689–97.
Crick, F. (1970). Diffusion in embryogenesis. *Nature* **225**: 420–2.
Cronquist, A. (1968). *The Evolution and Classification of Flowering Plants.* London & Edinburgh: Nelson.
Cronquist, A. (1981). *An Integrated System of Classification of Flowering Plants.* New York: Columbia University Press.
Crow, W. B. (1928). Symmetry in organisms. *Am. Nat.* **62**: 207–27.
Crowe, D. W. (1990). Mathematics and symmetry, a personal report. *Symmetry: Culture and Science* **1**(2): 139–53.
Cuénod, A. (1951). Du rôle de la feuille dans l'édification de la tige. *Bull. Soc. Nat. Tunisie* **4**: 3–15.
Cuénod, A. (1954). Les vrais spirales "foliaires" génératrices de la souche, de la tige et de la fleur. *Bull. Soc. Bot. Fr.* **101**: 207–13.
Cull, P. & Vogt, A. (1973). Mathematical analysis of the asymptotic behavior of the Leslie population matrix model. *Bull. Math. Biol.* **35**: 645–61.
Cusset, G. (1982). The conceptual bases of plant morphology. *Acta Biotheor.* **31A**: 8–86.
Cutter, E. G. (1956). Experimental and analytical studies of pteridophytes. XXXIII. The experimental induction of buds from leaf primordia in *Dryopteris aristata* Druce. *Am. Bot. (N.S.)* **20**: 143–68.
Cutter, E. G. (1959). On a theory of phyllotaxis and histogenesis. *Biol. Rev.* **34**: 243–63.
Cutter, E. G. (1963). Experimental modification of the pattern of organogenesis in *Hydrocharis. Nature* **198**: 504–5.
Cutter, E. G. (1964). Phyllotaxis and apical growth. *New Phytol.* **63**: 39–46.
Cutter, E. G. (1965). Recent experimental studies on the shoot apex and shoot morphogenesis. *Bot. Rev.* **31**: 7–113.

Cutter, E. G. (1966). Patterns of organogenesis in the shoot. In *Trends in Plant Morphogenesis,* ed. E. G. Cutter. London: Longmans, pp. 220–34.

Cutter, E. G. (1967). Morphogenesis and developmental potentialities of unequal buds. *Phytomorphology* **17**: 437–45.

Cutter, E. G. (1972a). A morphogeneticists's view of correlative inhibition in the shoot. In *The Dynamics of Meristem Cell Population,* eds. M. W. Miller & C. C. Kuehnert. New York & London: Plenum Press, pp. 51–73.

Cutter, E. G. (1972b). Regulation of branching in decussate species with unequal lateral buds. *Ann. Bot. (London)* **36**: 207–20.

Cutter, E. G. & Voeller, B. R. (1959). Changes in leaf arrangement in individual fern apices. *J. Linn. Bot.* **56**: 225–36.

Darvas, G. & Nagy, D. (1992). Symmetry of patterns and patterns of symmetry. *Symmetry: Culture and Science* **3**(1): 6–8.

Davenport, H. (1952). *The Higher Arithmetic.* London: Hutchinson.

Davies, P. A. (1937). Leaf arrangement in *Ailanthus altissima. Am. J. Bot.* **24**: 401–7.

Davies, P. A. (1939). Leaf position in *Ailanthus altissima* in relation to the Fibonacci series. *Am. J. Bot.* **26**: 67–74.

Davis, E. L. & Steeves, T. A. (1977). Experimental studies on the shoot apex of *Helianthus annuus:* The effect of surgical bisection on quiescent cells in the apex. *Can. J. Bot.* **55**: 606–14.

Davis, T. A. (1963). The dependence of yield on asymmetry in coconut palms. *J. Genetics* **58**: 186–215.

Davis, T. A. (1964). Possible geographical influence on asymmetry in coconut and other plants. Tech. Work. Paper #IX on Coconut, Colombo. *Proc. FAO* **2**: 59–69.

Davis, T. A. (1970). Fibonacci numbers for palm foliar spirals. *Acta. Bot. Neerl.* **19**: 249–56.

Davis, T. A. (1971). Why Fibonacci sequence for palm leaf spirals. *Fib. Q.* **9**: 237–44.

Davis, T. A. & Bose, T. K. (1971). Fibonacci systems in aroids. *Fib. Q.* **9**: 253–63.

Davis, T. A. & Davis, B. (1987). Association of coconut foliar spirality with latitude. *Mathematical Modelling* **8**: 730–733. (*Proc. 1st Int. Symp. on Phyllotaxis, U. of Berkeley,* July 1985; pp. 729–61 of *Proc. 5th Int. Conf. on Math. Modelling in Sci. and Tech.*)

Davis, T. A. & Mark, A. J. (1981). Clockwise and countercloskwise rotatory structures in certain New Zealand and Australian plants. *Proc. Ind. Natl. Sci. Acad.* **B47**: 175–84.

Davis, T. A. & Mathai, A. M. (1973). A mathematical explanation of the emergence of foliar spirals in palms. *Proc. Ind. Natl. Sci. Acad.* **39**: 194–202.

Delattre, P. (1987). Foreword. In *Une Approche Mathématique de la Biologie,* ed. R. V. Jean. Montréal: Gaétan Morin Publisher.

Delpino, F. (1883). *Teoria generale della fillotassi.* Genoa: Alti della R. Universita de Genova.

Demetrius, L. (1981). The Malthusian parameter and the effective rate of increase. *J. Theor. Biol.* **92**: 141–61.

Denffer, D. von, Schumacher, W., Magdefrau, K. & Ehrendorfer, F. (1971). *Strasburger's Textbook of Botany.* London & New York: Longman.

Depew, D. & Weber, B. (eds.) (1986). *Evolution at a Crossroads.* Boston: MIT Press.

Derome, J. R. (1977). Biological similarity and group theory. *J. Theor. Biol.* **65**: 369–78.

DeRosier, D. J. & Klug, A. (1972). Structure of the tubular variants of the head of bacteriophage T4 (polyheads). I. Arrangement of subunits in some classes of polyheads. *J. Mol. Biol.* **65**: 469–88.

Deschartres, R. (1948). Notes sur les hélices foliaires. *Rev. Scient. Bourbannais Centr. France* **1948**: 14–34.

Deschartres, R. (1954). Recherches sur la phyllotaxie du genre *Sedum*. *Rev. Gen. Bot.* **61**: 501–70.

Deschartres, R., Bugnon, F. & Loiseau, J. E. (1955). Dissociation fonctionnelle entre centres générateurs foliaires. *Bull. Soc. Bot. France* **102**: 481–7.

DeVries, H. (1899). On biastrepsis in its relation to cultivation. *Ann. Bot.* **13**: 395–420.

DeVries, H. (1920). Monographie der Zwangsdrehungen. In *Hugo de Vries Opera E Periodicis Collate*. Utrecht: V. A. Oosthoek, pp. 232–406. (Originally published in *Pringsheim's Jahrbücher für Wissenschaftliche Botanik* **23**: 13, 1892.)

Deysson, G. (1967). *Organisation et classification des plantes vasculaires*, vol. 2. Paris: Sedes.

Dickson, L. E. (1952). *History of the Theory of Numbers*, vol. 1. New York: Chelsea.

Dixon, R. (1981). The mathematical daisy. *New Scientist* **92**: 792–5.

Dixon, R. (1983). The mathematics and computer graphics of spirals in plants. *Leonardo* **16**(2): 86–90.

Dixon, R. (1989). Spiral phyllotaxis. *Computers Math. Applic.* **17**(4–6): 535–8. Also in *Symmetry 2: Unifying Human Understanding*, ed. I. Hargittai. Oxford: Pergamon Press.

Dixon, R. (1992). Green spirals. In *Spiral Symmetry*, eds. I. Hargittai & C. A. Pickover. Singapore: World Scientific, pp. 353–68.

Donald, C. M. (1968). The design of wheat ideotype. In *Proc. 3rd Int. Wheat Genetics Symp.*, eds. K. W. Finlay & K. W. Shepherd. Canberra: Butterworths, pp. 377–87.

Dormer, K. J. (1955). Mathematical aspects of plant development. *Discovery* **16**: 59–64.

Dormer, K. J. (1965a). Correlations in plant development: General and basic aspects. In *Encyclopedia of Plant Physiology*, ed. W. Ruhland. New York: Springer, pp. 452–91.

Dormer, K. J. (1965b). Self-regulatory phenomena in plant development, vol. 15(1). In *Handbuch der Pflanzenphysiologie*, ed. W. Ruhland. New York: Springer, pp. 479–91.

Dormer, K. J. (1972). *Shoot Organization in Vascular Plants*. Syracuse University Press.

Dormer, K. J. (1980). *Fundamental Tissue Geometry for Biologists*. Cambridge: Cambridge University Press.

Douady, S. & Couder, Y. (1992). Phyllotaxis as a physical self-organized process. In *Growth Patterns in Physical Sciences and Biology*, Proc. NATO ARW, Granada, Spain (7–11 Oct. 1991); *Phys. Rev. Letters* **68**(13): 2098–101.

Dutrochet, R. J. H. (1834). Observations sur les variations accidentelles du mode, suivant lequel les feuilles sont distribuées sur les tiges des végétaux. *Nouv. Ann. Mus. Hist. Nat.* **3**: 161–200.

Edwards, L. (1982). *The Field of Form*. Edinburgh: Floris Books.

Ehleringer, J. & Werk, K. S. (1986). Modification of solar-radiation absorption patterns and implications for carbon gain at the leaf level. In *On the Economy of Plant Form and Function*, ed. T. J. Givnish. Cambridge: Cambridge University Press, pp. 57–82.

Ellenberger, F. (1991). *Recherches d'Amateur sur la Phyllotaxie: La Mathématique Végétale.* Orsay, France: Centre Interdisciplinaire d'Etude de l'Evolution des Idées, des Sciences et des Techniques.
Endress, P. K. (1987). Floral phyllotaxis and floral evolution. *Botanische Jahrbücher für Systematik* **108**: 417–38.
Endress, P. K. (1989). Chaotic floral phyllotaxis and reduced perianth in *Achlys* (Berberidaceae). *Botanica Acta* **102**: 159–63.
Erickson, R. O. (1959). Patterns of cell growth and differentiation in plants. In *The Cell. Biochemistry, Physiology, Morphology,* vol. 1, eds. J. Brachet & A. E. Mirsky. New York: Academic Press, pp. 497–535.
Erickson, R. O. (1973a). Tubular packing of spheres in biological fine structures. *Science* **181**: 705–16.
Erickson, R. O. (1973b). Geometry of tubular ultrastructures. In *Basic Mechanisms in Plant Morphogenesis. Brookhaven Symp. Biol.* **25**: 111–28.
Erickson, R. O. (1976). Modeling of plant growth. *Ann. Rev. Plant Physiol.* **27**: 407–34.
Erickson, R. O. (1983). The geometry of phyllotaxis. In *The Growth and Functioning of Leaves,* ed. J. E. Dale & F. L. Milthorpe. Cambridge: Cambridge University Press, pp. 53–88.
Erickson, R. O. & Kuhn Silk, W. (1980). The kinematics of plant growth. *Sci. Am.* **242**: 102–13.
Erickson, R. O. & Meicenheimer, D. R. (1977). Photoperiod induced change in phyllotaxis in *Xanthium. Am. J. Bot.* **64**: 981–8.
Erickson, R. O. & Michelini, F. J. (1957). The plastochron index. *Am. J. Bot.* **44**: 297–305.
Esau, K. (1943), Vascular differentiation in the vegetative shoot of *Linum.* II. The first phloem and xylem. *Am. J. Bot.* **30**: 248–55.
Esau, K. (1954). Primary vascular differentiation in plants. *Biol. Rev. Camb. Phil. Soc.* **29**: 46–86.
Esau, K. (1965a). *Plant Anatomy,* 2nd ed. New York: Wiley.
Esau, K. (1965b). *Vascular Differentiation in Plants.* New York: Holt-Rinehart-Winston.
Esau, K. (1977). *Anatomy of Seed Plants,* 2nd ed. New York: Wiley.
Esau, K. & Kosakai, I. H. (1975). Leaf arrangement in *Nelumbo nucifera*: A re-examination of a unique phyllotaxy. *Phytomorphology* **25**: 100–13.
Evans, G. C. (1972). *The Quantitative Analysis of Plant Growth.* Oxford: Blackwell.
Fan, G. Y. & Cowley, J. M. (1985). Auto-correlation analysis of high resolution electron micrographs of near-amorphous thin films. *Ultramicroscopy* **17**: 345–56.
Ferré, J. B. & LeGuyader, H. (1984). The geometry of leaf morphogenesis: A theoretical proposition. *Acta Biotheor.* **33**: 85–132.
Fisher, J. B. (1986). Branching patterns and angles in trees. In *On the Economy of Plant Form and Function,* ed. T. J. Givnish. Cambridge: Cambridge University Press, pp. 493–524.
Florin, R. (1951). Evolution in cordaites and conifers. *Acta Hort. Berg.* **15**: 285–388.
Fomin, S. V. & Berkinblitt, M. B. (1975). *Problèmes Mathématiques en Biologie.* Moscow: Mir.
Fosket, E. B. (1968). The relation of age and bud break to the determination of phyllotaxis in *Catalpa speciosa. Am. J. Bot.* **55**: 894–9.
Foster, A. S. (1939). Problems of structure, growth and evolution in the shoot apex of seed plants. *Bot. Rev.* **5**: 454–70.

Fowler, D., Hanan, J. & Prusinkiewicz, P. (1989). Modelling spiral phyllotaxis. *Computers and Graphics* **13**: 291–6.
Fowler, D. R., Prusinkiewicz, P. & Battjes, J. (1992). A collision based model of spiral phyllotaxis. *Computers and Graphics* **26**: 361–8.
Fox, M. A. (1973). Some thoughts on the requirements for a theory of biology. *Bull. Math. Biol.* **35**: 11–19.
Frank, F. C. (1951). Crystal dislocations – elementary concepts and definitions. *Phil. Mag.* **42**: 809–19.
Franke, W. H. (1967). Das ordungsprinzip der Beruhrungszeilen. *Kakteen und andere Sukkulenten* **18**: 207–9.
Franquin, P. (1974). Un modèle théorique du développement de la structure de la plante. *Physiol. Veg.* **12**: 459–65.
Frey-Wyssling, A. (1954). Divergence in helical polypeptide chains and in phyllotaxis. *Nature* **173**: 596.
Frey-Wyssling, A. (1975). "Rechts" und "Links" im Pflanzenreich. *Biol. Unserer Zeit* **5**: 147–54.
Friedman, D. (1992). Determination of spiral symmetry in plants and polymers. In *Spiral Symmetry,* eds. I. Hargittai & C. A. Pickover. Singapore: World Scientific, pp. 251–79.
Frijters, D. (1978a). Principles of simulation of inflorescence development. *Ann. Bot.* **42**: 549–60.
Frijters, D. (1978b). Mechanisms of developmental integration for *Aster novae-angliae* L. and *Hieracium murorum* I. *Ann. Bot.* **42**: 561–75.
Frijters, D. & Lindenmayer, A. (1976). Developmental descriptions of branching patterns with paracladial relationships. In *Automata, Languages, Development,* eds. A. Lindenmayer & G. Rozenberg. Amsterdam: North Holland, pp. 57–73.
Fritsch, F. E. (1961). *The Structure and Reproduction of the Algae,* vol. II. Cambridge: Cambridge University Press.
Fritz, S. (1991). The mathematical gardener. *OMNI* **Fev.**: 65–71.
Fujita, T. (1937). Ueber die Reihe 2, 5, 7, 12,... in der schraubigen Blattstellung und die mathematische Betrachtung verschiedener Zahlenreihensysteme. *Bot. Mag. Tokyo* **51**: 480–9.
Fujita, T. (1938). Statistische Untersuchung über die Zahl konjugierten Parastichen bei den schraubigen Organstellungen. *Bot. Mag. Tokyo* **52**: 425–33.
Fujita, T. (1939). Statistische Untersuchungen über den Divergenzwinkel bei den schraubigen Organstellungen. *Bot. Mag. Tokyo* **53**: 194–9.
Fujita, T. (1942). Zurkenntnis der Organstellungen im Pflanzenreich. *Jap. J. Bot.* **12**: 1–55.
Fujita, T. (1964). Phyllotaxis of *Cuscuta. Bot. Mag. Tokyo* **77**: 73–6.
Gantmacher, F. R. (1959). *Theory of Matrices, 1 & 2.* New York: Chelsea.
Gardner, M. (1961). *The 2nd Scientific American Book of Mathematical Puzzles and Diversions.* New York: Simon and Schuster.
Gardner, M. (1982). *The Ambidextrous Universe,* 2nd ed. New York: Pelican.
Gauquelin, M. (1973). *Rythmes Biologiques, Rythmes Cosmiques.* Verviens: Marabout Université.
Gierer, A. (1981). Generation of biological patterns and form: Some physical, mathematical, and logical aspects. *Prog. Biophys. Molec. Biol.* **37**: 1–47.
Gierer, A. & Meinhardt, H. (1972). A theory of biological pattern formation. *Kybernetik* **12**: 30–9.
Gierer, A. & Meinhardt, H. (1974). Biological pattern formation involving

lateral inhibition. In *Lectures on Mathematics in the Life Sciences,* vol. 7. Providence, R.I.: American Mathematical Society, pp. 163–84.

Gifford, E. M. & Corson, C. E. (1971). The shoot apex in seed plants. *Bot. Rev. (Lancaster)* **37**: 143–229.

Gifford, E. M. & Tepper, H. B. (1962). Ontogenetic and histochemical changes in the vegetative shoot tip of *Chenopodium album. Am. J. Bot.* **49**: 902–11.

Girolami, G. (1953). Relation between phyllotaxis and primary vascular organization in *Linum. Am. J. Bot.* **40**: 618–25.

Givnish, T. J. (1984). Leaf and canopy adaptations in tropical forests. In *Physiological Ecology of Plants of the Wet Tropics,* eds. E. Medina, H. A. Mooney, & C. Vasquez-Yanez. The Hague: Dr. Junk, pp. 51–84.

Gladyshev, G. P. (1988). *Thermodynamics and Macrokinetics.* Moscow: Academy of Sciences. (In Russian.)

Glansdorff, P. & Prigogine, I. (1971). *Structure, Stabilité et Fluctuation.* Paris: Masson.

Goebel, K. (1969). *Organography of Plants.* New York: Hafner. (Facsimile of 1900 edition.)

Goel, N. S. & Thompson, R. L. (1986). Organization of biological systems; some principles and models. *Int. Rev. Cytol.* **103**: 1–88.

Goethe, J. W. von (1790). *Die Metamorphose der Pflanzen.* Leipzig.

Goldberg, S. (1961). *Introduction to Difference Equations with Illustrative Examples from Economics, Psychology and Sociology.* New York: Wiley.

Gomez-Campo, C. (1970). The direction of the phyllotactic helix in axillary shoots of six plant species. *Bot. Gaz.* **131**: 110–15.

Gomez-Campo. C. (1974). Phyllotactic patterns in *Bryophyllum tubiflorum, Harv. Bot. Gaz.* **135**: 49–58.

Goodal, C. R. (1985). Statistical and data digitization techniques applied to an analysis of leaf initiation in plants. In *Computer Sciences and Statistics, Proc. 16th Symp. on the Interface,* ed. L. Billard. New York: Elsevier.

Goodal, C. R. (1991). Eigenshape analysis of a cut-grow mapping for triangles and its application to phyllotaxis in plants. *SIAM J. Appl. Math.* **51**: 775–98.

Goodwin, B. C. & Brière, C. (1987). The concept of the morphogenetic field in plants. In *Le Développement des Végétaux, Aspects Théoriques et Synthétiques,* ed. H. LeGuyader. Paris: Masson et Cie, pp. 329–37.

Goodwin, B. C., Holder, N. & Wylie, C. C. (eds.) (1983). *Development and Evolution.* New York: Cambridge University Press.

Goodwin, B. C. & Saunders, P. T. (eds.) (1989). *Theoretical Biology: Epigenetic and Evolutionary Order from Complex Systems.* Edinburgh: Edinburgh University Press.

Gould, S. J. (1966). Allometry and size in ontogeny and phylogeny. *Biol. Rev.* **41**: 587–640.

Green, P. B. (1980). Organogenesis – a biophysical view. *Annual Rev. Plant Physiol.* **31**: 51–82.

Green, P. B. (1985). Surface of the shoot apex: A reinforcement-field theory for phyllotaxis. In *The Cell Surface in Plant Growth and Development,* eds. K. Roberts, A. W. B. Johnston, C. W. Lloyd, P. Shaw, & Woolhouse. Cambridge: Company of Biologists. *J. Cell Sci. (Suppl.)* **2**: 181–201.

Green, P. B. (1986). Plasticity in shoot development: A biophysical view. In *Plasticity in Plants. Symp. 40, Soc. Exp. Biol.,* eds. D. H. Jennings & A. J. Trewavas. Cambridge, U.K., pp. 211–32.

Green, P. B. (1987). Inheritance of pattern: Analysis from phenotype to gene. *Am. Zool.* **27**: 657–73.
Green, P. B. (1989). Shoot morphogenesis, vegetative through floral, from a biophysical perspective. In *Plant Reproduction: from Floral Induction to Pollination,* eds. E. Lord & G. Bernier. *Am. Soc. of Plant Physiologists, Symp. Series* **1**: 58–75.
Green, P. B. (1991). Morphogenesis. In *Plant Physiology. A Treatise,* vol. 10, eds. F. C. Stewart & R. G. Bidwell. New York: Academic Press, pp. 1–64.
Green, P. B. (1992). Pattern formation in shoots: A likely role for minimal energy configurations of the tunica. *Int. J. Plant Sci.* **153**(3): S59–75.
Green, P. B. (Unpublished). Phyllotaxis and flower formation – analysis as generative resonance in two dimensions.
Green, P. B. (Unpublished). Biomathématique de la géométrie de la tige. Paper presented at *54th ACFAS Congress, Montréal,* 1986.
Green, P. B. & Baxter, D. R. (1987). Phyllotactic patterns: Characterization by geometrical activity at the formative region. *J. Theor. Biol.* **128**: 387–95.
Gregory, F. G. (1968). Forms in plants. In *Aspects of Forms,* ed. L. L. Whyte. New York: American Elsevier Pub., pp. 57–76.
Gregory, R. A. & Romberger, J. A. (1972). The shoot apical ontogeny of the *Picea abies* seedling. I. Anatomy, apical dome diameter, and plastochron duration. II. Growth rates. *Am J. Bot.* **59**: 587–97, 598–606.
Greyson, R. I. & Walden, D. B. (1972). The ABPHYL syndrome in *Zea Mays.* I. Arrangement, number and size of leaves. *Am. J. Bot.* **59**: 466–72.
Greyson, R. I., Walden, D. B., Hume, J. A. & Erickson, R. O. (1978). The ABPHYL syndrome in *Zea Mays.* II. Patterns of leaf initiation and the shape of the shoot meristem. *Can. J. Bot.* **56**: 1545–50.
Grobstein, C. (1962). Levels and ontogeny, *Am. Sci.* **50**: 46–58.
Gu, Lianhong. (Unpublished). Unveiling phyllotaxis by computer simulation.
Gu, Lianhong. (Unpublished). A study on relationship between divergence angles and spatial distribution of primordia with computer. (In Chinese, with English summary.)
Guédès, M. (1979). *Morphology of Seed-Plants.* Vaduz: Cramer.
Guédès, M. & Dupuy, P. (1983). From decussation to distichy, with some comments on current theories of phyllotaxy. *Bot. J. Linn. Soc.* **87**: 1–12.
Guiasu, S. (1977). *Information Theory with Applications.* New York: McGraw-Hill.
Guinochet, M. (1965). *Notions Fondamentales de Botanique Générale.* Paris: Masson.
Gunckel, J. E. (1965). Modifications of plant growth and development induced by ionizing radiations. In *Encyclopedia of Plant Physiology,* vol. 2, ed. W. Ruhland. New York: Springer-Verlag, pp. 365–87.
Gunckel, J. E. & Wetmore, R. H. (1946). Studies of development in long shoots and short shoots of *Ginkgo biloba* L. II. Phyllotaxis and the organization of the primary vascular system; primary phloem and primary xylem. *Am. J. Bot.* **33**: 532–43.
Gunther, B. (1975). Dimensional analysis and theory of biological similarity. *Physiol. Rev.* **55**: 659–99.
Gunther, B. & Guerra, E. (1955). Biological similarities. *Acta Physiol. Latine Am.* **5**: 169–86.
Haccius, B. (1939). Untersuchungen über die Bedeutung der Distichie für das Verstandnis der zerstreuten Blattstellung bei den Dikotylen. *Bot. Archiv.* **40**: 58–150.

Haccius, B. (1942). Untersuchungen über die Blattstellung der Gattung Clematis. *Bot. Archiv.* **43**: 469–86.

Haccius, B. (1950). Weitere Untersuchungen zum Verstandnis der zerstreuten Blattstellung bei den Dikotylen. *Math. Naturwiss.* kl, **6**: 289–337.

Hackett, W. P. (1985). Juvenility, maturation, and rejuvenation in woody plants. *Hort. Rev.* **7**: 109–55.

Hackett, W. P., Cordero, R. E. & Srinivasan, C. (1987). Apical meristem characteristics and activity in relation to juvenility in *Hedera*. In *Manipulation of Flowering*, ed. J. G. Atherton. London: Butterworths, pp. 93–9.

Hagemann, W. (1984). Morphological aspects of leaf development in ferns and angiosperms. In *Contemporary Problems in Plant Anatomy*, eds. R. A. White & Dickison. New York: Academic Press, pp. 301–49.

Halberg, C. J. A. (1955). Spectral theory of linked operators in the l^p spaces. Doctoral Dissertation, UCLA.

Hallé, F., Oldeman, R. A. A. & Tomlinson, P. B. (1978). *Tropical Trees and Forests: An Architectural Analysis.* Berlin: Springer-Verlag.

Hallé, N. (1979a). Analyse du réseau phyllotaxique des ecussons du cône chez *Pinus. Adansonia, Ser. 2* **18**(4): 393–408.

Hallé, N. (1979b). Essai de phyllotaxie dynamique interprétée comme primitive. *Bull. Mus. Natl. Hist. Nat. Paris, 4e Ser.*, **1**, Section B, No. 2: 71–95.

Hallé, N. (1979c). Sur une nouvelle méthode descriptive du réseau phyllotaxique des ecussons du cône chez *Pinus* L. et son intérêt taxonomique. *C. R. Acad. Sci. Paris, Ser. D* **288**: 59–62.

Halperin, W. (1978). Organogenesis at the shoot apex. *Ann. Rev. Plant Physiol.* **29**: 239–62.

Hamermesh, M. (1962). *Group Theory and its Application to Physical Problems.* London: Addison-Wesley.

Hancock, H. (1964). *Development of the Minkowski Geometry of Numbers.* New York: Dover.

Harary, F. (1955). The number of linear, directed, rooted and connected graphs. *Trans. Am. Math. Soc.* **78**: 445–63.

Harary, F. (1960). Unsolved problems in the enumeration of graphs. *Publ. Math. Inst. Hung. Acad. Sci.* **5**: 63–95.

Hardwick, R. C. (1984). Sink development, phyllotaxy, and dry matter distribution in a large inflorescence. *Plant Growth Regulators* **2**: 393–405.

Hardwick, R. C. (1986). Construction kits for modular plants. *New Scientist* **10**: 39–42.

Hardy, G. H. & Wright, E. M. (1945). *An Introduction to the Theory of Numbers*, 2nd ed. London: Oxford University Press.

Hargittai, I., & Hargittai, M. (1987). *Symmetry through the Eyes of a Chemist.* New York: VCH Pub.

Hargittai, I. & Pickover, C. A. (eds.). (1992). *Spiral Symmetry.* Singapore: World Scientific.

Harris, T. E. (1963). *The Theory of Branching Processes.* Berlin: Springer-Verlag.

Harris, W. F. (1977). Disclinations. *Sci. Am.* **237**: 130–45.

Harris, W. F. & Erickson, R. O. (1980). Tubular arrays of spheres: Geometry, continuous and discontinuous contraction and the role of moving dislocations. *J. Theor. Biol.* **83**: 215–46.

Harris, W. F. & Scriven, L. E. (1970). Cylindrical crystals contractile mechanisms of bacteriophage and the possible role of dislocation in contraction. *J. Theor. Biol.* **27**(2): 233–57.

Harris, W. F. & Scriven, L. E. (1971). Moving dislocations in cyclindrical crystals cause waves bending. *J. Mechanochem. Cell Motility* **1**(1): 33–40.

Harrison, L. G. (1987). What is the status of diffusion–reaction theory thirty-four years after Turing? *J. Theor. Biol.* **125**: 369–84.

Harrison, L. G. (1992). *Kinetic Theory of Living Pattern*. Cambridge: Cambridge University Press.

Hawking, S. W. (1988). *A Brief History of Time. From the Big Bang to Black Holes*. New York & London: Bantam Books.

Haynes, K. E., Phillips, F. Y. & Mohrfald, J. W. (1980). The entropies: Some roots of ambiguity. *Socio-Econ. Plan. Sci.* **14**: 137–45.

Heimans, J. (1978). Problems of phyllotaxis. *Proc. Kon. Nedel. Akad. Wet.* **81C**: 91–8.

Hejnowicz, Z. (1973). *Anatomia Rozwojowa*. Warsaw: Drzew, Panstwowe, Wydawn-Naukowe.

Hejnowicz, Z. & Nabielski, J. (1979). Modelling of growth in the shoot apical dome. *Acta Soc. Bot. Poloniae* **48**: 423–41.

Hellendoorn, P. H. & Lindenmayer, A. (1974). Phyllotaxis in *Bryophyllum tubiflorum:* Morphogenetic studies and computer simulations. *Acta Bot. Neerl.* **23**: 473–92.

Heller, R. (1978). *Abrégé de Physiologie Végétale*. Paris: Masson.

Herman, G. T., Lindenmayer, A. & Rozenberg, G. (1975). Description of developmental languages using recurrence systems. *Math. Syst. Theory* **8**: 316–41.

Herman, G. T. & Rozenberg, G. (1975). *Developmental System and Languages*. New York: American Elsevier.

Herman, G. T. & Vitanyi, P. M. B. (1976). Growth functions associated with biological development. *Am. Math. Mon.* **83**: 1–16.

Hermant, A. (1946). Structures et formes naturelles, géométrie et architecture des plantes. *Techniques et Architecture* **VI**: 9–10.

Hermant, A. (1947). Croissance et topologie, architecture de plantes. *Techniques et Architecture* **VII**: 421–31.

Hernandez, L. F. (1988). Organ morphogenesis and phyllotaxis in the capitulum of sunflower (*Helianthus annuus* L.). Ph.D. Dissertation, University of New South Wales, Australia.

Hernandez, L. F. & Palmer, J. H. (1988a). Regeneration of the sunflower capitulum after cylindrical wounding of the receptacle. *Am. J. Bot.* **75**: 1253–61.

Hernandez, L. F. & Palmer, J. H. (1988b). A computer program to create the Fibonacci florets patterns of the sunflower head. *Proc. 12th Int. Sunflower Conf. 2, Novi Sad, Yugoslavia,* pp. 150–5.

Hersh, A. H. (1941). Allometric growth: The ontogenetic and phylogenic significance of differential rates of growth. *Growth* (suppl.) **5**: 113–45.

Hersh, A. H. & Anderson, B. G. (1941). Differential growth and morphological pattern in *Daphnia*. *Growth* **5**: 359–64.

Herz-Fischler, R. (1987). *A Mathematical History of Division in Extreme and Mean Ratio*. Waterloo, Canada: Wilfrid Laurier University Press.

Heyde, C. C. (1980). On a probabilistic analogue of the Fibonacci sequence. *J. Appl. Prob.* **17**: 1079–82.

Heyde, C. C. (1981). On Fibonacci (or lagged Bienaymé–Galton–Watson) branching processes. *J. Appl. Prob.* **18**: 583–91.

Hieser, C. B. (1976). *The Sunflower*. Norman: University of Oklahoma Press.

Hilbert, D. & Cohn-Vossen, S. (1952). *Geometry and the Imagination*. New York: Chelsea Pub.

Hirmer, M. (1922). *Zür Lösung der Problems der Blattstellungen.* Jena: Fisher.

Ho, I. C. & Peel, A. J. (1969). Transport of 14C-labelled assimilates and 32P-labelled phosphate in *Salix viminalis* in relation to phyllotaxis and leaf age. *Ann. Bot.* 33: 743–51.

Ho, M. W. (1989). A structuralism of process: Towards a post-Darwinian rational morphology. In *Dynamic Structures in Biology,* eds. B. C. Goodwin, A. Sibatani & G. Webster. Edinburgh: Edinburgh University Press, pp. 31–48.

Ho, M. W. & Fox, S. W. (eds.) (1988). *Evolutionary Processes and Metaphors.* New York: John Wiley and Sons.

Hofmeister, W. (1868). Allgemeine Morphologie des Gewachse. In *Handbuch der Physiologischen Botanik,* vol. 1. Leipzig: Engelmann, pp. 405–664.

Holland, J. M. (1972). *Studies in Structure.* New York: Macmillan.

Horridge, J. S. & Cockshull, K. E. (1979). Chrysanthemum shoot apex in relation to inflorescence initiation and development. *Ann. Bot.* **44**: 547–56.

Hunt, R. (1981). The fitted curve in plant growth studies. In *Mathematics and Plant Physiology,* eds. D. A. Rosc & D. A. Charles-Edwards. London: Academic Press, pp. 283–98.

Huntley, H. E. (1970). *The Divine Proportion. A Study in Mathematical Beauty.* New York: Dover.

Hutchinson, J. E. (1981). Fractals and self-similarity. *Indiana Univ. J. Math.* **30**: 713–47.

Huxley, J. S. (1932). *Problems of Relative Growth.* London: Methuen. (Reprinted in 1972, New York: Dover.)

Huxley, J. S., Needham, J. & Lerner, I. M. (1941). Terminology of relative growth rates. *Nature* **148**: 225.

Huyghes, R. (1971). *Formes et Forces (de l'Atome à Rembrandt).* Paris: Flammarion.

Illert, C. (1983). The mathematics of gnomonic seashells. *Math. Biosci.* **63**: 21–56.

Imai, Y. (1927). The right- and left-handedness of phyllotaxis. *Bot. Mag. Tokyo* **41**: 592–6.

Isenberg, C. (1978). *The Science of Soap Films and Soap Bubbles.* Clevedon: Tieto Ltd.

Ishida, Y. & Iyama, S. (1976). Observation of wedge disclinations and their behavior in a bubble raft crystal. *Acta Metal.* **24**: 417–23.

Iterson, G., van (1907). *Mathematische und mikroskopisch-anatomische Studien über Blattstellungen, nebst Betraschtungen über den Schalenbau der Miliolinen.* Jena: Gustav-Fischer-Verlag.

Iterson, G., van (1960). New studies on phyllotaxis. *Proc. Kon. Nederl. Akad. Wet.* **63C**: 137–50.

Jean, R. V. (1976a). Growth matrices in phyllotaxis. *Math. Biosci.* **32**: 65–76.

Jean, R. V. (1976b). La G-entropie en phyllotaxie. *Rev. Bio-Math.* **55**: 111–30.

Jean, R. V. (1978a). Growth and entropy: Phylogenism in phyllotaxis. *J. Theor. Biol.* **71**: 639–60.

Jean, R. V. (1978b). *Phytomathématique.* Québec: Les Presses de l'Université du Québec.

Jean, R. V. (1979a). Developmental algorithms in phyllotaxis and their implications: A contribution to the field of growth functions of L-systems by an application of Perron–Frobenius spectral theory. *J. Theor. Biol.* **76**: 1–30.

Jean, R. V. (1979b). A rigorous treatment of Richards' mathematical theory of phyllotaxis. *Math. Biosci.* **44**: 221–40.

Jean, R. V. (1979c). Some consequences of the hierarchical approach to phyllotaxis. *J. Theor. Biol.* **81**: 309–26.
Jean, R. V. (1979d). A mathematical structure for a theory of growth. *Math. Reports Acad. Sci. Can.* **1**(5): 237–40.
Jean, R. V. (1980a). La notion systémique de contenu d'information structurelle et ses dérivés. *Rev. Bio-Math.* **70**: 5–25.
Jean, R. V. (1980b). A systemic model of growth on botanometry. *J. Theor. Biol.* **87**: 569–84.
Jean, R. V. (1980c). Contribution à la théorie des fonctions de croissance des L-systèmes et à la biométrie végétale. *C. R. Acad. Sci. Paris (Sect. Anal. Math.)* **290A**: 949–52.
Jean, R. V. (1981a). The use of continued fractions in botany. *UMAP Module #571,* and in *UMAP Tools for Teaching 1985,* COMAP Inc., Arlington, Mass., 1986, pp. 91–120.
Jean, R. V. (1981b). An L-system approach to nonnegative matrices for the spectral analysis of discrete growth functions of populations. *Math. Biosci.* **55**: 155–68.
Jean, R. V. (1981c). A new approach to a problem of plant growth. In *Applied Systems and Cybernetics,* ed. G. E. Lasker. New York: Pergamon Press, pp. 1906–10.
Jean, R. V. (1981d). La modélisation en phyllotaxie, l'approche fonctionnelle. *Proc. First Int. Conf. Appl. Modelling and Simul.,* pp. 194–5.
Jean, R. V. (1982). The hierarchical control of phyllotaxis. *Ann. Bot. (London)* **49**: 747–60.
Jean, R. V. (1983a). Allometric relations in plant growth. *J. Math. Biol.* **18**: 189–200.
Jean, R. V. (1983b). Mathematical modeling in phyllotaxis: The state of the art. *Math. Biosci.* **64**: 1–27.
Jean, R. V. (1983c). *Croissance Végétale et Morphogénèse.* Paris: Masson et Cie; Montréal: Les Presses de l'Université du Québec.
Jean, R. V. (1983d). A fundamental problem in plant morphogenesis, from the standpoint of differential growth. In *Mathematical Modelling in Science and Technology,* eds. X. J. R. Avula, R. E. Kalman, A. I. Liapis & E. Y. Rodin. New York: Pergamon Press, pp. 774–7.
Jean, R. V. (1983e). Le phénomène de la phyllotaxie des plantes. *Ann. Sci. Nat. Bot.* **13**: 5, 45–6.
Jean, R. V. (1984a). *Mathematical Approach to Pattern and Form in Plant Growth.* New York: Wiley-Interscience. (Published in Chinese by Chinese Acad. Periodicals Pub. House, Beijing, 1990.)
Jean, R. V. (1984b). The Fibonacci sequence. *UMAP Journal* **5**(1): 23–47.
Jean, R. V. (1984c). Differential growth, Huxley's allometric formula, and sigmoid growth. *UMAP Module #635,* and in *UMAP Tools for Teaching 1983,* COMAP Inc., Arlington, Mass., 1985, pp. 415–46.
Jean, R. V. (1985a). A geometric representation of α-polypeptide chains revisited. *J. Theor. Biol.* **112**: 819–25.
Jean, R. V. (1985b). Modèles descriptifs et explicatifs des patterns observés chez les plantes. In *Quelques Recherches en Biomathématiques,* ed. R. V. Jean. Montréal: Cahier de l'ACFAS #26, pp. 35–44.
Jean, R. V. (1986a). A basic theorem on and a fundamental approach to pattern formation on plants. *Math. Biosci.* **79**: 127–54.
Jean, R. V. (1986b). An interpretation of Fujita's frequency diagrams in phyllotaxis. *Bull. Math. Biol.* **48**(1): 77–86.

Jean, R. V. (1986c). Predictions confronted with observations in phyllotaxis. *J. Infer. Deduct. Biol.* **1**(3): 1–19.
Jean, R. V. (1986d). Théorème central et développement récents dans l'étude des patterns observés chez les plantes. In *Biomathématiques*. Montréal: Cahier de l'ACFAS #45, pp. 79–88.
Jean, R. V. (1986e). A methodological paradigm in plant biology. *Math. Modelling* **7**: 49–60.
Jean, R. V. (1987a). A mathematical model and a method for the practical assessment of the phyllotactic patterns. *J. Theor. Biol.* **129**: 69–90.
Jean, R. V. (1987b). Towards a more universal model of phyllotaxis. *Math. Modelling* **8**: 734–9.
Jean, R. V. (1987c). About new developments in the systemic approach to pattern generation on plants. In *Cybernetics and Systems: The Way Ahead*, ed. J. Rose. Lancs.: Thales Pub., pp. 372–5.
Jean, R. V. (1987d). Modèles récents pour la phyllotaxie. In *Le Développement des Végétaux, Aspects Théoriques et Synthétiques*, ed. H. LeGuyader. Paris: Masson et Cie, pp. 359–74.
Jean, R. V. (1988a). Number-theoretic properties of two-dimensional lattices. *J. Number Theory* **29**: 206–23.
Jean, R. V. (1988b). Model of pattern generation on plants, based on the principle of minimal entropy production. In *Thermodynamics and Pattern Formation in Biology*, eds. I. Lamprecht & A. I. Zotin. West Berlin: W. de Gruyter Pub., pp. 249–64.
Jean, R. V. (1988c). Phyllotactic pattern generation: A conceptual model. *Ann. Bot. (London)* **61**: 293–303.
Jean, R. V. (1989a). Phyllotaxis: A reappraisal. *Can. J. Bot.* **67**: 3103–7.
Jean, R. V. (1989b). Concepts mathématiques pour l'etude du développement des pousses chez les plantes. In *Acta Quatro Congreso Int. de Biomatematica, Lima*, pp. 97–122.
Jean, R. V. (1990). A synergic approach to plant pattern generation. *Math. Biosci.* **98**: 13–47.
Jean, R. V. (1991). Hierarchy and autoevolutionism in a general system approach to plant pattern morphogenesis. *Int. J. General Systems* **18**: 201–12.
Jean, R. V. (1992a). Model testing in phyllotaxis. *J. Theor. Biol.* **156**(1): 41–52.
Jean, R. V. (1992b). Nomothetical modelling of sprial symmetry in biology. *Symmetry* **1**(1): 81–91; and in *Fivefold Symmetry*, ed. I. Hargittai. Singapore: World Scientific, pp. 505–28.
Jean, R. V. (1992c). On the origins of spiral symmetry in plants. In *Spiral Symmetry*, eds. I. Hargittai & C. A. Pickover. Singapore: World Scientific, pp. 323–51.
Jean, R. V. (1993a). Main results of a systemic model in plants. In *Recent Advances in Cybernetics and Systems*, eds. A. Ghosal and P. N. Murthy. India: McGraw-Hill, pp. 413–16.
Jean, R. V. (1993b). A theory and a model for pattern generation in plants. *J. Biol. Syst.* **1**(2): 159–86.
Jean, R. V. (In press). Modélisation de patterns chez les plantes. In *Dynamique et Morphologie*, ed. D. Barabé. Montreal: Orbis Publisher.
Jean, R. V. (Unpublished). Aspects mathématiques du problème des patterns à la surface des plantes et dans les bourgeons. Paper presented at *54th ACFAS Congress, Montréal*, 1986.
Jean, R. V. & Bicknell-Johnson, M. (1989). An adventure into applied mathematics with Fibonacci numbers. *School Science and Mathematics* **89**(6): 487–98.

Jean, R. V. & Schwabe, W. W. (1990). Shoot development of plants, *UMAP Module 702,* COMAP, Lexington, Mass.; and in *UMAP Tools for Teaching 1989,* pp. 169–208.
Jensen, L. C. W. (1968). Primary stem vascular patterns in three subfamilies of the Crassulaceae. *Am. J. Bot.* **55**: 553–63.
Jesuthasan, S. & Green, P. B. (1989). On the mechanism of decussate phyllotaxis: Biophysical studies on the tunica layer of *Vinca major. Am. J. Bot.* **76**: 1152–66.
Jolicoeur, P. (1963). The multivariate generalization of the allometry equation. *Biometrics* **19**: 497–9.
Jolicoeur, P. (1989). A simplified model for bivariate complex allometry. *J. Theor. Biol.* **140**: 41–9.
Kaplan, D. R. & Hagemann, W. (1991). The relationship of cell and organism in vascular plants. Are cells the building blocks of plant form? *Bioscience* **41**: 693–703.
Kappraff, J. (1992a). The spiral in nature, myth, and mathematics. In *Spiral Symmetry,* eds. I. Hargittai & C. A. Pickover. Singapore: World Scientific, pp. 1–46.
Kappraff, J. (1992b). The relations between mathematics and mysticism of the golden mean through history. In *Fivefold Symmetry,* ed. I. Hargittai. Singapore: World Scientific, pp. 33–65.
Katz, M. B. (1978). *Questions of Uniqueness and Resolution in Reconstruction from Projections, Lecture Notes in Biomathematics.* New York: Springer-Verlag.
Kavanagh, A. J. & Richards, O. W. (1942). Mathematical analysis of the relative growth of organisms. *Proc. Rochester Acad. Sci.* **8**: 150–74.
Kawaguchi, Y. (1982). A morphological study of the form of nature. *Computer Graphics* **16**(3): 223–32.
Kedves, M. (1990). Quasi-crystalloid basic molecular structure of the sporoderm. *Rev. Palaeobot. Palynol.* **64**: 181–6.
Kelley, A. & Groves, G. W. (1970). *Crystallography and Crystal Defects.* Bristol: W. Arrowsmith Ltd.
Kendrew, J. C. (1961). The three-dimensional structure of a protein molecule. *Sci. Am.* **Dec.**: 96–110.
Kerns, K. R., Collins, J. L., & Kim, H. (1936). Developmental studies of the pineapple, *Ananas comosus* (L.) Merr. I. Origin and growth of leaves and inflorescence. *New Phytol.* **35**: 305–17.
Khintchine, A. I. (1957). *Mathematical Foundations of Information Theory.* New York: Dover.
Khintchine, A. I. (1963). *Continued Fractions.* Groningen: Noordhoffeltd.
Kilmer, W. L. (1971). On growing pine cones and other Fibonacci fruits; McCulloch's localized algorithm. *Math. Biosci.* **11**: 53–7.
King, C. E., Gallaher, E. E. & Levin, D. A. (1975). Equilibrium diversity in plant pollinator systems. *J. Theor. Biol.* **53**: 263–75.
King, R. W. (1983). The shoot apex in transition: Flowers and other organs. In *The Growth and Functioning of Leaves,* eds. J. E. Dale & F. L. Milthorpe. Cambridge: Cambridge University Press, pp. 109–44.
Kirchoff, B. K. & Rutishauser, R. (1990). The phyllotaxis of *Costus* (Costaceae), *Bot. Gaz.* **151**: 88–105.
Klein, F. (1932). *Elementary Mathematics from an Advanced Standpoint.* New York: Macmillan.

Knuth, D. E. (1973). *The Art of Computer Programming,* vols. 1–3. Palo Alto: Addison-Wesley.

Koch, A. J. (1989). Des cristaux colloidaux à la phyllotaxie, deux exemples de réseaux cristallins en géométrie cylindrique. Doctoral Thesis, University of Lausanne.

Kofler, L. (1963). *Croissance et Développement des Plantes.* Paris: Gauthier-Villars.

Komai, T. (1951). Phyllotaxis-like arrangement of organs and zoöids in some Medusae. *Am. Naturalist* **85**(820): 75–6.

Komai, T. & Yamazi, I. (1945). Order found in the arrangement of organs and zoöids in some Medusae. *Annotationes Zoologicae Japonensis* **23**(1): 1–7.

Kresling, B. (1992). Origami as simulation method for nature's folding patterns. *Symmetry: Culture and Science* 3(1): 46–7.

Kruger, F. (1973). On the problem of summing up the allometric formula. *Rev. Bio.-Math.* **12**: 123–8.

Kubitzki, K. (1987). Origin and significance of trimerous flowers. *Taxon* **36**(1): 21–8.

Kuich, W. (1970). On the entropy of context-free languages. *Inf. Control* **16**: 173–200.

Kumazawa, M. & Kumazawa, M. (1971). Periodic variations of the divergence angle, internode length and leaf shape, revealed by correlogram analysis. *Phytomorphology* **21**: 376–89.

Lacalli, T. C. & Harrison, L. G. (1979). Turing's conditions and the analysis of morphogenetic models. *J. Theor. Biol.* **76**: 419–36.

Lacroix, C. & Sattler, R. (1988). Phyllotaxis theories and tepla-stamen superposition in *Basella rubra. Am. J. Bot.* **75**: 906–17.

Lacroix, N. H. J. (1985). Deux modèles mathématiques basés sur des principes générateurs. In *Quelques Recherches en Biomathématiques,* ed. R. V. Jean. Montréal: Cahier de l'ACFAS #26, pp. 45–54.

Lacroix, N. H. J. (1986). Formes de bourgeons floraux et géométrie projective. In *Biomathématiques,* ed. R. V. Jean. Montreal: Cahier de l'ACFAS #45, pp. 89–95.

Lambert, D. M. & Hughes, A. J. (1989). Keywords and concepts in structuralist and functionalist biology. In *Dynamic Structures in Biology,* eds. B. C. Goodwin, A. Sibatani & G. Webster. Edinburgh: Edinburgh University Press, pp. 62–76.

Larson, P. R. (1976). Development and organization of the secondary vessel system in *Populus grandidentata. Am. J. Bot.* **63**: 369–81.

Larson, P. R. (1977). Phyllotactic transitions in the vascular system of *Populus deltoïdes* Barts, as determined by ^{14}C labelling. *Planta* **134**: 241–9.

Larson, P. R. (1983). Primary vascularization and the siting of primordia. In *The Growth and Functioning of Leaves,* eds. J. E. Dale & F. L. Milthorpe. Cambridge: Cambridge University Press, pp. 25–51.

Lauffer, M. A. & Stevens, C. L. (1968). Structure of the tobacco mosaic virus particle; polymerisation of the tobacco mosaic virus protein. *Advan. Virus Res.* **13**: 1–63.

Lebeau, B., Jolicoeur, P., Pageau, G. & Crossman, E. J. (1986). Asymptotic growth, egg production and trivariate allometry in *Esox masquinongy* Mitchell. *Growth* **50**: 185–200.

LeGuyader, H. (1985). Les langages de la théorie en biologie. Doctoral Thesis, University of Rouen, France.

LeGuyader, H. (1988). La phyllotaxie ou le rêve du crystal vivant. In *Théories Biologiques – Éthique et Expérimentation en Médecine,* ed. H. Barreau. Paris: Éditions du CNRS, pp. 11–153.

Lehmann, N. L. & Sattler, R. (In press). Irregular floral development in *Calla palustris* L. (Araceae) and the concept of homeosis.

Lehninger, A. L. (1969). *Bioénergétique.* Paris: Ediscience.

Lehninger, A. L. (1982). *Principles of Biochemistry.* New York: Worth.

Leigh, E. G. (1972). The golden section and spiral leaf-arrangement. *Trans. Conn. Acad. Arts Sci.* **44**: 163–76.

Leins, P. & Erbar, C. (1985). Ein Beitrag zur Blutenentwicklung der Aristolochiaceen, einer Vermittlergruppe zu den Monokotylen. *Bot. Jahrb. Syst.* **107**: 343–68.

Leins, P. & Schwitalla, S. (1984). Studien an Cactaceen Bluten. I. Einige Bemerkungen zur Blutenentwicklung von Pereskia. *Beitr. Bio. Pflanzen* **60**: 313–23.

Leppik, E. E. (1961). Phyllotaxis, anthotaxis, and semataxis. *Acta Biotheor.* **14**: 1–28.

Leppik, E. E. (1970). Evolutionary differentiation of the flower head of the Compositae. II. *Ann. Bot. Fennici* **7**: 325–52.

Leppik, E. E. (1977). The evolution of capitulum types of the Compositae in the light of insect–flower interaction. In *The Biology and Chemistry of the Compositae,* eds. V. H. Heywood, J. B. Harborne & B. L. Turner. New York: Academic Press, pp. 61–90.

Lestiboudois, M. T. (1848). Phyllotaxie anatomique. *Ann. Sci. Nat.* **3**(1), #10: 15–105.

LeVèque, W. J. (1977). *Fundamentals of Number Theory.* Reading, Mass.: Addison-Wesley.

Levitov, L. S. (1990). Hamiltonian approach to phyllotaxis. *Landau Institute Preprint.*

Levitov, L. S. (1991a). Energetic approach to phyllotaxis. *Europhys. Lett.* **14**(6): 533–9.

Levitov, L. S. (1991b). Phyllotaxis of flux lattices in layered superconductors. *Phys. Rev. Lett.* **66**(2): 224–7.

Lewontin, R. C., Rose, S. & Kamin, L. J. (1984). *Biology, Ideology, and Human Nature. Not in Our Genes.* New York: Pantheon Books.

Li, Jing Gong (1986). On plant phyllotaxy. *J. Biomathematics* **1**: 1–8. (In Chinese with English summary.)

Lima-de-Faria, A. (1988). *Evolution without Selection.* New York: Elsevier.

Linden, F. M. J. van der (1990). Creating phyllotaxis, the dislodgement model. *Math. Biosci.* **100**(2): 161–200.

Lindenmayer, A. (1971). Developmental systems without cellular interaction: Their languages and grammars. *J. Theor. Biol.* **30**: 455–84.

Lindenmayer, A. (1975). Developmental algorithms for multicellular organisms: A survey of L-systems. *J. Theor. Biol.* **54**: 3–22.

Lindenmayer, A. (1977a). Theories and observations of developmental biology. In *Foundational Problems in Special Sciences,* eds. Butts & Hintikka. Dordrecht, Holland: Reidel, pp. 108–18.

Lindenmayer, A. (1977b). Paracladial relationships in leaves. *Ber. Dtsch. Bot. Ges.* **90**: 287–301.

Lindenmayer, A. (1978a). Algorithms for plant morphogenesis. In *Theoretical Plant Morphology,* ed. R. Sattler. The Hague: Leiden University Press, pp. 37–81. (Suppl. to *Acta Biotheoretica* **27**.)

Lindenmayer, A. (1978b). Growth functions of cell populations with lineage control. In *Biomathematics and Cell Kinetics,* eds. A. J. Valleron & P. D. M. MacDonald. New York: Elsevier, pp. 117–33.

Linford, M. B. (1983). Fruit quality studies. II. Eye number and eye weight. *Pineapple Quarterly* **3**(4): 185–95.

Lishchitovich, L. I. (1974). Mathematical modelling of stationary morphogenesis of roots in dicotyledonous plants. *Fiziologiya i Biochimiya Kulturnick Rastenii* **6**: 488–95. (In Russian with English summary.)

Loiseau, J. E. (1957). Evolution de la phyllotaxie chez *Tropaeolum majus* L. cultivé en chambres lumineuses conditionnées. *Bull Sc. Bourgogne* **18**: 57–9.

Loiseau, J. E. (1959). Observations et expérimentations sur la phyllotaxie et le fonctionnement du sommet végétatif chez quelques balsaminacées. *Ann. Sci. Nat. Bot.* **20**: 1–214.

Loiseau, J. E. (1965). Anisocladie hélicoïdale et hélices foliaires multiples chez quelques angiospermes. In *Travaux Dédiés à L. Plantefol.* Paris: Masson, pp. 367–89.

Loiseau, J. E. (1969). *La Phyllotaxie.* Paris: Masson et Cie.

Loiseau, J. E. (1976). Anisocladie hélicoïdale chez deux schrophylariacées: *Veronica Scutellata, Ilysanthes Parviflora. Phytomorphology* **26**: 23–35.

Loiseau, J. E. & Deschartres, R. (1961). Les phyllotaxies bijuguées. *Mem. Bull. Soc. Bot. Fr.* **108**: 105–16.

Loiseau, J. E. & Messadi, M. (1957). Observations sur des phyllotaxies à angles oscillants. *Bull. Sci. Bourgogne* **18**: 61–4.

Lovtrup, S. & Sydow, B. von (1974). D'Arcy Thompson's theorems and the shape of the molluscan shell. *Bull. Math. Biol.* **36**: 567–75.

Luck, J. & Luck, H. B. (1984). Un mécanisme générateur d'hélices phyllotaxiques. *Actes du 4e Séminaire de l'Ecole de Biologie Théorique, Solignac,* June 4–8, 1984. CNRS, France.

Luck, H. B. (1975). Elementary behavioral rules as a foundation for morphogenesis. *J. Theor. Biol.* **54**: 23–34.

Luck, H. B. & Luck, J. (1986). Unconventional leaves (an application of map 0L-systems to biology). In *The Book of L,* eds. G. Rozenberg & A. Salomaa. Berlin: Springer-Verlag, pp. 275–89.

Ludwig, F. (1896). Weiteres über Fibonaccicurven. *Botanisches Centralblatt* **68**: 7.

Lyndon, R. F. (1972). Leaf formation and growth at the shoot apical meristem. *Physiol. Veg.* **10**: 209–22.

Lyndon, R. F. (1976). The shoot apex. In *Cell Division in Higher Plants,* ed. M. M. Yeoman. London: Academic Press, pp. 285–314.

Lyndon, R. F. (1977). Interacting processes in vegetative development and in the transition to flowering at the shoot apex. In *Integration of Activity in the Higher Plant,* ed. D. H. Jennings. *Symp. Soc. Exper. Biol.* **31**: 221–50, Cambridge University Press.

Lyndon, R. F. (1978a). Flower development in *Silene*: Morphology and sequence of primordia. *Ann. Bot.* **42**: 1343–8.

Lyndon, R. F. (1978b). Phyllotaxis and the initiation of primordia during flower development in *Silene. Ann. Bot.* **42**: 1349–60.

Lyndon, R. F. (1979a). Rates of growth and primordial initiation during flower development in *Silene* at different temperatures. *Ann. Bot.* **43**: 539–51.

Lyndon, R. F. (1979b). A modification of flowering and phyllotaxis in *Silene. Ann. Bot.* **43**: 553–8.

Lyndon, R. F. (1990). *Plant Development: The Cellular Basis.* London: Unwin Hyman.
MacCurdy, E. (1955). *The Notebooks of Leonardo da Vinci.* New York: Braziller.
Mackay, A. L. (1980). The packing of three-dimensional spheres on the surface of a four-dimensional hypersphere. *J. Physics, A: Math. and Gen.* **13**: 3373–9.
Mackay, A. L. (1986). Generalised crystallography. *Comp. and Maths. with Appls.* **12B**: 21–37.
Mackay, A. L. (Unpublished). A time quasi-crystal.
Magnus, W. & Grossman, I. (1971). *Les Groupes et leurs Graphes.* Paris: Dunod.
Mahler, K. Cassels, J. W. S. & Ledermann, W. (1951). Farey section in $k(i)$ and $k(p)$. *Phil. Trans. Roy. Soc. London* **243A**: 585–628.
Majumdar, G. P. (1948). Leaf development at the growing apex and phyllotaxis in *Heracleum. Proc. Ind. Acad. Sci.* **28**: 83–98.
Maksymowych, R. (1973). *Analysis of Leaf Development.* Cambridge: Cambridge University Press.
Maksymowych, R., Cordero, R. E. & Erickson, R. O. (1976). Long-term developmental changes in *Xanthium* induced by gibberellic acid. *Am. J. Bot.* **68**: 1047–53.
Maksymowych, R. & Erickson, R. O. (1977). Phyllotactic change induced by gibberellic acid in *Xanthium* shoot apices. *Am. J. Bot.* **64**: 33–44.
Mandelbrot, B. B. (1982). *The Fractal Geometry of Nature.* San Francisco: W. H. Freeman.
Marc, J. & Hackett, W. P. (1991). Gibberellin-induced reorganization of spatial relationships of emerging leaf primordia at the shoot apical meristem in *Hedera helix* L. *Planta* **185**: 171–8.
Marc, J. & Palmer, J. H. (1981). Photoperiodic sensitivity of inflorescence initiation and development in sunflower. *Field Crops Res.* **4**: 155–64.
Marc, J. & Palmer, J. H. (1982). Changes in mitotic activity and cell size in the apical meristem of *Helianthus annuus* L. during the transition to flowering. *Am. J. Bot.* **69**: 768–75.
Marc, J. & Palmer, J. H. (1984). Variation in cell cycle time and nuclear DNA content in the apical meristem of *Helianthus annuus* L. during the transition to flowering. *Am. J. Bot.* **71**: 588–95.
Marvin, D. A. (1989). Dynamics of telescoping Inovirus: A mechanism for assembly at membrane adhesions. *Int. J. Biol. Macromol.* **11**: 159–64.
Marvin, D. A. (1990). Model-building studies of Inovirus: Genetic variations on a geometric theme. *Int. J. Biol. Macromol.* **13**: 125–38.
Marzec, C. (1987). Phyllotaxis as a dissipative structure. *Mathematical Modelling* **8**: 740–5. (*Proc. 1st Int. Symp. on Phyllotaxis, U. of Berkeley,* July 1985; pp. 729–61 of *Proc. 5th Int. Conf. on Math. Modelling in Sci. and Tech.*)
Marzec, C. & Kappraff, J. (1983). Properties of maximal spacing on a circle related to phyllotaxis and to the golden mean. *J. Theor. Biol.* **103**: 201–26.
Mathai, A. M. & Davis, T. A. (1974). Constructing the sunflower head. *Math. Biosci.* **20**: 117–33.
Maynard-Smith, J. (1971). *Mathematical Ideas in Biology.* Cambridge: Cambridge University Press.
Mayr, E. (1988). *Towards a New Philosophy of Biology.* Cambridge, Mass.: Harvard University Press.

McCully, M. E. & Dale, H. M. (1961). Variations in leaf number in *Hippuris*. A study of whorled phyllotaxis. *Can. J. Bot.* **39**: 611–25.
McDonald, E. A. (1975). Effects of primordial spacing in phyllotaxis. Ph.D. Dissertation, Wye College, University of London.
McIntire, G. I. (1987). The role of water in the regulation of plant development. *Can. J. Bot.* **65**: 1287–98.
Meakin, P. (1986). Computer simulation of growth and aggregation processes. In *On Growth and Form: Fractal and Non-Fractal Patterns in Physics*, eds. H. E. Stanley & N. Ostrowsky. Boston: Nijhoff, pp. 111–35.
Meicenheimer, R. D. (1979). Relationships between shoot growth and changing phyllotaxy of *Ranunculus*. *Am. J. Bot.* **66**: 557–69.
Meicenheimer, R. D. (1980). Growth characteristics of *Epilobium hirsutum* shoots exhibiting bijugate and spiral phyllotaxy. Ph.D. Dissertation, Department of Botany, Washington State University.
Meicenheimer, R. D. (1981). Changes in *Epilobium* phyllotaxy induced by N-1-naphthylphthalamic acid and α-4-chlorophenoxyisobutryic acid. *Am. J. Bot.* **68**: 1139–54.
Meicenheimer, R. D. (1982). Change in *Epilobium* phyllotaxy during reproductive transition. *Am. J. Bot.* **69**: 1108–18.
Meicenheimer, R. D. (1986). Role of Parenchyma in *Linum usitatissimum* leaf trace patterns. *Am. J. Bot.* **73**(12): 1649–64.
Meicenheimer, R. D. (1987a). Role of stem growth in *Linum usitatissimum* leaf patterns. *Am. J. Bot.* **74**: 857–67.
Meicenheimer, R. D. (1987b). Empirical model of stem growth and vasculature differentiation processes. *Mathematical Modelling* **8**: 746–50. (*Proc. 1st Int. Symp. on Phyllotaxis, U. of Berkeley*, July 1985; pp. 729–61 of *Proc. 5th Int. Conf. on Math Modelling in Sci. and Tech.*)
Meicenheimer, R. D. & Zagorska-Marek, B. (1989). Consideration of the geometry of phyllotactic triangular unit and discontinuous phyllotactic transitions. *J. Theor. Biol.* **139**: 359–68.
Meinhardt, H. (1974). The formation of morphogenetic gradients and fields. *Ber. Dtsch. Bot. Ges.* **87**: 101–8.
Meinhardt, H. (1976). Morphogenesis of lines and nets. *Differentiation* **6**: 117–23.
Meinhardt, H. (1982). *Models of Biological Pattern Formation*. New York: Academic Press.
Meinhardt, H. (1984). Models of pattern formation and their application to plant development. In *Positional Controls in Plant Development*, eds. P. W. Barlow & D. J. Carr. Cambridge: Cambridge University Press, pp. 1–32.
Meinhardt, H. & Gierer, A. (1974). Applications of a theory of biological pattern formation based on lateral inhibition. *J. Cell Sci.* **15**: 321–46.
Mesarovic, M. D. (1968). Systems theory and biology – view of a theoretician. In *Systems Theory and Biology*, ed. M. D. Mesarovic. New York: Springer-Verlag, pp. 59–87.
Mesarovic, M. D. (1972). A mathematical theory of general systems. In *Trends in General Systems Theory*, ed. G. Klir. New York: Wiley-Interscience, pp. 251–69.
Mesarovic, M. D., Macko, D. & Takahara, Y. (1970). *Theory of Hierarchical, Multilevel Systems*. New York: Academic Press.
Meyen, S. V. (1973). Plant morphology in its nomothetical aspects. *Bot. Rev.* **39**(3): 205–60.

Meyen, S. V. (1978). Nomothetical plant morphology and the nomothetical theory of evolution: The need for cross-pollination. *Theoretical Plant Morphology, Acta Biotheoretica* **27**: 21–36.
Michelini, F. J. (1958). The plastochron index in developmental studies of *Xanthium italicum* Moretti. *Am. J. Bot.* **45**: 525–33.
Millener, L. H. (1952). An experimental demonstration of the dependence of phyllotaxis on rate of growth. *Nature* **169**: 1052–3.
Milthorpe, F. L. (ed.). (1956). *The Growth of Leaves.* London: Butterworths.
Mingo-Castel, A. M., Gomez-Campo, C., Tortosa, M. E. & Pelacho, A. M. (1984). Hormonal effects on phyllotaxis of *Euphorbia lanthyres* L. *Bot. Mag. Tokyo* **97**: 171–8.
Mitchison, G. J. (1977). Phyllotaxis and the Fibonacci series. *Science* **196**: 270–5.
Mohr, H. (1982). Principles in plant morphogenesis. In *Axioms and Principles of Plant Construction,* ed. R. Sattler. Boston: Nyhoff and Junk, pp. 93–111.
Moncur, M. W. (1981). *Floral Initiation in Field Crops.* Melbourne, Australia: SCIRO.
Monroy Ata, A. (1989). Le réseau hydrique du xylème comme un possible système de coordination fonctionnelle de la plante entière. In *Installation de Plantes Pérennes de la Zone Aride Soumises à des Contraintes Hydriques Contrôlées et à des Coupes.* Doctoral Dissertation, Université des Sciences et des Techniques du Languedoc, Montpellier, France.
Morowitz, H. J. (1968). *Energy Flow in Biology.* New York: Academic Press.
Morowitz, H. (1972). *Entropy for Biologists – An Introduction to Thermodynamics,* 3rd ed. New York: Academic Press.
Moseley, H. (1842). On conchyliometry. *Philos. Mag. Ser. 3* **21**(138): 300–5.
Mowshowitz, A. (1967). Entropy and the complexity of graphs. *Technical Report Concomp,* University of Michigan. (Doctoral dissertation published as four articles in *Bull. Math. Biophys.* **30**, 1968.)
Murray, B. J., Mauk, C. & Nooden, L. D. (1982). Restricted vascular pipelines and orthostichies in plants. *What's New in Plant Physiology* **13**(9): 33–6.
Murray, J. (1988). Les taches du léopard. *Pour la Science* **May**: 78–87.
Nabarro, F. R. N. (1967). *Theory of Crystal Dislocations.* Oxford: Clarendon Press.
Nägeli, C. (1858). Das Wachstum des Stammes und der Wurzel bei den Gefäfasspflanzen und die Anordnung der Gefäasstrāange im Stengel. *Beitr. Wiss. Bot.* **1**: 1–156.
Nagy, D. (1987). Ideal and fuzzy symmetries: From the hard approach to the soft one. In *Symmetry in a Cultural Context,* ed. D. Nagy. Tempe: Arizona State University Press, pp. 1–6.
Namboodiri, K. K. & Beck, C. B. (1968). A comparative study of the primary vascular system of conifers. *Am. J. Bot.* **55**: 447–72.
Naumann, C. F. (1845). *Ueber den Quincunx als Grundgesetz der Blattstellung vieler Pflanzen.* Dresden & Leipzig: Arnoldische Buchhandlung.
Needham, A. R. (1991). A study of sunflower quasicrystallography. Master of Science Thesis, University of Melbourne, Australia.
Needham, A. R., Rouse, J. L. & Bursill, L. A. (In press). Chirality and phyllotaxis of *Helianthus tuberosus*: A multihead sunflower.
Nelson, D. R. (1986). Quasicrystals. *Sci. Amer.* **255**(2): 42–51.
Nicolis, G. & Prigogine, I. (1977). *Self-Organization in Non-Equilibrium Systems.* New York: Wiley.

Niklas, K. J. (1986). Computer-simulated plant evolution. *Sci. Amer.* **254**: 78–86. Also L'évolution des plantes simulée par ordinateur. *Pour la Science* **May**: 34–44.
Niklas, K. J. (1988). The role of phyllotactic pattern as a "developmental constraint" on the interception of light by leaf surfaces. *Evolution* **42**(1): 1-16.
Nobel, P. S. (1986). Form and orientation in relation to PAR interception by cacti and agaves. In *On the Economy of Plant Form and Function,* ed. T. J. Givnish. Cambridge: Cambridge University Press, pp. 83–103.
Northrop, E. P. (1975). *Riddles in Mathematics, a Book in Paradoxes.* New York: Krieger.
Nougarède, A. (1967). Experimental cytology of the shoot apical cells during vegetative growth and flowering. *Int. Rev. Cytol.* **21**: 203–351.
Nougarède, A. (1971). Méristèmes. In *Encyclopedia Universalis* **10**, 1st ed., Paris, pp. 808–18.
Novak, F. A. (1966). *The Pictorial Encyclopedia of Plants and Flowers,* ed. J. G. Barton. New York: Crown Publishers Inc.
Occelli, R. (1985). Transitions ordre–désordre dans les structures convectives bi-dimensionelles. Doctoral Thesis, University of Provence, Marseille.
O'Grady, R. T. & Brooks, D. R. (1988). Teleology and biology. In *Entropy, Information, and Evolution, New Perspectives on Physical and Biological Evolutions,* eds. B. H. Weber, D. J. Depew & J. D. Smith. Cambridge, Mass.: MIT Press.
Olds, C. D. (1963). *Continued Fractions.* Toronto: Random House.
O'Neil, T. (1961). Primary vascular organization of *Lupinus* shoot. *Bot. Gaz.* **123**: 1–9.
Palmer, J. H. & Hernandez, F. L. (1988a). Organization of florets and seed rows in the sunflower capitulum. *Proc. 7th Australian Sunflower Workshop, Echuca,* pp. 31–4.
Palmer, J. H. & Hernandez, F. L. (1988b). Techniques to change the number of floret and seed rows in the sunflower capitulum. *Proc. 12th Int. Sunflower Conf. 2, Novi Sad, Yugoslavia,* pp. 156–7.
Palmer, J. H. & Marc, J. (1982). Wound-induced initiation of involucral bracts and florets in the developing sunflower inflorescence. *Plant and Cell Physiol.* **23**: 1401–9.
Palmer, J. H. & Steer, B. T. (1985). The generative area as the site of floret initiation in the sunflower capitulum and its integration to predict floret number. *Field Crops Res.* **11**: 1–12.
Papentin, F. (1980). On order and complexity. I. General considerations. *J. Theor. Biol.* **87**: 421–56.
Pattee, H. H. (1969a). How does a molecule become a message? *Developmental Biol. Suppl.* **3**: 1–16.
Pattee, H. H. (1969b). Physical conditions for primitive functional hierarchies. In *Hierarchical Structures,* eds. L. L. Whyte, A. G. Wilson & D. Wilson. New York: American Elsevier, pp. 161–78.
Pattee, H. H. (1970). The problem of biological hierarchy. In *Towards a Theoretical Biology,* vol. 3, ed. C. H. Waddington. Edinburgh: Edinburgh University Press, pp. 117–36.
Pattee, H. H. (1971). Can life explain quantum mechanics? In *Quantum Theory and Beyond,* ed. T. Bastin. Cambridge: Cambridge University Press, pp. 307–21.
Pattee, H. H. (ed.) (1974). *Hierarchy Theory, the Challenge of Complex Systems.* New York: Braziller.

Pauling, L. & Corey, R. B. (1951). The structure of synthetic polypeptides. *Proc. Natl. Acad. Sci. USA* **37**: 241–50.
Pauling, L. & Corey, R. B. (1953). Compound helical configurations of polypeptide chains: Structure of proteins of the α-keratin type. *Nature* **171**: 59.
Paz, A., and Solomaa, A. (1973). Integral sequential word functions and growth equivalence of Lindenmayer systems. *Inf. Control* **23**: 313–43.
Pearsall, W. H. (1927). Growth studies. VI. On the relative sizes of growing plant organs. *Ann. Bot.* **41**: 549–56.
Petukhov, S. V. (1981). *Biomechanic, Bionic and Symmetry.* Moscow: USSR Academy of Sciences. (In Russian.)
Petukhov, S. V. (1988). *Geometries of Living Nature and Algorithms of Self-Organization.* In the series Mathematics and Cybernetics #6, Moscow: Znanie. (In Russian.)
Peusner, L. (1974). *Concepts in Bioenergetics.* Englewood Cliffs, N.J.: Prentice-Hall.
Phelouzat, R. (1955). Disposition des feuilles dans les premiers stades du développement chez quelques dicotylédones. *Rev. Gén. Bot.* **62**: 454–97.
Philipson, W. R. (1935). A grass with spiral phyllotaxis *Micraira subolifoila. Kiev Bull. Add. Ser.* **1935**: 324–6.
Philipson, W. R. (1949). The ontogeny of the shoot apex in dicotyledons. *Biol. Rev.* **24**: 21–50.
Philipson, W. R. (1963). Vascular patterns in dicotyledons. *Bot. Rev.* **29**: 382–404.
Piélou, E. C. (1969). *Introduction to Mathematical Ecology.* New York: Wiley-Interscience.
Pilet, P. E. (1967). *L'Energétique Végétale.* Paris: Les Presses Universitaires de France.
Pippenger, N. (1978). La théorie de la complexité. *Pour la Science* **10**: 86–95.
Plantefol, L. (1946a). Sur les méthodes en phyllotaxie. *C. R. Acad. Sci. Paris* **222**: 1508–10.
Plantefol, L. (1946b). Fondements d'une théorie phyllotaxique nouvelle. I. Historique et critique. II. La phyllotaxie des monocotylédones. *Ann. Sci. Nat. Bot., Sér. II* **7**: 153–229.
Plantefol, L. (1947). Fondements d'une théorie phyllotaxique nouvelle. III. La phyllotaxie des dicotylédones. IV. Généralisations et conclusions. *Ann. Sci. Nat. Bot., Sér. II* **8**: 1–71.
Plantefol, L. (1948). *La Théorie des Hélices Foliaires Multiples.* Paris: Masson.
Plantefol, L. (1949). A new theory of phyllotaxis. *Nature* **163**: 331–2.
Plantefol, L. (1950). La phyllotaxie. *Année Biol.* **54**: 447–60.
Plantefol, L. (1951). Phyllotaxie et point végétatif. *Scientia* **86**: 91–8.
Plantefol, L. (1956). Sur les variations phyllotaxiques de *Stapelia hirsuta. C. R. Acad. Sci. Paris* **243**: 916–19.
Plantefol, L. (1963). Rapport sur la tératologie des organes végétatifs. *Mem. Soc. Bot. Fr.* **42**: 5–16.
Poethig, R. S. (1990). Phase change and the regulation of shoot morphogenesis in plants. *Science* **250**: 923–30.
Pollard, J. H. (1973). *Mathematical Models for Growth of Human Populations.* Cambridge: Cambridge University Press.
Popham, R. A. & Chan, A. P. (1952). Origin and development of the receptacle of *Chrysanthemum morifolium. Am. J. Bot.* **39**: 329–39.
Popper, K. (1968). *The Logic of Scientific Discovery.* New York: Harper & Row.

Priestley, J. H. & Scott, L. I. (1933). Phyllotaxis in the dicotyledon from the standpoint of developmental anatomy. *Biol. Rev. (London)* **8**: 241–68.

Priestley, J. H. & Scott, L. I. (1936). The vascular anatomy of *Helianthus annuus* L. *Proc. Leeds Phil. Lit. Soc.* **3**: 159–73.

Priestley, J. H. Scott, L. I. & Gillett, E. C. (1935). The development of the shoot in *Alstroemeria* and the unit of shoot growth in monocotyledons. *Ann. Bot.* **49**: 161–79.

Priestley, J. H., Scott, L. I. & Mattinson, K. M. (1937). Dicotyledon phyllotaxis from the standpoint of development. *Proc. Leeds Phil. Lit. Soc.* **3**: 380–8.

Prigogine, I. & Glansdorff, P. (1971). *Structure, Stabilité et Fluctuation*. Paris: Masson.

Prusinkiewicz, P. & Hanan, J. (1989). Lindenmayer systems, fractals and plants. *Lecture Notes in Biomathematics* **79**. New York: Springer-Verlag.

Prusinkiewicz, P. & Lindenmayer, A. (1990). *The Algorithmic Beauty of Plants*. New York: Springer-Verlag.

Radl, E. (1930). *History of Biological Theories*. Oxford: Oxford University Press.

Rashevsky, N. (1954). Topology and life: In search of general mathematical principles in biology and sociology. *Bull. Math. Biophys.* **16**: 317–48.

Rashevsky, N. (1955). Life, information theory and topology. *Bull. Math. Biophys.* **17**: 229–35.

Rashevsky, N. (1960). *Mathematical Biophysics, Physico-Mathematical Foundations of Biology*, vols. 1 & 2. New York: Dover.

Rashevsky, N. (1961). *Mathematical Principles in Biology and their Applications*. Springfield, Ill.: Thomas.

Rashevsky, N. (1965). Models and mathematical principles in biology. In *Theoretical and Mathematical Biology*, eds. T. H. Watermann & H. J. Morowitz. Toronto: Blaisdel, pp. 36–54.

Rashevsky, N. (1968). Organismic sets. II. Some general considerations. *Bull. Math. Biophys.* **30**: 163–74.

Rasmussen, N. (1992). Studies on the determination of organ pattern and organ identity in flower development. Ph.D. Dissertation, Stanford University, Stanford, California.

Rees, A. R. (1964). The apical organization and phyllotaxis of the oil palm. *Ann. Bot. (N.S.)* **28**: 57–69.

Reeves, E. C. R. & Huxley, J. S. (1972). Some problems in the study of allometric growth. In *Problems of Relative Growth*, ed. J. S. Huxley. New York: Dover, pp. 267–303.

Reffye, P. de, Edelin, C. & Jaeger, M. (1989). La modélisation de la croissance des plantes. *La Recherche* **20**(207): 159–68.

Reinberg, A. (1977). *Des Rhythmes Biologiques à la Chronobiologie*. Paris: Gauthier-Villars.

Rényi, A. (1961). On measures of entropy and information. *Proc. 4th Berkeley Symposium on Mathematical and Statistical Problems*, vol. 1. Berkeley: University of California Press, pp. 547–61.

Reverberi, D. (1985). Clockspring model of sprouting and blossoming. *J. Appl. Math. and Phys.* **36**(5): 743–56.

Ricard, J. (1960). *La Croissance des Végétaux*, Collection Que sais-je? Paris: Presses Univ. de France.

Ricciardi, L. M. (1977). Diffusion processes and related topics. *Lecture Notes in Biomathematics* **14**. New York: Springer-Verlag.

Richards, F. J. (1948). The geometry of phyllotaxis and its origin. *Symp. Soc. Exper. Biol.* **2**. Cambridge: Cambridge University Press, pp. 217–45.

Richards, F. J. (1951). Phyllotaxis: Its quantitative expression and relation to growth in the apex. *Philo. Trans. Roy. Soc. London* **235B**: 509–64.

Richards, F. J. (1955). Ueber einige Fragen betreffend die Messung von Blattstellung. *Planta* **45**: 198–207.

Richards, F. J. (1956). Spatial and temporal correlations involved in leaf pattern production at the apex. In *The Growth of Leaves,* ed. F. L. Milthorpe. London: Butterworths, pp. 66–76.

Richards, F. J. (1959). A flexible growth function for empirical use. *J. Exp. Bot.* **10**: 290–300.

Richards, F. J. (1969). The quantitative analysis of growth. In *Plant Physiology, a Treatise,* vol. 5A, ed. F. C. Steward. New York: Academic Press, pp. 3–76.

Richards, F. J. & Schwabe, W. W. (1969). Phyllotaxis: A problem of growth and form. In *Plant Physiology, a Treatise,* vol. 5A, ed. F. C. Steward. New York: Academic Press, pp. 79–116.

Richards, O. W. & Kavanagh, A. J. (1945). The analysis of growing form. In *Essays on Growth and Form,* eds. W. E. Le Gros Clark & P. B. Medawar. London: Oxford University Press, pp. 188–230.

Richter, P. H. & Schranner, R. (1978). Leaf arrangement, geometry, morphogenesis, and classification. *Naturwissenschaften* **65**: 319–27.

Ridley, J. N. (1982a). Packing efficiency in sunflower heads. *Math. Biosci.* **58**: 129–39.

Ridley, J. N. (1982b). Computer simulation of contact pressure in capitula. *J. Theor. Biol.* **95**: 1–11.

Ridley, J. N. (1986). Ideal phyllotaxis on general surfaces of revolution. *Math. Biosci.* **79**: 1–24.

Ridley, J. N. (1987). Descriptive phyllotaxis on surfaces with circular symmetry. *Mathematical Modelling* **8**: 751–5. (*Proc. 1st Int. Symp. on Phyllotaxis, U. of Berkeley,* July 1985; pp. 729–61 of *Proc. 5th Int. Conf. on Math. Modelling in Sci. and Tech.*)

Rigaut, J. P. (1987). Fractals, semi-fractals et biométrie. In *Fractals, Dimensions Non-Entières et Applications,* ed. G. Sherbit. Paris: Masson et Cie, pp. 231–81.

Rijven, A. H. G. C. (1969). Randomness in the genesis of phyllotaxis. II. *New Phytol.* **68**: 377–86.

Rivier, N. (1986). A botanical quasicrystal. *J. Phys. France, Colloque C3, Suppl.* 7(47): 299–309.

Rivier, N. (1988). Crystallography of spiral lattices. *Mod. Phys. Letters B* **2**: 953–60.

Rivier, N. (1990). The structure and dynamics of patterns of Bénard convection cells. *Proc. IUTAM Symp. on Fluid Mechanics of Stirring and Mixing, La Jolla, California,* August, 1990.

Rivier, N., Koch, A. J. & Rothen, F. (1991). Crystallography of composite flowers: Mode locking and dynamical maps. In *Proc. Cargese Symp. on Biologically Inspired Physics,* September, 1990, ed. S. Leibler. New York: Plenum.

Rivier, N., Occelli, R., Pantaloni, J. & Lissowski, A. (1984). Structure of Bénard convection cells, phyllotaxis and crystallography in cylindrical symmetry. *J. Phys.* **45**: 49–63.

Roach, W. A. (1939). Plant injection as a physiological method. *Ann. Bot. (N.S.)* **3**: 155–226.
Robbins, J. (1957). Gibberellic acid and the reversal of adult *Herdera* to a juvenile state. *Am. J. Bot.* **44**: 743–6.
Roberts, D. W. (1977). A contact pressure model for semidecussate and related phyllotaxis. *J. Theor. Biol.* **68**: 583–97.
Roberts, D. W. (1978). The origin of Fibonacci phyllotaxis – an analysis of Adler's contact pressure model and Mitchison's expanding apex model. *J. Theor. Biol.* **74**: 217–33.
Roberts, D. W. (1984). A chemical contact pressure of phyllotaxis. *J. Theor. Biol.* **108**: 481–90.
Roberts, D. W. (1987). The chemical contact pressure model for phyllotaxis: Application to phyllotaxis changes in seedlings and to anomalous phyllotaxis. *J. Theor. Biol.* **125**: 141–61.
Roberts, D. W. (Unpublished). Phyllotaxis. A reappraisal of Church's theory.
Roche, C. (1949). Étude phyllotaxique sur *Hedera helix*. *Rev. Gén. Bot.* **56**: 49–73.
Roche, D. (1970). Manifestations de l'anisocladie hélicoïdale après une variation phyllotaxique provoquée chez la stellaire (*Stellaria media* L. Vill.). *Mém. Bull. Soc. Bot. France* **1970**: 205–14.
Roche, D. (1971). Suppression d'une hélice foliaire par action de l'acide phénylborique sur la stellaire (*Stellaria media* L. Vill.). *C. R. Acad. Sci. Paris* **273**: 2522–5.
Rogler, C. E. & Hackett, W. P. (1975). Phase change in *Hedera helix:* Induction of the mature to juvenile phase by gibberellin A_3. *Physiol. Plant.* **34**: 141–7.
Rohrbach, P. (1867). Ueber Pycnophyllum Remy nebst Bemerkungen über die Blattstellung der Caryophylleen. *Bot. Zeitung (Berlin)* **25**: 297–300.
Romberger, J. A. & Gregory, R. A. (1977). The shoot apical ontogeny of *Picea abies* seedling. III. Some age-related aspects of morphogenesis. *Am. J. Bot.* **64**: 622–30.
Rosen, R. (1960). A quantum-theoretic approach to genetic problems. *Bull. Math. Biophys.* **26**: 227–55.
Rosen, R. (1967). *Optimality Principles in Biology.* London: Butterworth.
Rosen, R. (1972). *Foundations of Mathematical Biology.* New York: Academic Press.
Rosen, R. (1977). The generation and recognition of patterns in biological systems. In *Mathematics and the Life Sciences, Lecture Notes in Biomathematics Series,* ed. S. Levin. New York: Springer-Verlag, pp. 222–341.
Rosen, R. (1978). *Fundamentals of Measurement and Representation of Natural Systems.* New York: North Holland.
Rothen, F. & Koch, A. J. (1989a). Phyllotaxis, or the properties of spiral lattices. I. Shape invariance under compression. *J. Phys. France* **50**: 633–57.
Rothen, F. & Koch, A. J. (1989b). Phyllotaxis, or the properties of spiral lattices. II. Packing of circles along logarithmic spirals. *J. Phys. France* **50**: 1603–21.
Rothen, F. & Pieranski, P. (1986). Les cristaux colloidaux. *La Recherche* **175**(17): 312–21.
Rouse, J. L., Needham, A. & Bursill, L. A. (In press). Phyllotaxis of the sunflower, *Helianthus tuberosus* L. *J. Theor. Biol.*

Rouse Ball, W. W. & Coxeter, H. S. M. (1987). *Mathematical Recreations and Essays*. New York: Dover.
Rozenberg, G. (1976). Bibliography of L-systems. In *Automata, Languages, Development*, eds. A. Lindenmayer & G. Rozenberg. Amsterdam: North Holland, pp. 351–66.
Rozenberg, G. & Lindenmayer, A. (1973). Developmental systems with locally catenative formulas. *Acta Informatica* **2**: 214–48.
Rozenberg, G. & Salomaa, A. (eds.) (1986). *The Book of L*. Berlin: Springer-Verlag.
Runion, E. G. (1972). *The Golden Section and Related Curiosa*. Glenview, Ill.: Scott, Foresman & Co.
Russin, W. A., Clayton, M. K. & Durbin, R. D. (1991). Modelling non-exponential growth of a short cycling plant – *Tagetes erecta* (Asterraceae). *Can. J. Bot.* **69**: 316–20.
Rutishauser, R. (1981). Blattstellung und Sprossentwicklung bei Blütenpflanzen. *Dissertationes Botanicae* **62**. Vaduz: Cramer.
Rutishauser, R. (1982). Der Plastochronquotient als Teil einer quantitativen Blattstellungsanalyse bei Samenpflanzen. *Beitr. Biol. Pflanzen* **57**: 323–57.
Rutishauser, R. (1986). Phyllotactic patterns in phyllodinous acacias (*Acacia* subg. *Heterophyllum*) – promising aspects for systemics. *Bull. Int. Group Study Mimosoideae* **14**: 77–108.
Rutishauser, R. (1989). A dynamic multidisciplinary approach to floral morphology. *Prog. in Botany* **51**: 54–69.
Rutishauser, R. & Sattler, R. (1985). Complementarity and heuristic value of contrasting models in structural botany. I. General considerations. *Bot. Jahrb. Syst.* **107**(1–4): 415–55.
Rutishauser, R. & Sattler, R. (1987). Complementarity and heuristic value of contrasting models in structural botany. II. Case study of leaf whorls: *Equisetum* and *Ceratophyllum*. *Bot. Jahrb. Syst.* **109**: 227–55.
Rutishauser, R. & Sattler, R. (1989). Complementarity and heuristic value of contrasting models in structural botany. III. Case study on shoot-like leaves and leaf-like shoots in *Utricularia macrorhiza* and *U. purpurea* (Lentibulariaceae). *Bot. Jahrb. Syst.* **111**: 121–37.
Ryan, G. W., Rouse, J. L. & Bursill, L. A. (1991). Quantitative analysis of sunflower seed packing. *J. Theor. Biol.* **147**: 303–28.
Sachs, J. (1882). *Text-Book of Botany*, 2nd ed. London & New York: Oxford University Press.
Sachs, T. (1991). *Pattern Formation in Plant Tissues*. Cambridge: Cambridge University Press.
Sakaguchi, S., Hogetsu T. & Hara, N. (1988). Arrangement of cortical microtubules in the shoot apex of *Vinca major* L. *Planta* **175**: 403–11.
Salomaa, A. & Soittola, M. (1978). *Automata-Theoretic Aspects of Formal Power Series*. New York: Springer-Verlag.
Salthe, S. N. (1985). *Evolving Hierarchical Systems – Their Structure and Representation*. New York: Columbia University Press.
Sattler, R. (1966). Towards a more adequate approach to comparative morphology. *Phytomorphomology* **16**: 417–29.
Sattler, R. (1973). *Organogenesis of Flowers*. Toronto: Toronto University Press
Sattler, R. (1974a). A new conception of the shoot of higher plants. *J. Theor. Biol.* **47**: 367–82.
Sattler, R. (1974b). Essentialism in plant morphology. *Proc. 14th Int. Congr. Hist. Sci.* **3**: 464–7.

Sattler, R. (1978a). "Fusion" and "continuity" in floral morphology. *Notes Roy. Bot. Gard. Edinburgh* **36**: 397–405.
Sattler, R. (1978b). What is theoretical plant morphology? In *Theoretical Plant Morphology,* ed. R. Sattler. The Hague: Leiden University Press, pp. 5–20. (Suppl. to *Acta Biotheoretica* **27**).
Sattler, R. (ed.) (1982). *Axioms and Principles of Plant Construction.* The Hague: Nijhoff/Junk.
Sattler, R. (1984). Homology – a continuing challenge. *Syst. Bot.* **9**: 382–94.
Sattler, R. (1986). *Biophilosophy.* New York: Springer-Verlag.
Sattler, R. (1988a). A dynamic multidimensional approach to floral morphology. In *Aspects of Floral Development,* eds. P. Leins, S. C. Tucker & P. K. Endress. Berlin: Cramer, pp. 1–6.
Sattler, R. (1988b). Homeosis in plants. *Am. J. Bot.* **75**: 1606–17.
Sattler, R. (1990). Towards a more dynamic plant morphology. *Acta Biotheor.* **38**: 303–15.
Sattler, R. (1992). Process morphology: Structural dynamics in development and evolution. *Can. J. Bot.* **70**: 708–14.
Sattler, R., Luckert, D. & Rutishauser, R. (1988). Symmetry in plants: Leaf and stipule development in *Acacia longipedunculata. Can. J. Bot.* **66**: 1270–84.
Sattler, R. & Rutishauser, R. (1990). Structural and dynamic descriptions of the development of *Utricularia foliosa* and *U. australis. Can. J. Bot.* **68**: 1989–2003.
Sattler, R. & Rutishauser, R. (1992). Partial homology of pinnate leaves and shoots orientation of leaflet inception. *Bot. Jahrb. Syst.* **114**(1): 61–79.
Sattler, R. & Singh, W. (1978). Floral organogenis of *Echinodorus amazonicus* Rataj and floral construction of the Alismatales. *Bot. J. Linn. Soc.* **77**: 141–56.
Sawada, Y. & Honjo, H. (1986). Mais d'où vient donc la forme des dendrites? *La Recherche* **175**(17): 522–4.
Schaffalitzky de Muckadell, M. (1959). Investigations on ageing of apical meristems in woody plants and its importance in silviculture. *Forstl. Forsgsv. Danm.* **25**: 310–455.
Schaffner, J. H. (1926). The change of opposite to alternate phyllotaxy and repeated rejuvenations in hemp by means of photoperidocity. *Ecology* **7**: 315–25.
Schaffner, J. H. (1927). Spiral shoots of *Equisetum. Am. Fern J.* **17**: 43–7.
Schaffner, J. H. (1938). Spiral systems in the vascular plants. *Bull. Torrey Bot. Club* **65**(8): 507–29.
Schimper, C. F. (1830). Beschreibung des Symphytum Zeyheri und seiner zwei deutschen Verwandten der *S. Bulborum* Schimp. und *S. tuberosum* Jacq. *Geiger's Mag. für Pharm.* **29**: 1–92.
Schimper, C. F. (1836). Geometrische Anordnung der um eine Axe periferischen Blattgebilde. *Verhandl. Schez. Naturf. Ges.* **21**: 113–17.
Schmidt, B. L. & Millington, W. F. (1968). Regulation of leaf shape in *Proserpinaca palustris. Bull. Torrey Bot. Club* **95**: 264–86.
Schmucker, T. (1933). Zur Entwicklungsphysiologie der schraubigen Blattstellung. *Planta* **19**: 139–53.
Schneiter, A. A. & Miller, J. F. (1981). Description of sunflower growth stages. *Crop Science* **21**: 901–3.
Scholz, H. J. (1985). Phyllotactic iterations. *Ber. Bunsenges Phys. Chem.* **89**: 699–703.

Schoute, J. C. (1913). Beitrage zur Blattstellunglehre. I. Die Theorie. *Rec. Trav. Bot. Néerl.* **10**: 153–339.
Schoute, J. C. (1935). Note on some properties of decussate foliage leaves which are connected with space-filling in the bud. *Rec. Trav. Bot. Néerl.* **32**: 317–22.
Schoute, J. C. (1936). On whorled phyllotaxis. III. True and false whorls. *Rev. Trav. Bot. Néerl.* **33**: 670–87.
Schoute, J. C. (1938). On whorled phyllotaxis. IV. Early binding whorls. *Rec. Trav. Bot. Néerl.* **35**: 415–558.
Schrödinger, E. (1962). *What is Life?* Cambridge: Cambridge University Press.
Schuepp, O. (1921). Zur Theorie der Blattstellung. *Ber. Deutsch. Got. Ges.* **39**: 249–57.
Schuepp, O. (1934). Untersuchungen und Konstruktionen zur Theorie der zweizeilig dorsiventralen Blattstellung. *Jahrb. Wiss. Bot.* **80**: 36–73.
Schuepp, O. (1936). Konstruktionen zur Theorie der schiefen Quirle. Modelle zur Blattstellungstheorie. *Jahrb. Wiss. Bot.* **82**: 555–80.
Schuepp, O. (1946). Geometrische Betrachtungen über Wachstum und Formwechsel. *Ber. Schweiz. Bot. Ges.* **56**: 629–55.
Schuepp, O. (1959). Konstruktionen zur Theorie der Blattstellung. *Denkschr. Schweiz. Naturf. Ges.* **82**: 592–629.
Schuepp, O. (1963). Mathematisches und Botanisches über Allometrie. *Verhandl. Naturf. Ges. Basel* **74**: 69–105.
Schuepp, O. (1966). *Meristeme*. Basel & Stuttgart: Birkhäuser-Verlag.
Schuepp, O. (1969). Morphological concepts: Their meaning in the ideal morphology and a comparative ontogeny. *Amer. J. Bot.* **56**: 799–804.
Schuster, H. G. (1984). *Deterministic Chaos. An Introduction*. Weinheim: Physic-Verlag.
Schwabe, W. W. (1958). Some effects of environment and hormone treatment on reproductive morphogenesis in the *Chrysanthemum*. *J. Linn. Soc. Lond.* **56**: 254–61.
Schwabe, W. W. (1963). Morphogenetic responses to climate. In *Environmental Control of Plant Growth,* ed. L. T. Evans. New York: Academic Press, pp. 311–36.
Schwabe, W. W. (1971). Chemical modification of phyllotaxis and its implication. In *Control Mechanisms of Growth and Differentiation,* eds. D. C. Davis & M. Balls. Cambridge: Cambridge University Press, pp. 301–22. (*Symp. Soc. Exp. Biol.* **25**.)
Schwabe, W. W. (1979). Organogenesis at the plant apex with special reference to the transition to flowering. *British P. Growth Reg. Group, Monograph* **3**: 75–86.
Schwabe, W. W. (1984). Phyllotaxis. In *Positional Control in Plant Development,* eds. P. W. Barlow & D. J. Carr. Cambridge: Cambridge University Press, pp. 403–40.
Schwabe, W. W. & Clewer, A. G. (1984). Phyllotaxis – a simple computer model based on the theory of a polarly-translocated inhibitor. *J. Theor. Biol.* **109**: 595–619.
Schwendener, S. (1878). *Mechanische Theorie der Blattstellungen*. Leipzig: Engelmann.
Schwendener, S. (1883). Zur Theorie der Blattstellungen. *Sitzungsber. Akad. Wiss. Berl.,* pp. 741–73.
Schwendener, S. (1909). *Theorie der Blattstellungen, Mechanische Probleme der Botanik*. Leipzig: Englemann.

Selvam, M. A. (1990a). A cell dynamical system model for turbulent shear flows in the planetary atmospheric boundary layer. *Proc. Ninth Symp. on Turbulence and Diffusion,* May 1990, Roskilde, Denmark. Pub. Am. Meteorological Soc., Boston, Mass., pp. 262–5.

Selvam, M. A. (1990b). Deterministic chaos, fractals and quantum-like mechanics in atmospheric flows. *Can. J. Physics* **68**: 831–41.

Shannon, C. E. (1948). A mathematical theory of communication. *Bell Syst. Tech. J.* **27**: 379–656.

Shannon, C. E. & Weaver, N. (1967). *The Mathematical Theory of Communication,* 11th ed. Urbana: University of Illinois Press.

Shao, S. (1988). Fuzzy self-organizing controller and its application for dynamic processes. *Fuzzy Sets and Systems* **26**: 151–64.

Sharp, W. E. (1971). An analysis of the laws of stream order for Fibonacci drainage patterns. *Water Resources Research* **7**(6): 1548–57.

Sharp, W. E. (1972). Fibonacci drainage patterns. *Fib. Q.* **10**: 643–55.

Shaver, D. L. (1967). Decussate phyllotaxy in maize. *Maize Genet. Coop. News Letter* **41**: 33–4.

Shool, D. A. (1954). Regularities in growth curves, including rhythm and allometry. In *Dynamics of Growth Processes,* ed. J. E. Boel. Princeton, N.J.: Princeton University Press, pp. 224–41.

Sifton, H. W. (1944). Developmental morphology of vascular plant growth. *New Phytol.* **43**: 87–129.

Silk, K. W. & Erickson, R. O. (1979). Kinematics of plant growth. *J. Theor. Biol.* **76**: 481–501.

Simon, H. A. (1962). The architecture of complexity. *Proc. Am. Phil. Soc.* **106**: 467–82.

Simon, H. A. (1977). *Models of Discovery.* Boston: Reidel.

Simpson, G. G. (1967). *The Major Features of Evolution.* New York: Columbia University Press.

Sinnott, E. W. (1960). *Plant Morphogenesis.* New York: McGraw-Hill.

Sinnott, E. W. (1966). The geometry of life. In *Trends in Plant Morphogenesis,* ed. E. G. Cutter. London: Longmans, pp. 88–93.

Skipworth, J. P. (1962). The primary vascular system and phyllotaxis in *Hectorella caespitosa* Hook f. *New Zealand J. of Science* **5**: 253–8.

Slover, J. de (1958). Le sens longitudinal de la différentiation du procambium, du xylène et du phloème chez *Coleus, Ligustrum Anagallis* et *Taxus. La Cellule* **59**: 55–202.

Smith, A. R. (1984). Plants, fractals, and formal languages. *Computer Graphics* **18**(3): 1–10.

Smith, B. W. (1941). The phyllotaxis of *Custus,* from the standpoint of development. *Proc. Leeds Phil. Soc.* **4**: 42–63.

Smith, R. (1980). Rethinking allometry. *J. Theor. Biol.* **87**: 97–111.

Snow, M. (1955). Spirodistichy re-interpreted. *Phil. Trans. Roy. Soc. London* **239B**: 45–88.

Snow, M. & Snow, R. (1931). Experiments on phyllotaxis. I. The effect of isolating a primordium. *Phil. Trans. Roy. Soc. London* **221B**: 1–43.

Snow, M. & Snow, R. (1933). Experiments on phyllotaxis. II. The effect of displacing a primordium. *Phil. Trans. Roy. Soc. London* **222B**: 353–400.

Snow, M. & Snow, R. (1934). The interpretation of phyllotaxis. *Biol. Rev., Cambridge Phil. Soc.* **9**: 132–7.

Snow, M. & Snow, R. (1935). Experiments on phyllotaxis. III. Diagonal splits through decussate apices. *Phil. Trans. Roy. Soc. London* **225B**: 63–94.

Snow, M. & Snow, R. (1937). Auxin and leaf formation. *New Phytol.* **36**: 1–18.
Snow, M. & Snow, R. (1947–1948). On the determination of leaves. *New Phytol.* **46**: 5–19; *Soc. Exp. Biol. Symp.* **2**: 263–75.
Snow, M. & Snow, R. (1952). Minimum areas and leaf determination. *Proc. Roy. Soc. London* **139B**: 545–66.
Snow, M. & Snow, R. (1956). Regulation of sizes of leaf primordia by growing-point of stem apex. *Proc. Roy. Soc. London* **144B**: 222–9.
Snow, M. & Snow, R. (1959). Regulation of sizes of leaf primordia by older leaves. *Proc. Roy. Soc. London* **151B**: 39–47.
Snow, M. & Snow, R. (1962). A theory of the regulation of phyllotaxis based on *Lupinus albus*. *Phil. Trans. Roy. Soc. London* **244B**: 483–513.
Snow, R. (1942). Further experiments on whorled phyllotaxis. *New Phytol.* **41**: 108–24.
Snow, R. (1948). A new theory of leaf formation. *Nature* **162**: 798.
Snow, R. (1949). A new theory of phyllotaxis. *Nature* **163**: 332.
Snow, R. (1951). Experiments on bijugate apices. *Phil. Trans Roy. Soc. London* **235B**: 291–310.
Snow, R. (1952). On the shoot apex and phyllotaxis of *Costus*. *New Phytol.* **51**: 359–63.
Snow, R. (1954). Phyllotaxis of flowering teasels. *New Phytol.* **53**: 99–107.
Snow, R. (1955). Problems of phyllotaxis and leaf determination. *Endeavour* **14**: 190–9.
Snow, R. (1958). Phyllotaxis of *Kniphofia* and *Lilium candidum*. *New Phytol.* **57**: 160–7.
Snow, R. (1965). The causes of the bud eccentricity and the large divergence angles between leaves in Cucurbitaceae. *Philos. Trans., Ser. B.* **250**: 53–77.
Sober, E. (1984). *The Nature of Selection: Evolutionary Theory in Philosophical Focus*. Cambridge, Mass.: MIT Press.
Soberon, J. M. & Delrio, M. C. (1981). The dynamics of a plant–pollinator interaction. *J. Theor. Biol.* **91**: 363–78.
Soma, K. (1958). Morphogenesis in the shoot apex of *Euphorbia lathyris* L. *J. Fac. Sci. U. Tokyo III* **7**: 199–256.
Steeves, T. A., Hicks, M. A., Taylor, J. M. & Rennie, P. (1969). Analytical studies on the shoot apex of *Helianthus annuus*. *Can. J. Bot.* **47**: 1367–75.
Steeves, T. A. & Sussex, I. M. (1989). *Patterns in Plant Development*, 2nd ed. New York: Cambridge University Press.
Stein, K. (1982). Kawaguchi's spirals. *OMNI* **5**(2): 110–15.
Steinhaus, H. (1960). *Mathematical Snapshots*. New York: Oxford University Press.
Sterling, C. (1945). Growth and vascular development in the shoot apex of *Sequoia sempervirens* (Lamb) Endl. II. Vascular development in relation to phyllotaxis. *Am. J. Bot.* **32**: 380–6.
Stevens, P. S. (1974). *Patterns in Nature*. Boston: Little, Brown & Co. (French edition by Seuil, Paris, under the title *Les Formes dans la Nature*, 1978.)
Stevens, P. S. (1980). *Handbook of Regular Patterns*. Cambridge, Mass.: MIT Press.
Steward, F. C. (1968). *Growth and Organization in Plants*. London: Addison-Wesley.
Stewart, B. M. (1966). *Theory of Numbers*, 2nd ed. New York: Macmillan.
Stewart, W. N. (1964). An upward outlook in plant morphology. *Phytomorphology* **14**: 120–34.
Stoll, R. R. (1952). *Linear Algebra and Matrix Theory*. New York: McGraw-Hill.

Sweet, S. S. (1980). Allometric inference in morphology. *Am. Zool.* **20**: 643–52.

Sykes, Z. M. (1969). On discrete stable population theory. *Biometrics* **25**: 285–93.

Szabo, Z. (1930). A Dipsacaceak Viragzatanak Fejlodestani Ertelmezese (Entwicklungsegeschichtliche Deutung des Blutenstandes der Dipsacaceen), *A Szent Istvan Akademia Mennyisegtan Tereszettudomanyi Osztalyanak Felolvasasai* **2**: 3–72.

Szilard, R. (1974). *Theory and Analysis of Plants. Classical and Numerical Methods.* Englewood Cliffs, N.J.: Prentice-Hall.

Tait, P. G. (1872). On phyllotaxis. *Proc. Roy. Soc. Edinburgh* **7**: 391–4.

Tavcar, A. (1941). The inheritance of the number and position of the leaves of *Zea mays* L. *Bull. Int. Acad. Yougoslave Sci., Cl. Sci. Math.* **34**: 1–90.

Taylor, A. (1967). *Introduction to Functional Analysis.* New York: Wiley.

Tennekone, K., Dayatilaka, R. K. D. & Ariyaratne, S. (1982). Right–left symmetry in phyllotaxy and imbrication of flowers of *Hibiscus furcartus* L. *Ann. Bot. (London)* **50**: 397–400.

Teissier, G. (1948). La relation d'allométrie, sa signification statistique et biologique. *Biometrics* **4**: 14–53.

Thom, R. (1975). *Structural Stability and Morphogenesis.* Reading, Mass.: Benjamin.

Thomas, R. L. (1975). Orthostichy, parastichy and plastochrone ratio in a central theory of phyllotaxis. *Ann. Bot.* **39**: 455–89.

Thomas, R. L. & Cannell, M. R. (1980). The generative spiral in phyllotaxis theory. *Ann. Bot.* **45**: 237–49.

Thomas, R. L., Chan, K. W. & Easau, P. T. (1969). Phyllotaxis in the oil palm: Arrangement of fronds on the trunk of mature palms. *Ann. Bot. (London)* **33**: 1001–8.

Thomas, R. L., Chan, K. W. & Ng, S. C. (1970). Phyllotaxis in the oil palm: Arrangement of male/female spikelets on the inflorescence stalk. *Ann. Bot. (London)* **34**: 93–105.

Thomas, R. L., Ng, S. C. & Wong, C. C. (1970). Phyllotaxis in the oil palm: Arrangement of male/female florets along the spikelets. *Ann. Bot.* **34**: 107–15.

Thompson, D. W. (1917). *On Growth and Form.* Cambridge: Cambridge University Press. (Reprinted in 1942 and 1968; abridged edition by J. T. Bonner, 1971.)

Thompson, K. S. (1984). Reductionism and other isms in biology. *Am. Sci.* **72**: 388–90.

Thornley, J. H. M. (1975a). Phyllotaxis. I: A mechanistic model. *Ann. Bot. (London)* **39**: 491–507.

Thornley, J. H. M. (1975b). Phyllotaxis. II: A description in terms of intersecting logarithmic spirals. *Ann. Bot.* **39**: 509–24.

Thornley, J. H. M. (1976). *Mathematical Models in Plant Physiology.* New York: Academic Press.

Thornley, J. H. M. (1977). A model of apical bifurcation applicable to trees and other organisms. *J. Theor. Biol.* **64**: 165–76.

Thornley, J. H. M. & Cockshull, K. E. (1980). A catastrophe model for the switch from vegetative to reproductive growth in the shoot apex. *Ann. Bot.* **46**: 333–41.

Thornley, J. H. M. & Johnson, I. R. (1990). *Plant and Crop Modelling.* Oxford: Clarendon Press.

Thrall, R. M., Mortimer, J. A., Rebman, K. R. & Baum, R. F. (eds.) (1967). *Some Mathematical Models in Biology,* Rep. #40241-R-7, University of Michigan Press.

Tieghem, P. van (1874). Remarques sur la disposition des feuilles dites opposées et verticillées. *Bull. Soc. Bot. France* **21**: 360–3.

Ting, T. (1963). Translocation of assimilates of leaf of main stem in relation to the phyllotaxis in cotton plant. *Acta. Biol. Exp. Sin.* **8**: 656–63.

Tomlinson, P. B. (1982). Change and design in the construction of plants. *Acta Biotheor.* **31A**: 162–83.

Tomlinson, P. B. & Wheat, D. W. (1979). Bijugate phyllotaxis in Rhizophoreae. *Bot. J. Linn. Soc.* **78**: 317–21.

Tort, M. (1969). Modifications phyllotaxiques provoquées par des traitements thermiques et par l'acide gibbérellique chez le crosne du Japon, *Stachys sieboldii* Miq. *Mém. Bull. Soc. Bot. Fr.* **1969**: 179–95.

Tort, M. & Loiseau, J. E. (1967). Modifications phyllotaxiques provoquées chez le *Phlox Drumondii* Hook. (nouvelles observations). *Mém. Bull. Soc. Bot. Fr.* **1967**: 31–49.

Troll, W. (1937). *Vergleichende Morphologie der hoheren Pflanzen Vegetations-Organe,* vol. 1. Berlin: Borntraeger.

Troll, W. (1938). Ueber die zerstreute Blattstellung bei den Dikotylen. *Chron. Bot.* **4**: 39–40.

Trucco, E. (1956a). A note on the information content of graphs. *Bull. Math. Biophys.* **18**: 129–35.

Trucco, E. (1956b). On the information content of graphs: Compound symbols; different states for each point. *Bull. Math. Biophys.* **18**: 237–53.

Truemper, K. (1978). Algebraic characterizations of unimodar matrices. *SIAM J. Appl. Math.* **35**: 328–32.

Tucker, S. C. (1961). Phyllotaxis and vascular organization of the carples in *Michelia fuscata. Am. J. Bot.* **48**: 60–71.

Tucker, S. C. (1962). Ontogeny and phyllotaxis of the terminal vegetative shoots of *Michelia fuscata. Am. J. Bot.* **49**: 722–37.

Tucker, S. C. (1990). Loss of floral organs in *Ateleia* (Leguminosae: Papilionoideae: Sophoreae). *Am. J. Bot.* **77**: 750–61.

Turing, A. M. (1952). The chemical basis of morphogenesis. *Phil. Trans. Roy. Soc. (London)* **B237**: 37–52.

Varga, R. S. (1962). *Matrix Iterative Analysis.* Englewood Cliffs, N.J.: Prentice-Hall.

Veen, A. H. (1973). A computer model for phyllotaxis based on diffusion of an inhibitor on a cylindrical surface. Part of a thesis presented to the Moore School of Electrical Engineering, University of Pennsylvania, Philadelphia.

Veen, A. H. & Lindenmayer, A. (1977). Diffusion mechanism for phyllotaxis, theoretical, physico-chemical and computer study. *Plant Physiol.* **60**: 127–39.

Veh, R. von (1931) Untersuchungen und Betrachtungen zum Blattstellungsproblem. *Flora* **125**: 83–154.

Velenovsky, J. (1905). *Vergleichende Morphologie der Pflanzen.* Prague.

Vieth, J. (1963). Étude de quelques inflorescences anormales de scabieuses. *Mem. Bull. Soc. Bot. Fr.*, pp. 36–45.

Vieth, J. (1964). Le capitule de *Dipsacus* représente-*t*-il un système bigugué? *Mem. Bull. Soc. Bot. Fr.*, pp. 38–47.

Vieth, J. (1965). Etude morphologique et anatomique de morphoses induites par voie chimique sur quelques dipsacacées. Doctor of Science Thesis, University of Dijon.

Vieth, J. (1971). Persistance de quelques chimiomorphoses induites sur *Dipsacus silvestris* Mill. *Ann. ACFAS* **38**: 36.

Vieth, J. & Arnal, C. (1961). Signification de quelques aberrations phyllotaxiques induites chez les dipsacaées. *Bull. Soc. Bot. Fr.* **108**: 92–101.

Vincent, J. R. & Tomlinson, P. B. (1983). Architecture and phyllotaxis of *Anisophyllea disticha* (Rhizophoraceae). *Gard. Bull. Sing.* **36**: 3–18.

Vinogradov, I. M. (1954). *Elements of Number Theory.* New York: Dover.

Vitanyi, P. M. B. (1980). *Lindenmayer Systems: Structure, Languages and Growth Functions.* Amsterdam: Math. Center Tracts.

Voeller, B. R. & Cutter, E. G. (1959). Experimental and analytical studies of Pteridophytes. XXXVIII. Some observations in spiral and bijugate phyllotaxis in *Dryopteris aristata* Druce. *Ann. Bot. (N.S.)* **23**: 391–6.

Vogel, H. (1979). A better way to construct the sunflower head. *Math. Biosci.* **44**: 179–89.

Vogel, T. (1956). *Physique Mathématique Classique.* Paris: Colin.

Waddington, C. H. (ed.) (1968–1972). *Towards a Theoretical Biology.* Edinburgh: Edinburgh University Press, pp. 1–4.

Waldrop, M. M. (1992). Artificial life's rich harvest. *Science* **257**: 1040–2.

Wardlaw, C. W. (1949a). Experimental and analytical studies of Pteridophytes. XIV. Leaf formation and phyllotaxis in *Dryopteris aristata* Druce. *Ann. Bot. (N.S.)* **13**: 164–98.

Wardlaw, C. W. (1949b). Phyllotaxis and organogenesis in ferns. *Nature* 164: 167–9.

Wardlaw, C. W. (1949c). Experiments on organogenesis in ferns. *Growth* (suppl.) **13**: 93–131.

Wardlaw, C. W. (1952). *Phylogeny and Morphogenesis.* London: Macmillan.

Wardlaw, C. W. (1955). Evidence relating to the diffusion-reaction theory of morphogenesis. *New Phytol.* **54**: 39–48.

Wardlaw, C. W. (1956). The inception of leaf primordia. In *The Growth of Leaves,* ed. F. L. Milthorpe. London: Butterworths, pp. 53–65.

Wardlaw, C. W. (1957). On the organization and reactivity of the shoot apex in vascular plants. *Am. J. Bot.* **44**: 176–85.

Wardlaw, C. W. (1963). Experimental investigations of floral morphogenesis in *Petasites hybridus. Nature* **198**: 560–1.

Wardlaw, C. W. (1965a). *Organization and Evolution in Plants.* London: Longman.

Wardlaw, C. W. (1965b). The organization of the shoot apex. In *Encyclopedia of Plant Physiology,* vol. I, 15, ed. W. Ruhland. New York: Springer-Verlag, pp. 424–42, 443–51, 966–1076.

Wardlaw, C. W. (1966). Leaves and buds: Mechanisms of local induction in plant growth. In *Cell Differentiation and Morphogenesis.* Amsterdam: North Holland, pp. 96–119.

Wardlaw, C. W. (1968a). *Morphogenesis in Plants. A Comparative Study.* London: Methuen.

Wardlaw, C. W. (1968b). *Essays on Form in Plants.* Manchester: Manchester University Press.

Wardlaw, C. W. (1970). *Cellular Differentiation in Plants and Other Essays.* Manchester: Manchester University Press.

Wardlaw, C. W. & Cutter, E. G. (1956). Experimental and analytical studies of Pteridophytes. XXXI. The effect of shallow incisions on organogenesis in *Dryopteris aristata* Druce. *Ann. Bot.* **20**: 39–57.

Wareing, P. F. & Phillips, I. D. J. (1978). *The Control of Growth and Differentiation in Plants.* New York: Pergamon Press.

Warfield, J. N. (1987). Thinking about systems. *Syst. Res.* **4**(1): 227–34.
Waterman, T. H. (1968). Systems theory and biology – view of a biologist. In *Systems Theory and Biology,* ed. M. D. Mesarovic. New York: Springer-Verlag, pp. 1–37.
Waterman, T. H. & Morowitz, H. J. (eds.) (1965). *Theoretical and Mathematical Biology.* New York: Blaisdel.
Watson, M. A. & Casper, B. B. (1984). Morphogenetic constraints on patterns of carbon distribution in plants. *Ann. Rev. Ecol. Syst.* **15**: 233–58.
Weber, D. F. & Weatherwax, P. C. (1966). A plant with opposite leaves. *Maize Genet. Coop. News Letter* **40**: 49.
Weisse, A. (1894). Neue Beitrage zur mechanischen Blattstellunglehre. *Jahrb. Wiss. Bot.* **26**: 236–94.
Weisse, A. (1897). Die Zahl der Randbluthen an Compositenkopfchen in ihrer Beziehung zur Blattstellung und Ernahrung. *Jahrb. Wiss. Bot.* **30**: 453–83.
Weisse, A. (1904). Untersuchungen über die Blattstellung an Cacteen und anderen Stamin-Succulenten. *Jahrb. Wiss. Bot.* **34**: 343–423.
Weisse, A. (1935). Zur Frage der spiraligen Blattstellungen. *Ber. Deutsch. Bot. Ges.* **53**: 438–65.
West, B. J. & Goldberger, A. L. (1987). Physiology in fractal dimensions. *Am. Sci.* **75**: 354–65.
Wetmore, R. H. (1943). Leaf stem relationship in the vascular plants. *Torreya* **43**: 16–28.
Wetmore, R. H. (1956). Growth and development in the shoot system of plants. *Soc. Dev. Growth. Symp.* **14**: 173–90.
Weyl, H. (1952). *Symmetry.* Princeton, N.J.: Princeton University Press.
Whaley, G. W. (1961). Growth as a general process. In *Encyclopedia of Plant Physiology,* vol. 14, ed. W. Ruhland. New York: Springer-Verlag, pp. 71–112.
White, A. T. (1973). *Graphs, Groups and Surface.* Amsterdam: North-Holland.
Whitehead, A. N. (1925). *The Principles of Natural Knowledge.* Cambridge: Cambridge University Press.
Whyte, L. L., Wilson, A. G. & Wilson, D. (eds.) (1969). *Hierarchical Structures.* New York: American Elsevier.
Wicken, J. S. (1988). Thermodynamics, evolution, and emergence: Ingredients for a new synthesis. In *Entropy, Information, and Evolution, New Perspectives on Physical and Biological Evolutions,* eds. B. H. Weber, D. J. Depew & J. D. Smith. Cambridge, Mass.: MIT Press, pp. 139–69.
Wicks, K. (1992). Spiral-based self-similar sets. In *Spiral Symmetry,* eds. I. Hargittai & C. A. Pickover. Singapore: World Scientific, pp. 107–21.
Wiener, N. (1948). *Cybernetics or Control and Communication in the Animal and the Machine.* Paris: Hermann et Cie.
Wiesner, J. (1907). *Der Lichtgenuss der Pflanzen.* Leipzig: Engelmann.
Williams, R. F. (1975). *The Shoot Apex and Leaf Growth: A Study in Quantitative Biology.* Cambridge: Cambridge University Press.
Williams, R. F. & Brittain, E. G. (1984). A geometrical model of phyllotaxis. *Austr. J. Bot.* **32**: 43–72.
Williams, R. F. & Metcalf, R. A. (1985). The genesis of form in Casuarinaceae. *Aust. J. Bot.* **33**: 563–78.
Willis, J. C. (1973). *A Dictionary of Flowering Plants and Ferns,* 8th ed. Revised by H. K. Airy-Shaw. Cambridge: Cambridge University Press.
Wilson, C. L. (1953). The telome theory. *Bot. Rev.* **19**: 417–37.

Winfree, A. T. (1980). *The Geometry of Biological Time.* New York: Springer-Verlag.

Winkler, H. (1901). Untersuchungen zur Theorie der Blattstellungen. I. *Jahrb. Wiss. Bot.* **36**: 1–79.

Winkler, H. (1903). Untersuchungen zur Theorie der Blattstellungen. II. *Jahrb. Wiss. Bot.* **38**: 501–44.

Wolpert, L. (1971). Positional information and pattern formation. *Curr. Topics Develop. Biol.* **6**: 183–224.

Woodger, J. H. (1967). *Biological Principles.* New York: Humanities Press. (Reissued with a new introduction.)

Wright, C. (1873). On the uses and origin of the arrangements of leaves in plants. *Mem. Am. Acad. Arts Sci.* **9**: 379–415.

XuDong, F. & Bursill, L. A. (In press). Spiral packing of contact discs with logarithmic growth laws. *Modern Phys. Letters B.*

XuDong, F., Bursill, L. A. & JuLin, P. (1988). Fourier transforms and structural analysis of spiral lattices. *Int. J. Modern Phys. B* **2**(1): 131–46.

XuDong, F., Bursill, L. A. & JuLin, P. (1989). Packing of equal discs on a parabolic spiral lattice. *Modern Phys. Letters B* **3**(2): 119–24.

XuDong, F., JuLin, P. & Bursill, L. A. (1988). Algorithm for the determination of spiral lattice chirality. *Int. J. Modern Phys. Letters B* **2**(1): 121–9.

Yanofsky, M. F., Ma, H., Bowman, J. L., Drews, G. N., Feldman, K. A. & Meyerowitz, E. M. (1990). The protein encoded by the *Arabidopsis* homeotic gene *agamous* resembles transcription factors. *Nature* **346**: 35–9.

Yegappan. T. M., Paton, D. M., Gates, C. T. & Mueller, W. J. (1980). Water stress in sunflower (*Helianthus annuus,* L.). I. Effect on plant development. *Ann. Bot. (London)* **46**: 61–70.

Young, D. A. (1978). On the diffusion theory of phyllotaxis. *J. Theor. Biol.* **71**: 421–32.

Young, D. A. (1987). Some thoughts on phyllotaxis. *Mathematical Modelling* **8**: 729. (*Proc. 1st Int. Symp. on Phyllotaxis, U. of Berkeley,* July 1985; pp. 729–61 of *Proc. 5th Int. Conf. on Math. Modelling in Sci. and Tech.*)

Young, D. A. & Corey, E. M. (1990). Lattice models of biological growth. *Technical Paper,* Lawrence Livermore Nat. Lab., University of California, Livermore.

Zagorska-Marek, B. (1985). Phyllotactic patterns and transitions in *Abies balsamea. Can. J. Bot.* **63**: 1844–54.

Zagorska-Marek, B. (1987). Phyllotaxis triangular unit; phyllotactic transitions as the consequences of the apical wedge disclinations in a crystal-like pattern of the units. *Acta Soc. Bot. Poloniae* **56**(2): 229–55.

Zhitkov, V. S. (1983). Size of leaf base as a criterion of classification of the forms of phyllotaxis and of metamerism pattern of the shoot of flowering plants. *J. Obchtch. Biol.* **44**(6): 802–22. (In Russian, with English summary.)

Zimmermann, W. (1953). Main results of the telome theory. *Paleobotanist* **1**: 456–70.

Zimmermann, W. (1959). *Die Phylogenie des Pflanzen,* 2nd. ed. Stuttgart: Fischer.

Author index

Subject index